Human Characteristics

Human Characteristics: Evolutionary Perspectives on Human Mind and Kind

Edited by

Henrik Høgh-Olesen, Jan Tønnesvang and Preben Bertelsen

CAMBRIDGE SCHOLARS

PUBLISHING

Human Characteristics: Evolutionary Perspectives on Human Mind and Kind,
Edited by Henrik Høgh-Olesen, Jan Tønnesvang and Preben Bertelsen

This book first published 2009

Cambridge Scholars Publishing

12 Back Chapman Street, Newcastle upon Tyne, NE6 2XX, UK

British Library Cataloguing in Publication Data
A catalogue record for this book is available from the British Library

ISBN (10): 1-4438-0213-1, ISBN (13): 978-1-4438-0213-0

TABLE OF CONTENTS

INTRODUCTION

HUMAN CHARACTERISTICS - SETTING THE STAGE

HENRIK HØGH-OLESEN

Every once in a while, we as scientists have to reconsider the perennial questions concerning human nature: What are the special human behaviours, social practices, and psychological structures that make us particularly human? Or, as the French crooner Charles Aznavour puts it, on one of my mother-in-law's old records: *"Tell me if you can, what makes man a man?"*

The answers given to this question may differ considerably, and each century or decennium may have its favourites. But one thing seems inevitable. Whenever a new distinction is launched, it stirs up fuss and provokes many people, both inside and outside the scientific community, to take a stand. Thus, when the Swedish botanist Carl von Linné, in the 10th edition of his *"Systema Naturae"* in 1758, placed humans and apes in the same zoological category of *"Primates"*, an outcry of indignation was heard all over Europe. Man was "imago Dei": The only living creature made in the image of God. And little did it help that the word "primate" literally indicated that we were among the first, foremost and highest in rank of all God's creatures (which was Linné's way of sugaring a bitter pill). The Rubicon between man and beast had been crossed, and – primate or not – from now on man was just another mammal with a natural history.

When, a hundred years later, Darwin consolidated this view in *"On the Origin of Species"* (1859) and *"The Descent of Man"* (1874), the reception was just as hostile, and the rave goes on as the latest roars from the supporters of intelligent design indicate.

So obviously, these matters most certainly matter to us. And indeed, this perpetual quest for origin, meaning, and ontological status – this existential need to know who we are – may in itself be one of the human particulars that we have set out to discover. Man, "the existential man-hunter", so to speak, constantly tries to track down his own footprints in order to catch up psychologically with himself and his origins.

It is also significant that the closer we get to our own time, the more answers are suggested and the quicker they change. Today, relevant data and theoretical deliberations from a broad range of sciences are pilling up, and so an overview is needed. To facilitate this breadth of view, a large inter-disciplinary conference entitled 'Human Mind – Human Kind' with participants from more than 20 nations was held at the University of Aarhus, Denmark in August 2007. More than 100 experts within the field of evolution and cognition presented their latest research. Forty-five papers were later submitted for publication, and after a review process, twenty of these found their way to this book, securing a volume of both quality and thematic coherence.

The contributions fall into three well defined but interconnected sections.

Section 1: *Evolution and Cognition – comparative and developmental perspectives* begins with Stange´s interesting work on art creation and appreciation in humans and animals. Aesthetic preferences may be functional to others than humans, but does non-human animals show any signs of aesthetic sensibility or artistic creativity? The evidence is scanty and it is difficult to know for certain ; however, Stange discusses what data we have, and ends his chapter with an actual sketching of an experimental procedure, that might help us answer some of these fascinating questions.

In chapter 2, Nadal, Capó, Peters & Cela-Conde continue this discussion in a review, that fuses results from empirical aesthetics, neuroimaging and comparative neuroscience, and substantiate the idea that aesthetic preference – rather than a content-specific modular mechanism – seems to be the result of several cognitive and affective processes (a "mosaic evolution"), some of which we have inherited from our primate ancestors.

The phenomenon of art itself is indicative of a complex inner world; but do other animals also have complex mental worlds, and how are we going to address this question empirically?

One way to approach the question is by investigating animals' abilities to travel mentally in time, and to recollect and plan for future states. Oswath has conducted a series of experiments on chimpanzees and orangutans that would seem to suggest, that humans are not alone "in the mental world of possible futures", as he so eloquently puts it in chapter 3.

Kingo & Krøjgaard (chapter 4), further develop the focus on complex symbolic and mental capacities by directing the attention towards language and tool use, two domains in which humans seem to be far superior, when comparatively viewed. There may be many reasons for this superiority, but one easily overlooked explanation offered here, is the human ability to extract and understand the functional characteristics of physical objects, and it is argued that this skill serves as a basic prerequisite for advanced language and tool use.

Krøjgaard (chapter 5) expands this line of thinking in a thorough article on object individuation in humans and apes. Basic aspects of object individuation may also be prominent in some of our primate relatives, but again humans excel and display a far more elaborate understanding of cultural artefacts than any of these. Additional skills, far beyond simple object tracking here and now, may account for this surplus, and the attention is directed towards the human ability to abstract spatio-temporal threads of specific objects. Finally, Nørager (chapter 6) closes the theme on tools, objects and cognitive capacities in an article that shows how our knowledge of primate cognition and tool use can be applied in making modern computerised technology more intuitive and user-friendly.

The applied field of psychotherapy also seems to gain important insights from animal comparisons. Descartes´ *Meditations*, and the thought processes produced in modern depth psychology, may be prime examples of a type of reflexive thinking only found in humans, but in chapter 7, the psychiatrist Brickmann turns our attention towards a continuum of succorance, running from arched back nursing in rats, and grooming in non-human primates, to supportive elements of psychotherapy in humans, hiding underneath all the "mentalizing".

Closing the section – and serving as a bridge to the social perspectives in section 2 - Smedt, de Cruz & Braeckman conduct an examination of human social cognition from an archaeological and comparative perspective, which suggests that the human brain is not simply an enlarged chimpanzee brain. Instead natural selection seems to have favoured a different social cognition in both species.

In Section 2*: Human Sociality, Morality & Religiosity,* scholars from within psychology, philosophy, ethics and the scientific study of religion, look at the interconnected vessels of sociality, morality and religiosity from an evolutionary perspective.

The section opens with a longitudinal study by Takahashi, Inoue, Yamakawa & Shibata on the development of close social relationships among young Japanese children, that introduce light and shade into the classic attachment theory by showing, that children as young as 3 years old, construct their own frameworks of social relationships consisting of *multiple* significant others, including non-kin, and this makes evolutionary sense.

To many (see Høgh-Olesen chapter 20), third-party sanctioning and norm enforcement is the moral Rubicon between humans and apes. In chapter 10, Ingram, Piazza, & Bering explore the possibility that human gossiping, and the phenomenon of absent third-party punishment, has generated the evolution of a set of unique cognitive structures (which enable people to model and manipulate their reputation in the minds of others), capable of handling this adaptive challenge. The model presented is supported by an observational study on tattling in two preschool samples, and an experimental study of giving, under threat of gossip, in a dictator game.

The theme on human morality is further developed in chapter 11, where de Jelle discusses empathy (affective and cognitive) as the trigger of altruistic motivation, and in chapter 12 and 13, where the philosophers Toft and Nissen take a closer look at some of the foundational problems related to an evolutionary approach on human altruism and morality. Toft by addressing the problem of altruism towards distant strangers in need, a phenomenon that according to him cannot be sufficiently accounted for by evolutionary theory ranging from reciprocal altruism to gene-culture co-evolution, and Nissen by investigating what conclusions to draw in philosophical ethics from the fact, that morality is an adaptation. An investigation that leads Nissen towards a notion of objective morality, that is strong enough to support moral realism.

Human sociality is complex. As humans we engage in large scale, high-risk cooperation with non-kin and strangers. We glue these complex social organisations together by strange symbolic embellishments and religious beliefs, cults and rituals, unparalleled in the rest of the animal world, and sometimes we even sacrifice our own life to defend or revenge an actual or symbolic threat to these ideological and collaborative mega-units (being a Dane, the "cartoon-crises" is still fresh in memory).

These phenomena are puzzling indeed. Liberoth (chapter 14) explores how far evolutionary theory will get us in understanding the evolution of martyrdom and religious self-sacrifice, while Levi (chapter 15) examines the nature and functions of religious cognition and behaviour and argues, that ritual behaviour and norms are among the forces that render human traditions evolutionary stable.

Petersen & Kennair (chapter 16) conclude the section with a bridge-building article that shows how evolutionary psychology and political and social science may cross-fertilize each other and join hands in the understanding of such complex social phenomena as cooperation, coalition formation, norm-sanctioning and public opinion.

The domain summarized under the heading of Section 3: *Human Sexuality and Mating Strategies,* has been a growing field in many years, in journals like "*Human Nature*" and "*Evolution and Human Behavior*", and three such studies have found their way into this volume.

Until now mating studies have largely ignored the fact, that mating decisions may be heavily influenced by parents and other kin, and that parents and offspring may have conflicting opinions, regarding what constitutes an ideal mate. Park, Dubbs & Buunk throw light on this subject in a new study presented in chapter 17, while Hald & Høgh-Olesen study the different mating strategies applied by women and men, by letting eight confederates approach strangers of the opposite gender on a university campus, asking one of the following three questions: 1)"Would you go on a date with me?", 2)"Would you come to my place tonight?", 3)"Would you go to bed with me tonight?". The gender-specific answers to these questions can be found in chapter 18.

Finally, Klavina, Buunk & Park, chapter 19, investigate the relationship between jealousy towards rivals from different ethnic groups and the perceived characteristics and stereotypes of these groups. An important aspect of intergroup conflict is competition for mates, especially among men, and because different outgroups pose different levels of threat, the group membership of rivals may be a characteristic that evokes jealousy. Data supporting this prediction is presented.

A book entitled *Human Characteristics* is more or less under an obligation to synthesize perspectives into a more inclusive answer to what exactly makes us human, and at least to outline the species-specific behaviours, social practices and psychological processes that appear to be uniquely human. The present author takes on this commitment in chapter 20, summing up and supplementing the perspectives presented so far. Four central fields of activity related to: 1) complex symbolic behaviours, 2) tool making and tool use, 3) culture and social transmission, 4) sociality and morality, are surveyed and comparatively analysed for similarities and differences. Supplementary data, from a broad range of sciences, are brought in, introducing light and shade into the picture.

The atmosphere at the conference was open-minded and characterized by inter-disciplinary curiosity, and so are the following articles. They naturally apply their own evolutionary and/or comparative approach without engaging in aggressive theoretical demarcations.

In my own field of science, psychology, we traditionally speak of the major paradigmatic "*forces*" that have dominated the field. Behaviourism, psychoanalysis, and humanistic psychology constitute the three first forces, and now, after some 30 years interregnum, a fourth unifying force, which introduces an evolutionary and cognitive angle to the study of man, has emerged strong and vital.

Naturally, there are differences between socio-biology, human behavioural ecology, gene-culture co evolution, evolutionary psychology, etc., but equally there were differences between Watsonian, Palovian, and Skinnerian behaviourism, and between humanistic and existential psychology. Therefore, let us not indulge in the narcissism of the little difference, as Freud put it. The different 'regiments' may wave different colours, but it is still the same force united in a common evolutionary approach.

SECTION I:

EVOLUTION AND COGNITION – COMPARATIVE AND DEVELOPMENTAL PERSPECTIVES

CHAPTER ONE

ART AND APPRECIATION: UNIQUELY HUMAN?

KEN STANGE

Abstract: Close observation of other species has resulted in the realization that most of the characteristics once thought to be unique to Homo sapiens are also found in other species, albeit in a less developed state. This paper addresses recent claims that even artistic creativity may not be unique to human beings. While rejecting the often cited and dubious evidence of animal art admired by humans, it offers an alternative approach which involves clearly operationally defining art, creativity, and aesthetic sensibility in terms of the unique characteristics associated with each. On the basis of these definitions, some evidence is offered to support the idea of an aesthetic sensibility in other species, and one feasible empirical investigation is proposed

What Makes Homo sapiens Unique? The Usual Suspects

Science has a justified reputation for attacking our self-esteem as a species. The first major attack was the Copernican Revolution which moved us and our home from the centre of the universe—and now cosmologists have relegated us to the far suburbs in but one insignificant galaxy among millions scattered over distances we can't even comprehend. Then a blow was struck even closer to the bone by what Daniel Dennett (1995) calls Darwin's "dangerous idea" of natural selection. Once we accept this humbling cornerstone of biology, we are tempted to reach out for something, anything, we can claim as uniquely ours—at least in the small corner of the universe to which we have access. Can we at least claim our species is qualitatively different from all the other life forms of which we have knowledge? Descartes infamously considered all non-human creatures as mere automatons, and many are afraid that recent neuroscience discoveries are implying that we too are automatons. Is there any justification remaining for thinking Homo sapiens differ *qualitatively* from other species?

The specific candidates traditionally offered up as evidence of our special status have been systematically eliminated—or at least many feel they have. The four major specific candidates are language, tool-use, cognitive skills at problem solving, and transgenerational transmission of culture. Of course one can choose to operationally define each of these so stringently that only *Homo sapiens* makes the cut, but to do so seems a mere *ex post facto* attempt at saving face.

For example, the degree to which the great apes can understand syntax, or whether they do so at all, is often acrimoniously debated, with researchers such as Francine Patterson, working with the gorilla Koko (Patterson & Linden, 1988), and Sue Savage-Rumbaugh, working with the bonobo Kanzi (Savage-Rumbaugh, Shanker, & Taylor, 2001), claiming to have evidence of sophisticated language abilities including syntactical usage in their protégés, while highly respected linguistic experts such as Stephen Pinker and Noam Chomsky view such claims as exaggerated projections of the researchers' expectations and involvement with their animals. "Possessing a language is the quintessentially human trait: all normal humans speak, no nonhuman animal does." (Pinker, 1995, p. 135).

Nevertheless, it is fair to say that if one doesn't set the bar too high, most researchers have to concede something that could be called 'language ability' in other species. Furthermore, regarding the other three former candidates for human uniqueness, there is plenty of evidence of some degree of cognitive skill at problem solving in many species (Wasserman, and Zentall, 2006), indisputable evidence of animal use of tools, even in birds (Bluff, Weir, Rutz, Wimpenny, and Kacelnik, 2007), and general acceptance of transgenerational transmission of acquired social or adaptive skills in non-human primates (Laland, and Hoppitt, 2003).

However, in more general terms, humankind does seem to be distinct from other species in having science, religion, and art. However science, in the modern sense, was not a characteristic of our species until relatively recent times. And science, in the broad sense of making observations and generalizing from them certainly does exist in primitive form in other species. The strict behaviourist's attempts to explain away apparent empirical reasoning by other creatures has been replaced by widespread acknowledgement of observational learning in animals (Bandura, 1977). Religion can be viewed as a combination of superstitious behaviour and ritualized behaviour, both of which have certainly been observed in other species (Dawkins, 2006). If one accepts this loose definition of science and this interpretation of religion, only art remains as the last bastion of 'hope' for any claim to the qualitative uniqueness of our species.

Some feel that this bastion is being battered by evidence that chimps and elephants and a few other species can, when given the necessary tools, create admirable paintings that are indistinguishable even by art critics from the work of contemporary artists. Whenever a painting by an elephant or a chimp sells for a substantial sum, it makes the news. For example, a painting by Kamala, an elephant at the Calgary Zoo in Canada, recently sold for $1,175. This news is greeted with glee both by those who wish to denigrate contemporary artists and by those who wish to elevate our estimation of animals to a higher plane. However, neither of these groups could be called unbiased evaluators of the significance of such news.

Yet it is a serious philosophical and empirical question whether or not other species demonstrate what could justifiably be called an aesthetic sense, and it is worthwhile to question whether or not aesthetically pleasing paintings created by other species are really valid evidence of such an aesthetic sense—and if they aren't, what really would constitute such evidence. These are the questions addressed in this paper.

To Claim Art as Uniquely Human Requires Defining Art

It is at first surprising that both philosophers of aesthetics and those engaged in doing research in empirical aesthetics rarely address the question of exactly what art is. The primary reason for this is that art, even more than science and religion, is notoriously resistant to definition. This problem stems from the incredible diversity of art forms which seem to have so very little in common. What does found art, such as Marcel Duchamp's urinal have in common with a Rembrandt painting or a Bach fugue or a Bergman film or the unrecorded 1913 *Rites of Spring* ballet performance that caused Stravinsky to flee the concert hall or the Köln Cathedral or the ritual masks of the Dani tribe of Papua New Guinea or the sonnets of Shakespeare or the draping of the Pont Neuf in Paris by Christo?

There is a simple solution to this problem: focus not on the thing called art by someone, not on the object or performance, but rather on the response to it (Stange, in prep). All of the above examples induce in some people what could be called an "aesthetic response", which is precisely what inspires someone to call something 'art'. There is far less confusing diversity in the "aesthetic response" than in what induces it: some of us are or could be 'moved' or 'touched' by all of the above examples, but our experience in each case is surprising similar. This shifting of emphasis from cause to effect may seem like begging the question, not very different from saying intelligence is what IQ tests measure just to avoid confronting what intelligence really is or what people mean when they use the word. However, defining art as what produces an aesthetic response is different. Like being in love, anyone who has had the experience knows what it is. And, from a more scientific perspective, one can cite specific unique characteristics associated with the experience.

There are three empirical criteria that seem to consistently distinguish the aesthetic experience:

(1) It involves an intense emotional response to simulations or imitations of 'real-life' events or things. Furthermore, some of these events or things are ones we would find aversive if indeed they were real.

(2) It results in a pleasurable cognitive response to relationships just for their own sake, independent of any apparent utility.

(3) It produces pleasure from pure perception. The perceptual experience seems to be an end in itself.

The seemingly reasonable objection to this definition of art is that one has such aesthetic responses to things we don't call art; e.g. a stunning landscape or the face of a beautiful woman. This objection misses the point that this is simply a working or operational definition of art. All such definitions are admittedly arbitrary, but nonetheless are considered good and useful if they match up with our general conception of what is being defined. We don't normally label as art what has been created by chance or by nature. So by eliminating such unintentional causes of the aesthetic response, one arrives at a reasonable working definition for art: art is what produces an aesthetic response and is *not* a product of random or natural events. A painting of a landscape that effects an aesthetic response in someone *is* usually called art. A landscape that effects an aesthetic response in someone *is not* usually labelled art. I realize that even this working definition can occasionally be problematic, but it is far less so than any based on the cause of the experience rather than the experience itself.

However, the first step to answering any question about artistic creativity in another species is to search for evidence of an aesthetic response to *anything*, art or nature. Before one can even consider the possibility of non-human artistic creation, one has to establish evidence of aesthetic appreciation. So now here is a closer look at the three criteria just mentioned.

First to be considered is the paradoxical emotional response we have to what we know are mere simulations or imitations of something, our strong emotional response to what are clearly not "real-life" events. We seek out such experiences even when they would be aversive if real. It makes sense to cry at the death of a loved one. However, we also cry at the death of Romeo and Juliet. Our adrenaline levels surge if we encounter a bear while walking through the woods, but our sympathetic nervous system also is activated by seeing a character in a film suddenly encounter a big, bad bruin. Again there may be a reasonable evolutionary explanation for this. It may be vestigial and residual and no longer particularly adaptive, something like our easily triggered fear of snakes even if most of our species now live where there are no venomous snakes. Or it may serve some function such as training our sense of empathy or rehearsing for dealing with real life events. Nevertheless, it is unique to the aesthetic experience.

Then there is the cognitive response. Much of aesthetic appreciation is largely rational. There is a pleasure inherent in seeing new relationships. A pun or any joke involves surprising us with an unconsidered relationship. A Shakespearean play is all about the complex inter-relationship of fictional individuals. A Bach fugue is about the intertwining of highly abstract contrapuntal and polyphonic sound sequences. Some art is called 'cerebral' because the pleasure one gets from it is relatively free of emotion. For example, the murder mysteries of Agatha Christie don't particularly excite our emotions, for we rarely mourn for the victim, but we derive pleasure from trying to untangle the motivational and situational relationships that led to the crime and finding the overall pattern that points to the perpetrator. There is no question that seeing relationships has tremendous survival value, so of the three markers of an aesthetic response this one has the most obvious evolutionary value.

Finally, one of the most striking things about the aesthetic experience is the pleasure we derive from the pure perception of an object or event. It is entirely understandable that a well "plated' meal should stimulate our pleasure centres. It makes sense to salivate at the sight of an appetizing meal. The pleasure we derive from perceiving an appetizing dish motivates us to consume the food, which has obvious survival value. It is no accident that some of the so-called "pleasure centres" in the brain such as the hypothalamus are also those that regulate homeostasis and drive us to eat when hungry and drink when thirsty, as electrical stimulation of this structure has repeatedly demonstrated (Bozarth, 1994). However it does not make sense that we derive profound pleasure from a beautiful still life painting of food when we are well aware that the canvas is inedible—and we do so even when we're not hungry. When one considers the great pleasure derived from something like an abstract piece of music, the mystery is even more overwhelming. Evolutionary psychologists and those researching empirical aesthetics are working hard to explain this phenomenon, but that is not what is at issue here. Suffice it to say that this is an important criterion that distinguishes the aesthetic experience.

To summarize, the aesthetic response is distinguished by emotional, cognitive, and perceptual experiences we find pleasurable and seek out, even though they have no apparent immediate relevance or practical value. And so the critical questions are whether other species also seek out such experiences and whether we can find evidence of this by applying these criteria.

Two Relevant Paradoxes

Before proceeding to the thematic question of whether art is unique to humankind, and how one could possibly answer that question, two apparent paradoxes have to be considered because both are extremely relevant. The first has to do with the famous Turing Test for artificial intelligence, and the second is what I call the "Creative/Critical Paradox". Both directly relate to any possible test of the uniqueness of art to our species.

The so-called Turing Test is the classic empirical test for artificial intelligence proposed in 1950 by the brilliant mathematician Alan Turing, one of the fathers of computer science. The protocol for the test is as follows. A human 'judge' sits at a computer keyboard terminal and communicates by typing messages to two sources located in closed rooms hidden from his view. In one room is a human being who reads and responds to the messages sent from the judge. In the other room there is a computer, allegedly possessing artificial intelligence, which also receives and responds to the messages and questions sent by the judge. If, after extensive questioning and interaction with both the computer and the human, the judge is not able to tell which room contains the human correspondent and which the computer, on a better than chance basis, the computer is said to have passed the test—and can be credited with intelligence equivalent to that of a human being.

The classic refutation of the validity of this test is Searle's "Chinese Room" thought experiment. Searle (1980) suggests that if you imagine yourself a monolingual English speaker "locked in a room, and given a large batch of Chinese writing" plus "a second batch of Chinese script" and "a set of rules" in English "for correlating the second batch with the first batch", and then a judge who is fluent in Chinese sends you messages, you will be able to 'reply' in a way that convinces the Chinese judge that you really know Chinese. You will have passed a Turing Test for Chinese linguistic intelligence without having that attribute.

I find Searle's argument persuasive. I even hold the unorthodox view that Alan Turing wasn't entirely serious and that he actually presented his so-called "test" more as a tongue-in-cheek critique of excessive trust in operational definitions than as a sincerely intended test by which artificial intelligence could be demonstrated. But whatever Turing's motives, his test paradigm and Searle's criticism are just as relevant to the alleged demonstration of artificial creativity as they are to artificial intelligence.

If an independent judge cannot distinguish artworks created by a computer from those created by a human artist, the computer would pass the Turing Test for artificial creativity. Now replace the computer with, for example, an elephant. Imagine you have this elephant in a (big) room painting pictures, while in another room you have an abstract expressionist artist also painting. If, after a substantial number of paintings have been created by both artists, the works are presented to a judge, and that judge cannot consistently distinguish the paintings of the human from those of the elephant, is it not reasonable to say the elephant has passed a variation of the Turing Test modified to determine animal creativity? And, despite the elephant passing the test, isn't the conclusion of animal creativity dubious?

The second apparent paradox is what I call the "Creative/Critical Paradox". It is common to view being critical as easy and being 'creative' as difficult. "Everyone is a critic!" is a common lament, especially by creative people. It is easy to find fault with ideas, but difficult to find new ideas, or so goes the folk wisdom. While there is an element of truth to this common perception of the nature of creativity, it can be misleading.

Creativity has two components: the production of something new and the evaluation of what has been produced. The fact is that production actually is the relatively easy part, while evaluation is the hard part. All good writers know writing is mostly revision. Ideas are a dime a dozen. Good ideas are rare. What distinguishes creativity is the ability to distinguish the wheat from the chaff. I have written software—which I call *Ghostwriter*—that randomly creates a virtually infinite number of syntactically correct sentences. Every once in a great while, like with the proverbial many monkeys at many typewriters with much time on their hands, a sentence sometimes appears that is stunningly beautiful, even profoundly insightful. Whatever creativity I have as a writer resides in my ability to detect these rare gems. The computer program can take care of the production part of creativity, but only a human can take care of the evaluative part.

The creation of art is primarily a matter of evaluating, filtering and then revising. It may be that we often do this evaluation mentally, before physically creating something, before writing a sentence or putting a brush stroke down on canvas. Nevertheless, it is *judgment* that is the critical and distinguishing component of creativity, of the creation of significant art *and* science.

Creativity, like all behaviour, can be schematized as consisting of input, processing, and output. The log-jam in AI development hasn't been at the processing or output parts; i.e., at the productive part of this circuit. It has been at the input part because input involves evaluation. Computers can do logical analysis and output the results. Computers can create images and sound sequences and present them. However, computers *cannot evaluate* their output because they cannot, when their output is sent back as input, *recognize* and *appreciate*—or at least so it seems so far.

If things that produce an aesthetic response result from random events (e.g., the reflections in an oil spill after a rainfall), from evolutionary natural selection (e.g., the male peacock's glorious tail), and from insentient computer algorithms (e.g., a sentence 'written' by my *Ghostwriter* program), should these things be considered creative and assumed to have an aesthetic sense? This is a rhetorical question, for obviously we do not credit happenstance, nature and algorithms with that attribute we call creativity or aesthetic sense.

Aesthetic evaluation depends on appreciation. So these two apparent paradoxes suggest that understanding the nature of appreciation is the key to understanding creativity.

So How To Detect An Aesthetic Response? By Art Appreciation?

To reiterate and summarize, the first place to look for any substantive evidence of an aesthetic sense in other species is not in what they may 'create' (and certainly not the creations judged by human—anthropocentric—standards), for there really is no way to determine if the work is intentional and driven by any aesthetic motives. Nor is it reasonable to focus on creative production as any kind of evidence, for an aesthetically pleasing thing can be created even by random events: What matters is the critical and evaluative aspect of creativity. Thus the logical thing to look for is some evidence of aesthetic *appreciation*, some evidence of an aesthetic response.

The following are the three aforementioned empirical criteria associated with an aesthetic response: 1) an emotional response independent of 'real-life' events; 2) a cognitive response to relationships independent of practical application; and 3) pleasure in perception independent of utility. Human beings clearly search out experiences that induce these responses. Can we possibly determine if other species exhibit these behaviours, behaviours we consistently associate with the aesthetic response?

Emotional responses independent of 'real-life' events? The difficulty in 'reading' animal emotions makes this very problematic, and the question of whether an animal can even distinguish reality from illusion is not easily answered. The closest thing to an empirical investigation of this is the research done with mirrors where there is some evidence great apes (and perhaps elephants and dolphins) recognized themselves as themselves in mirrors (de Waal, 2007). With species that don't show this ability, the observed reaction to simulations of real stimuli seems at first to be accepting the simulations as real stimuli, followed in some species by an indifference to the stimulus that seems to indicate a rejection of it as a mere illusion. An example is how one of my dogs responded to the appearance of canines in television shows. Initially, the sounds of barking and the images (which it should be noted dogs do not perceive as fluid movements because of a different flicker frequency threshold) caused my dog Nick to bark in social response and even look behind the television monitor. Eventually, however, he failed to respond with anything more than a glance at the tube when dogs appeared or barked in some show, even when the barks clearly were ones of distress or aggression.

Cognitive responses to relationships independent of practical application? Again this seems impossible to ascertain. How could we possibly determine if an animal is seeking out intellectual stimulation and challenge? So clearly this, too, seems outside the realm of empirical investigation.

Pleasure in perception independent of utility? Fortunately, this marker *does* seem to be measurable in other species, because it can be inferred if a creature repeatedly seeks out certain perceptual experiences that seem to have no obvious utility. Such behaviour would at least *suggest* a primitive aesthetic sense and *is* testable.

There are three behaviours that could be reasonably considered indicative of pleasure in perception independent of utility, and so perhaps of an aesthetic 'drive':

(1) The first of these is exploratory behaviour: the seeking out of novel experiences for their own sake.

(2) Secondly, there is the seeking of non-sexual and non-utilitarian sensual pleasures.

(3) Finally, the expression of clearly defined preferences for some stimuli over others, without any obvious immediate or evolutionary benefit, would be evidence of purely aesthetic motivation.

In searching for these behavioural markers in another species, I have chosen as my example *canis lupus familiaris*: the domestic dog. The reason for this choice is that our knowledge of the behaviour of dogs is far greater than that of any other species high enough on the evolutionary tree to possibly manifest the criteria behaviours. The domestic dog's intelligence has been estimated (Coren, 1994) to be that of a two-year old human child, and although the great apes exist on a more proximate branch of the evolutionary tree, and are usually assumed to be even closer to us in cognitive ability and behavioural traits, our knowledge of them is far less extensive than of the domestic dog with whom we have empathetically cohabited and observed for at least fourteen thousand years.

So do dogs show exploratory behaviour, the seeking out of experience for its own sake? Obviously they do, and of course exploratory behaviour is also widely observed in many other species. Its evolutionary function is obvious. When sated laboratory rats are placed in a novel maze they don't just lie down and sleep. They spend their time exploring; and if reintroduced to the maze when hungry, they learn the location of the reward faster than rats that hadn't previously had the opportunity to explore the maze (Tolman, 1948).

Do dogs seek non-sexual and non-utilitarian sensual pleasures? Again, obviously they do. There is no survival benefit to being petted, but any dog owner knows their aptly name 'pet' seems to have a biological need for it; and unlike most drives to fulfil a need, the need seems insatiable. A piece of canine wisdom (from which we could learn) is that "when they stop petting you, move on!" Unless we stop stroking them, most dogs seem willing to sit and be petted forever. Grooming behaviour in the primates may seem similar and is usually assumed to serve the useful functions of social bonding and hygiene, but neither seems a particularly likely explanation for the doggie drive to be petted. It is true that some research has indicated cats rub each other's faces as part of social bonding and that humans mimicking this behaviour with their hands produce "affiliative responses" in their feline pets. (Schmied et al., 2008) But dogs do not rub each other's backs, and dog trainers and owners know that the canines especially prefer being petted in places where they would not normally be stimulated—and certainly not by others of their species. The most parsimonious explanation of why dogs seek out petting is simply—it just feels good. (Of course, why something feels good which has no utilitarian or apparent evolutionary explanation is a central question in both aesthetics and general evolutionary theory, not to mention neuroscience.)

Nevertheless, it certainly seems that canines do show clearly defined preferences for some stimuli over others without any—at least *obvious*—utilitarian or evolutionary benefit. As every dog owner knows, dogs definitely do have preferences. They have preferred parts of their bodies to be rubbed, preferred places to rest, preferred foods, preferred scents, preferred toys, preferred human companions, etc. Many of their preferences seem to make no sense to us and seem entirely arbitrary and idiosyncratic. (Why my dog Maggie much prefers her fuzzy lion toy to the fuzzy bear toy is a mystery to me.) The meaning and function of preferences is a complex topic of great interest in the field of empirical aesthetics.

One of the ongoing projects of evolutionary psychologists is trying to explain wide-spread human preferences that are now actually maladaptive. For example, a preference for—and thus over-indulgence in—sweet and salty foods is accepted as a major contributor to a variety of medical problems in developed countries. We only have five taste receptors on our tongues and two of these are specialized for inorganic salts and for sugars. Salt is essential to the mammalian diet and once was scarce, so it isn't difficult to understand why we evolved special receptors to detect foods that contain it—and why consumption of it is so pleasing. Similarly, a preference for sweet foods makes perfect sense for two reasons: sugars are one of the most concentrated sources of quick energy and few poisonous plants taste sweet. When salt and high caloric foods were scarce, there was no need for an off switch for our drives to consume them.

When we move deeper into the area of aesthetic preferences the source of these preferences is less obvious, but nevertheless is sometimes discernable or inferable. One example of this is the widespread preference, both in pictures and in the design of parks, for landscapes that resemble the East African savannah which have places to hide safely, yet offer vistas of the surrounding area (Orians and Heerwagen, 1992). The typical urban park is designed as a variety of open and wooded spaces. Also both parks and popular landscape paintings usually contain bodies of water.

Thus it may be that many of our aesthetic preferences for some stimuli over others once had some practical function that has faded away over time and now seems mysterious and inexplicable. Why we prefer the smell of roses over the smell of marigolds may never be explained. Why our dogs couldn't care less about the smell of roses or marigolds, but definitely love the smell of dead and rotting fish, even to the extent of rolling on them to scent themselves, also may never be explained. The suggestion that they do this to disguise their own scent so they can sneak up on prey seems dubious. "Don't worry, pal, about that big, hairy mammal that seems to be stalking us: my nose tells me it is just a big fish out of water." I am being facetious, of course, and certainly if the prey were only relying on olfactory warnings, there may be some substance to explaining this behaviour as olfactory masking—although the usual prey species of wolves are mammals, usually more likely to be alerted by their visual systems' movement detectors than olfactory cues. (One must be careful about facile explanations that simply fit with one's adopted theoretical framework, e.g., evolutionary biology, and miss more parsimonious explanations. Could not dogs just *like* the smell of rotting fish?)

The virtually indisputable point is that dogs (and of course many animals) show strong preferences for certain—at least *apparently*—non-utilitarian stimuli, which they repeatedly seek out, just as we do when we go to a manicured park or an art gallery. With other species it may be easier to suggest some plausible past or present utility to the preference, but that does not make it qualitatively different from our preferences. For all we know, there very well may be an evolutionary explanation for many humans' love of baroque music.

A Modest Research Proposal

So let us say an elephant is given brushes and paints and creates a painting. How could one determine if the work is the result of some aesthetic sense or merely a random event? Certainly not by having human beings judge it, for that is incredibly and naively anthropocentric. Just as dogs have very, very different ideas of what smells good than we do, so presumably the elephant artist will have a very different idea of what looks good. They don't even have the same visual perceptual apparatus as we do: their visual acuity is mediocre; they are dichromats, and what colour they do see is similar to what a human with red-green colour blindness would see.

One has to keep in mind that while we may find something so beautiful and aesthetically pleasing as to want to hang it on the wall, it may be considered repulsive by another species. Because of their aesthetic appeal I have downloaded and saved images of the many beautifully pigmented poisonous frogs that live in the tropics, but it's reasonable to assume that—unlike me—would-be predators find these frogs very 'ugly' because their gaudy colouration is a warning against stopping for a nibble.

Relevant, too, is the lesson to be learned from studies of face recognition in chimps, where initially their ability was judged to be poor. However, the faces presented were human faces, and subsequent studies have shown that even young chimps have excellent face recognition *for their own species*, just not for human faces (Myowa-Yamakoshi, et al, 2005). ("All dem damn humans look alike!") Even sheep have demonstrated finely tuned ability to recognize other individuals of their species, even in photographs (Kendrick et al, 2001).

So if, as I have argued, the one readily measurable indicator of aesthetic sense is aesthetic preference, the following (admittedly quirky) experiment could cast some light on the question of animal creativity and aesthetic sense.

Give a number of elephants (or chimps) paints and canvas, and have them create a number of paintings. Then have an accomplished human artist study these works, and then produce an equivalent number of paintings in the same 'style'. Finally collect a set of paintings by artists working in various styles—of course matched in terms of size and other superficial parameters.

Set up some standard preference paradigm such as has been used in empirical aesthetics research since its inception. Just one of many possible methodologies, widely used for studies of preferential-looking in neonatal humans and animals, is eye and head-tracking of moving stimuli.

Have the animal artists 'evaluate' the works, as operationally defined by, for example, viewing time, to the four stimulus categories: 1) their own productions; 2) other conspecific works; 3) the human stylistic imitations of their works; and 4) a random sample of human art works. Achieving any statistical significance between these four conditions in such a study would be interesting, would be of value in understanding aesthetic preference in non-human species, and could offer some modest supporting evidence for the proposition that abstract, apparently non-utilitarian aesthetic preference isn't uniquely human.

If, for example, a clear preference—or lack of interest in—was only shown for the subject's own productions this would suggest that memory and familiarity were the determining variables. If, as another possibility, a significant preference for both the subject's own work and other conspecific creations was demonstrated, this would seem to indicate some of degree of aesthetic 'judgment' based on that species' perceptual preferences. Of course, neither of these possible outcomes would conclusively demonstrate that the animal is a 'creative artist'! However, it would certainly offer some modest supporting evidence for the proposition that abstract, apparently non-utilitarian aesthetic preference isn't uniquely human. Even if the animal subjects showed a clear preference for the human works over their own creations, perhaps because of novelty, this would be suggestive of aesthetic discrimination.

Conclusion

There is no question that art is one of the greatest achievements of humankind and that no other species has accomplished anything approaching what we have wrought. No whale song approaches the complex beauty of a Bach fugue. No elephant's painting can be compared to the ceiling of the Sistine Chapel. Nevertheless, we exist along the continuum of evolution, and we should be cautious about assuming that huge differences in accomplishment mean that an evolutionary quantum leap has occurred. None of the great apes are ASL signing (or typing) complex periodic sentences, but there is evidence of primitive language ability. We are profoundly different from the other fauna on this planet, but we have repeatedly found that on a very basic level we have more in common with them than previously believed.

The question of whether there is any primitive aesthetic sense in other species, and thus some precursor of human artistic creativity, isn't a trivial one. The recent trend to promote and sell paintings by animals is not motivated by any sincere attempt to answer this question, nor does human evaluation of these works really contribute anything to our understanding. But it has had the positive effect of raising the question. The first meaningful step to a serious scientific investigation of this would be to determine the characteristics that are uniquely associated with an aesthetic sense and aesthetic appreciation, which is itself a worthwhile endeavour. The next logical second step would be to design ways of empirically testing for those characteristics.

References

Bandura, A. (1977). *Social learning theory.* Englewood Cliffs, NJ: Prentice Hall.

Bluff, L.A., Weir, A.A.S., Rutz, C., Wimpenny, J.H., & Kacelnik, A. (2007). Tool-related cognition in New Caledonian crows. *Comparative Cognition & Behaviour Reviews, 2*, 1-25.

Bozarth, M.A. (1994). Pleasure systems in the brain. In D.M. Warburton (Ed.), *Pleasure: The politics and the reality*. New York: John Wiley & Sons.

Coren, S. (1994). The intelligence of dogs: Canine consciousness and capabilities. New York: The Free Press/Macmillan.

Dawkins, R. (2006). *The god delusion.* New York: Houghton Mifflin.

Dennett, D. (1995). *Darwin's dangerous idea.* New York: Simon & Schuster.

Kendrick, K., da Costa, A., Leigh, A., Hinton, M., & Peirce, J. (2001). Sheep don't forget a face. *Nature, 414 (6860),* 165.

Laland, K.N. & Hoppitt, W. (2003). Do animals have culture? *Evolutionary Anthropology: Issues, News, and Reviews, 12(3)*, 150-159.

Myowa-Yamakoshi, M.,Yamaguchi, M.K., Tomonaga, M.,Tanaka, M., Matsuzawa, T. (2005). Development of face recognition in infant chimpanzees. *Cognitive Development*, *20(1),* 49-63.

Orians, G. H. and Heerwagen, J. H. (1992). Evolved responses to landscapes. In J. Barkow, L. Cosmides & J. Tooby (Eds.), *The adapted mind.* New York: Oxford University Press, 558-572.

Patterson, F. & Linden E. (1988). *The education of Koko.* New York: Henry Holt & Co.

Pinker, S. (1995). Language Acquisition. In L.R. Gleitman & M. Liberman (Eds.), *An invitation to cognitive science.* Cambridge, MA: MIT Press, 135-281.

Savage-Rumbaugh, E.S., Shanker, S.G., & Taylor T.J. (2001). *Apes, language, and the human mind.* New York: Oxford University Press.

Schmied, C., Waiblinger S., Scharl T., Leisch F., & Boivin, X. (2008). Stroking of different body regions by a human: Effects on behaviour and heart rate of dairy cows. *Applied Animal Behaviour Science, 109 (1)*, 25-38.

Searle, J. (1980). Minds, brains, and programs. *Behavioral and Brain Sciences*, *3*, 417-424.

Stange, K. (in prep.). Secret agents: Creativity in the arts and sciences.

Tolman, E. C. (1948). Cognitive maps in rats and men. *The Psychological Review, 55(4),* 189-208.

Turing, A.M. (1950). Computing machinery and intelligence. *A Quarterly Review of Psychology and Philosophy, 59(236)*, 433.

de Waal, F. B.M. (2007). *On the possibility of empathy in other animals.* Paper presented at Human Mind – Human Kind: Inter-Disciplinary Conference on Human Characteristics, University of Aarhus, Denmark.

Wasserman, E.A., & Zentall, T.R. (Eds.) (2006). *Comparative Cognition: Experimental Explorations of Animal Intelligence.* New York: Oxford University Press.

CHAPTER TWO

THE MOSAIC EVOLUTION OF AESTHETIC PREFERENCE.

MARCOS NADAL, MIQUEL CAPO, AVANTI PETERS, CAMILO JOSE CELA-CONDE

Abstract: Evolutionary psychology has provided a fruitful framework for the development of several hypotheses regarding the evolution of aesthetic preference. These accounts generally assume that aesthetic preference is the product of content-specific information-processing mechanisms that evolved in a certain environment to solve particular adaptive problems. In this study, we review results from the fields of empirical aesthetics, neuroimaging, and comparative neuroscience and discuss their implications for approaches to the evolution of aesthetic preference. This review suggests that aesthetic preference is the result of several cognitive and affective processes associated with activity in diverse brain regions, none of which seems to play an exclusive role in the aesthetic experience. Furthermore, we argue that aesthetic preference is the result of a kind of mosaic evolution; whereas some of these underlying processes and their neural correlates must have appeared at some point in the human lineage, others seem to have been inherited from earlier primate ancestors.

Introduction

Our closest living relatives, chimpanzees and gorillas, are capable of carrying out pictorial activities, such as applying paints on canvas (Lenain, 1997). However, this behaviour has only been observed in captive enculturated apes, never in wild animals living in their natural environment. Yet, aesthetic preference – the interest in the beauty of the object produced by oneself or another individual – seems to be completely out of reach of apes and other animals that have been trained to paint. Hence, the capacity to appreciate the beauty of certain forms, colours or objects is usually considered to have appeared at some point during the evolution of human beings, after the divergence of the human and chimpanzee lineages, close to seven million years ago. This discontinuity between human beings, who are capable of viewing the world in terms of beauty and ugliness, and all other animal species, which we assume to be incapable of perceiving the world in these terms, represents one of the greatest challenges to understanding the evolution of this phenomenon. In addition, the earliest archaeological remains that indisputably reveal the existence of aesthetic preference in our ancestors are very recent in evolutionary terms, and they are undoubtedly the product of fully-fledged "aesthetic minds".

The difficulties posed by the lack of solid material evidence pose considerable challenges to provide a sound explanation of the evolution of aesthetic preference. There are, though, at least three reasons that make this such an interesting topic. First, given that there is no obvious continuity with non-human animal behaviour, the capacity to appreciate beauty has traditionally been considered one of the features that identify the human species, distinguishing it from its closest living and extinct relatives. In this sense, it defines and demarcates the boundary of "what being human is". Second, aesthetic preference is not only related to the creation and admiration of some of the most extraordinary manifestations of human culture, such as art and architecture. It is also manifested in many of our everyday activities, such as choosing what clothes to wear, which car to buy, and how to decorate our homes. Finally, we will only have a limited understanding of the capacity to appreciate beauty, or any other cognitive phenomenon, until we have a clear picture of its phylogenetic dimension.

Hence, aesthetic preference – the capacity to value beauty – appears to be a unique phenomenon in the natural world that permeates the life of humans, from the ornamentation of one's own body to the creation of great works of art. In this study, we review some of the hypotheses that have been proposed to explain its evolution, and we point out some of their fundamental postulations. Thereafter we review results from psychological, neuroimaging, and comparative neuroscience studies that we believe can shed light on the evolution of aesthetic preference. Finally, we sketch a new proposal that is coherent with these results and compare this proposal with existing evolutionary approaches.

Evolutionary Psychology of Aesthetic Preference

By providing an explanatory mechanism of the evolution of physical and functional traits, Darwin (1859/1991) opened the door to the foundational ideas on the evolution of aesthetic preference. A good early example of this is Clay's early exploration of possible answers to the question, "Why should emotional response to harmony or rhythm as such, whether in sound, colour or form, have a survival value?" (Clay, 1908; p. 287). He believed that modern humans inherited the ability to appreciate beauty from their earlier ancestors, which were endowed with a more rudimentary sense of beauty. The main selective advantage conferred by this capacity was, according to Clay, the possibility of distinguishing suitable from unsuitable environments:

"It does not seem unreasonable to suppose that the instinctive pleasure in harmony is due to the impelling need for suitability to environment; and that to any organism the power of feeling the first sign however faint that it was out of touch with its surroundings, instead of merely proving it by living or dying, would have an inestimable value in the struggle for existence, so that such powers would be quickly increased and developed" (Clay, 1908; p. 288-9).

Grant Allen (1880) presented a different proposal, arguing that the starting point for the study of "the primitive source of the appreciation of beauty" (Allen, 1880; p. 30), should be the observation of other animals' behaviour. He noted that most mammals and birds show some sensitivity to natural beauty. However, this sensitivity is restricted in range to sounds made by their own species and the forms and colours of their decorated mates. Reformulating Darwin's (1874/1998) own vision, Allen (1880) suggested that when the first humans appeared, after the differentiation of their lineage, they possessed only an elementary sensibility for the beauty of form, symmetry, and colour. The fully conscious manifestation of this capacity would occur solely in relation to physical features of their conspecifics of the opposite sex. The expression of this sensibility in relation to objects, such as flowers, fruits, and feathers, would have been very limited. Only with the continuing evolution of human beings did this primitive conception of beauty broaden to include the sensibility for natural and cultural elements (Allen, 1880).

The virtual disappearance of Darwinian thinking in all fields of psychology and other social sciences at the beginning of the twentieth century (Plotkin, 2004), also brought studies into the evolution of aesthetic preference to a full stop. In fact, researchers only resumed researching the topic when Wilson's (1975) Sociobiology and Eibl-Eibesfeldt's (1988, 1989) Human Ethology were introduced. However, the work carried out during the past two decades by evolutionary psychologists has proven to be the most fruitful, thought-provoking and recognized. Building on the early work, a large number of the proposals presented by these authors stressed the adaptive value of recognizing suitable environments or mates (Smith, 2005).

Orians (2001) argued that aesthetic experiences have been moulded by natural selection through the adaptive advantages conferred by emotional responses during decision-making and problem-solving: "Survival is enhanced by making better decisions about avoiding environmental hazards, where to find and how to choose food, places in which to live, and associates for various activities, including reproduction" (Orians, 2001; p. 25). Current aesthetic experiences involve emotional responses that have been shaped for generations because of the advantages they conferred in determining which elements in the environment require attention and the appropriate responses to them. From this perspective, natural selection would have endowed humans with a series of specific adaptations to assess the suitability of certain landscape features, animals, fruits, or natural indicators of the necessity to modify certain behaviours. These mechanisms would aid in the solution of various adaptive problems, such as the identification of safe and resource-rich places in which to settle, animals that may constitute a source of food or danger, fruits and other nutrient-rich vegetables, or the first indicators of rain, the sunset or the sunrise (Sánchez, 2005).

Given that our Pleistocene ancestors were hunter-gatherers living in groups that moved and resettled frequently through savanna-like landscapes, it is assumed that habitat selection was especially important to them: "Our aesthetic reactions to landscapes may have derived, in part, from an evolved psychology that functioned to help hunter-gatherers make better decisions about where to move, where to settle, and what activities to follow in various localities" (Orians & Heerwagen, 1992; p. 557). Kaplan (1992) argued that selective pressures in early humans favoured exploration and gathering information about the environment while not moving too far from what was familiar, and emphasized that selection must have moulded this capability to assess environments to be fast and unconscious. As noted by Kaplan (1987; p. 25), the preference for natural sceneries, environments, and landscapes is not a special case of aesthetics, but more likely the basis on which some of the more traditional aesthetic domains are anchored.

Alternatively, Darwin (1874/1998) and Allen's (1880) notion that the evolution of the appreciation of beauty is related with mate choice was recently reintroduced and updated by Miller (2001). He views art and aesthetics as biological adaptations acquired throughout human evolution:

"It evolved through sexual selection to serve the same courtship functions as almost all other examples of organic beauty and complex behavioural signals observable in nature. Such ornamentation often evolves as a reliable, costly indicator of the signaller's good health, good brain, and good genes. This leads to the further proposal that many design features of art function as indicators of the artist's virtuosity, creativity, intelligence, conscientiousness, and other important heritable mental and physical traits. This 'aesthetic fitness' view suggests that aesthetic judgment is a natural part of mate choice and social cognition, in which an art-work is viewed as the extended phenotype of the artist" (Miller, 2001).

He suggests that our aesthetic preferences evolved favouring works of art that could only have been created by high-fitness artists. Thus, we are inclined to consider people who are able to produce high qualitative work as attractive due to our evolved preferences for what is difficult, rare, skilful and costly. This serves as an indicator of such qualities as health, energy, creativity, access to rare materials, good learning abilities, intelligence and coordination, among others (Miller, 2001). From this point of view, evolution shaped our aesthetic preference to distinguish difficult from easy, rare from common, skilful from careless, and costly from cheap, by means of a domain-specific adaptive mechanism.

Although these approaches posit diverse selective advantages driving the evolution of aesthetic preference, their notions of aesthetic preference and its evolution exhibit certain common features. These features are most clear in approaches framed within what has been called narrow evolutionary psychology (see Bechtel, 2002), which follows the reasoning and methods as developed by Tooby and Cosmides (1992). Whether aesthetic preference is considered to have provided greater advantages in recognizing suitable environments or suitable mates, it is usually considered to be a single integral capacity. It has been conceived as a distinct mechanism that evolved because it solved a specific adaptive problem. As Kogan (1994) states, "Over evolutionary time, we should have genetically acquired a set of beauty detectors as an evolved psychological mechanism" (Kogan, 1994; p. 144). In fact, most evolutionary psychologists favour some kind of modular conception of cognitive processes: "From an evolutionary perspective, the human cognitive architecture is far more likely to resemble a confederation of hundreds or thousands of functionally dedicated computers, designed to solve problems endemic to the Pleistocene, than it is to resemble a single general-purpose computer equipped with a small number of general-purpose procedures such as association formation, categorization, or production-rule formation" (Tooby & Cosmides, 1995; p. 1189). Therefore, aesthetic preference is conceived as a distinct cognitive mechanism dedicated to the performance of a specific function. It is fair to say that this represents the views of most evolutionary approaches to aesthetic preference, though they may vary in the way they characterize modules (Barrett & Kurzban, 2006).

A second common assumption is to consider that aesthetic preference, together with many other cognitive mechanisms, appeared in response to the Pleistocene savanna-like environments of our hunter-gatherer ancestors. This kind of environment is sometimes generically referred to as the environment of evolutionary adaptedness. The comprehension of the evolution of aesthetic preference requires examining the adaptive problems posed by this kind of environment: "Thus, to map the structures of our cognitive devices, we need to understand the structures of the problems they solve, and the problem-relevant parts of the hunter-gatherer world" (Tooby & Cosmides, 1995; p. 1194). Hence, "If art evolved in our lineage over the last one or two million years, there is little reason to expect proto-art abilities in living non-human primates such as chimpanzees, which split off from us at least 5 million years ago" (Miller, 2001).

Tooby and Cosmides (2001) summarized the general view of evolutionary psychologists with great clarity: "Our species-typical neural architecture is equipped with motivational and cognitive programs that appear to be specially designed to input fictional experiences and engage in other artistic activities" (Tooby & Cosmides, 2001; p. 10). Although they consider the possibility that aesthetic preference, together with other aesthetic and artistic capacities, is an accidental result of brain and cognitive mechanisms adapted to perform different functions, they argued that aesthetic preference was directly selected for the specific functions and adaptive value this capacity conferred our hunter-gatherer ancestors: "We expect that humans have evolved motivational systems (or systems of aesthetic preference) that are designed to find rewarding the kinds of actions and experiences that would have been adaptive for our ancestors".

In contrast to this view of the evolution of aesthetic preference, we will review literature that suggests that aesthetic preference is best understood as the result of interaction among multiple component processes, none of which is exclusively dedicated. Furthermore, we will argue that these components seem to have different evolutionary histories. While some seem to be the result of evolutionary episodes which occurred throughout human evolution, others seem to be inherited from distant primate ancestors. We believe that the human capacity to appreciate beauty cannot be explained by a single adaptive function or selective pressure. On the contrary, we will conclude that aesthetic preference is the result of a mosaic of evolutionary processes. Multiple selective pressures influenced different component processes, conferring diverse selective advantages. It is possible, moreover, that this might have occurred at various moments throughout our species' evolution, both before and after humans and chimpanzees split. It is not unconceivable that humans share some of the cognitive and neural underpinnings of aesthetic preference with other primates, thus indicating it may in fact predate humans themselves.

Cognitive and Affective Processes Underlying Aesthetic Preference

Although the idea that some cognitive and neural mechanisms underlying human capacities, such as aesthetic preference, may be shared with non-human primates initially seemed far-fetched, it is currently being explored in language and morality. Some of the cognitive processes involved in language comprehension and acquisition, presumed to be specifically human traits, have also been identified in monkeys (for reviews of this research see Tincoff & Hauser, 2005; Weiss & Newport, 2006). Similarly, Flack and de Waal's (2000) division of human morality into four building blocks (sympathy related traits, norm related traits, reciprocity, and getting along) allowed them to identify their possible evolutionary roots in our primate relatives. This suggests that not all the constituent cognitive operations subservient to human morality and language appeared after the human and chimpanzee lineages diverged. On the contrary, it implies that they appeared long before humans, and that human language and morality evolved, at least in part, by recruiting pre-existing building blocks.

In this section we will review recent psychological models of aesthetic preference based on experimental data, together with neuroimaging studies, which suggest that conceiving aesthetic preference as a cognitive process, a computational programme, or a single mechanism, is a misleading over-simplification. To consider language, moral reasoning, or aesthetic preference, as single and unitary cognitive processes may suggest that each of these cognitive faculties constitutes a single and distinct piece of computing machinery, resulting from one or very few evolutionary episodes and selective pressures. However, as we just mentioned, viewing cognitive mechanisms as the result of the modification and novel combination of previously existing subcomponents has proved very fruitful to understanding their structure and evolution (Marcus, 2004). Evolutionary approaches to human behaviour and cognition must not lose sight of the fact that admiring the beauty of a sculpture or a painting, or creating a piece of art, are the result of the interplay of different cognitive processes, probably none of which are exclusive to the task. This has been highlighted by recent models of aesthetic experience that integrated the results of numerous studies that explored the influences of different factors on aesthetic preference (Chatterjee, 2003; Leder, Belke, Oeberst, & Augustin, 2004).

Leder and colleagues' (2004) model proposes that cognitive processes involved in aesthetic preference occur in five stages. The first of these stages includes processes related to perceptual analysis, such as organization, grouping, symmetry analysis, as well as the processes related to complexity and other perceptual variables that are known to affect aesthetic preference. The analysis of the stimuli on the grounds of familiarity, prototypicality and meaning is performed in the second stage, related to the implicit and automatic integration of information with pre-existing memory structures. An explicit classification is performed in the third phase, which includes cognitive operations related to the style and the content of the stimulus. This is followed by the specific art- and self-related interpretations, under the cognitive mastering stage. Finally, the model produces two different outputs: A cognitive state, the outcome of earlier cognitive stages, and an affective state, a result of continuous interactions between the aforementioned processes and diverse affective mechanisms. The cognitive state is the grounds for the aesthetic judgment, while the aesthetic emotion is grounded on the affective state.

Chatterjee's (2003) proposal presents a slightly different view. He suggests that, in the case of visual arts, during the initial stage, early visual processes divide the stimulus into simple components, such as colour and form, which are extracted and analyzed in different brain regions. In a subsequent stage, intermediate visual processes group certain elements and segregate others to form coherent representations. In late visual stages, certain regions of the stimulus are selected for processing in greater detail. At this moment, information stored in memory becomes active, objects are recognized and associated with their meanings. This visual analysis elicits emotions associated with the aesthetic experience and provides the foundations to formulate aesthetic judgment. This model also includes feedback of information, via attentional processes, from late visual levels and affective systems to early visual processing stages.

Whereas Chatterjee's (2003) proposal can be considered a neuroscientific model of aesthetic preference for a broad range of visual objects, Leder and colleagues' (2004) proposal was conceived as an information-processing model of aesthetic judgment of visual works of art (Vartanian & Nadal, 2007). Despite these conceptual differences, both models consider that aesthetic preference involves diverse cognitive and affective mechanisms. They both acknowledge the importance of early and late visual processes in the generation of an emotional response and arriving at a decision. These models also consider the influence of complexity, order, grouping and many other variables familiar to experimental aestheticians, as well as the interaction between affective and cognitive processes such as the activation of memories and the search for the stimulus' meaning. Additionally, both models suggest two different outputs: An emotional response or aesthetic emotion versus a decision or aesthetic judgment. Hence, it is currently very clear that this capacity relies on a variety of cognitive and affective processes. Some of these are related to perceptual analysis, others to recognition and related mnemonic processes, and some to decision-making, or analysis of the affective value of the stimulus.

With the increase of our understanding of these constituent processes, together with advances in the methodology and precision of neuroimaging techniques, researchers have become interested in describing the underlying neural mechanisms related to cognitive and affective processes involved in aesthetic preference. There are currently four published neuroimaging studies dealing with this issue. Kawabata and Zeki (2004) used functional magnetic resonance imaging (fMRI) to register the brain activity in participants while they rated the beauty of a series of visual stimuli. The results of this study revealed that the activity in the orbitofrontal cortex was greater for stimuli classified as beautiful, whereas activity in the motor cortex was greater for stimuli rated as ugly. Also by means of fMRI, Vartanian and Goel (2004) found that activity in the caudate nucleus, the anterior cingulate gyrus and occipital gyri increased with the preference of the participants for the presented stimuli. Jacobsen, Schubotz, Höfel, and von Cramon (2006), who used the same neuroimaging technique, identified a relation between aesthetic preference and neural activity in the temporal pole, the lateral prefrontal cortex and the frontal pole. Finally, using magnetoencephalography (MEG), Cela-Conde, Marty, Maestú, Ortiz, Munar, Fernández *et al.* (2004) found an increase in the activity of the left dorsolateral prefrontal cortex between 400 and 1000 milliseconds after the presentation of stimuli rated as beautiful by their participants.

It might seem surprising that four studies designed with the same objective – to identify the neural correlates of aesthetic preference– would lead to such diverse results. However, it was not claimed in any of the studies that the identified regions acted in isolation. In fact, the divergence in their results represents a clear testimony to the true complexity of processes underlying aesthetic preference. This capacity actually relies on the coordinated function of multiple cognitive and affective systems. We have previously suggested that certain aspects of the experimental designs and procedures might have led each of the four studies to register only a partial image of the complex underpinnings of aesthetic preference (Nadal, Munar, Capó, Rosselló, & Cela-Conde, 2008). Based on the results produced in the four studies and current knowledge of the neural correlates of cognitive and affective processes, we suggested that affective processes involved in aesthetic preference seem to be mediated by certain regions of the orbitofrontal cortex (Kawabata & Zeki, 2004), the caudate nucleus, and the anterior cingulate cortex (Vartanian & Goel, 2004). The recognition of the visual stimulus and the attribution of meaning in aesthetic preference seem to be related to activity in the temporal pole (Jacobsen *et al.*, 2006). The actual decisions required by the experimental settings are apparently mediated by the lateral prefrontal cortex and the frontal pole (Cela-Conde *et al.*, 2004; Jacobsen *et al.*, 2006). Finally, there is evidence of top-down enhancement of early visual processes carried out in the occipital cortex (Vartanian & Goel, 2004).

Clearly, psychological models of aesthetic preference that account for a broad range of experimental findings and neuroimaging studies converge on the notion that aesthetic preference is not a single cognitive process, nor does it rely on a single, unitary mechanism. They rather indicate that it is the result of several cognitive and affective processes related to different aspects of the stimulus. Moreover, neither psychological experiments nor neuroimaging studies have uncovered evidence for cognitive or neural mechanisms especially devoted to, or exclusively involved in aesthetic preference.

The Evolution of the Neural Correlates of Aesthetic Preference

Neuroimaging experiments have revealed that a broad network of brain regions support the multiple cognitive and affective processes that are involved in aesthetic preference. It is very possible that not all the neural structures involved in aesthetic preference, and the functions they perform, have undergone the same degree of transformation since the appearance of the human lineage. We have contended that together with the comparative method, knowledge of the neural correlates of aesthetic preference can be used to determine whether brain regions involved in aesthetic preference show any kind of distinctive features in humans, or whether, on the contrary, they remained unchanged since the divergence of our species from its closest living relatives (Nadal, Capó, Munar, Marty, & Cela-Conde, in press).

The comparative study of species provides a means to clarify the phylogenetic relations among them and to shed light on the evolutionary history of the traits that characterize them. The comparative method relies on two fundamental methodological tools: The distinction between homology and homoplasy, and the principle of parsimony. A trait is said to be homologous when it appears in two species because it was inherited from a common ancestor, such as the five digits in different species of mammals. Conversely, a homoplasy is a trait exhibited by two species as the result of independent evolutionary processes, like the wings of sparrows and bats. The principle of parsimony is primarily used as a means to organize evolutionary lineages of closely related species, especially in reference to traits that are not reflected in fossil remains. This principle states that if a trait appears in two closely related species, such as human beings and chimpanzees, or human beings and Old World monkeys, it can be assumed to be a homology. That is to say, the presence of the same character in these species is more likely to be the result of the inheritance from a common ancestor than of its independent evolution in both species. Implementing the principle of parsimony, in this section we will review findings from comparisons between human and non-human primates' brain structures and processes shown to be involved in aesthetic preference by the aforementioned neuroimaging experiments. We need to point out a caveat before we proceed. As Sejnowski and Churchland (1989) have noted, the brain is organized in several hierarchical levels, including systems, maps, networks, individual neurons, synapses and molecules. As with any other cognitive operation, there is no way of determining which level of analysis is the most relevant to the study of aesthetic preference. Furthermore, the evolution of such a capacity may result from alterations in any set of these levels. However, available knowledge of the neural underpinnings of aesthetic preference is limited to the level of systems. The little knowledge we have about human brain evolution makes it difficult to present meaningful hypotheses about modifications at most of the other levels. Hence, our analysis will be restricted to the higher levels in the organizational hierarchy of the brain.

Visual Processing System

As we saw above, the study performed by Vartanian and Goel (2004) revealed that visual processing activity in the occipital gyri was greater when participants awarded high preference ratings. Several studies have shown that throughout human evolution, brain regions involved in visual processing have expanded to a lesser degree than the whole brain has (Rilling, 2006; Schoenemann, 2006), suggesting a trade-off towards cognitive processes which are not directly bound to sensory information. Turning to a finer level of analysis, the retinotopic organization and functions of brain areas involved in early visual processing, known as V1 and V2, are conserved to a great extent in human beings (Orban, Van Essen, & Vanduffel, 2004). However, certain aspects of the human area V1 are derived. For instance, Preuss and Coleman (2002) found that humans differ from monkeys in certain features related to the cortical representation of the magnocellular system. It seems that some of these features occurred initially in the common ancestors of African apes and humans, while others appeared to have evolved exclusively along our lineage. Given that the magnocellular visual processing stream is involved in the analysis of luminance contrasts, movement, perspective, the relative size of objects and depth perception, it is reasonable to assume that the changes that appeared in this stream had an impact on these perceptual processes.

Whereas homology in monkeys and humans is relatively easy to study in primary visual areas, this becomes increasingly difficult as one moves to higher levels of the visual system. Various studies reviewed by Orban and colleagues (2004) suggest that the ventral and dorsal visual information streams have been transformed to different extents throughout human evolution. Specifically, the areas that constitute the ventral stream, related to the representation and categorization of objects, have experienced a smaller expansion than those that are part of the dorsal stream, involved in the representation of space and the analysis of visual information to organize action (Orban *et al.*, 2004). Barton (2006) noted that the fact that parietal areas of the dorsal stream receive information only from the magnocellular system, supports the aforementioned notion that cortical representations of information fed by the magnocellular system have been the target of special modifications throughout human evolution. The studies that have shown brain activity in homologous brain regions in humans and monkeys during the perception of symmetry (Sasaki, Vanduffel, Knutsen, Tyler, & Tootell, 2005), the representation of visual objects (Munakata, Santos, Spelke, Hauser, & O'Reilly, 2001), and their classification (Sigala, Gabbiani, & Logothetis, 2002), support the idea that the ventral visual processing stream is relatively conserved in human beings.

Temporal Poles

Jacobsen and colleagues' (2006) neuroimaging study mentioned above revealed that the task of rating the beauty of geometric visual stimuli is related to greater activity in the left temporal pole than the task of rating other visual features. The authors suggested that the function of this region was to generate an affective and semantic context, based on previous experiences, with which to frame decisions on the beauty of visual stimuli.

Rilling and Seligman (2002) compared diverse aspects of the temporal lobes of a broad sample of primates, including humans. Their results revealed that the temporal lobe has grown in surface, volume and white matter proportion throughout human evolution. This suggests that, to a certain extent, connectivity patterns have been reorganized since the appearance of the human lineage (Schenker, Desgouttes, & Semendeferi, 2005). Rilling (2006; Rilling & Seligman, 2002) has suggested that this reorganization, which is evident mainly in the left hemisphere, could be related to the appearance and expansion of regions of the temporal pole supporting linguistic functions. Specifically, the ventral visual processing stream seems to have shifted further ventrally to allow the expansion of speech and language related areas on the lateral surface of the temporal lobe, which seem to bear a close phylogenetic relation to those underlying the processing of species-specific calls in monkeys (Gil-da-Costa & Hauser, 2006).

In relation specifically to the temporal pole itself, there is evidence indicating that most of the functions it supports in our species are primitive. It is known that in humans this region is involved in the use of prior experiences to generate a semantic and affective context, which enhances the interpretation of the information being processed. Studies in comparative neuroanatomy have provided evidence that this region performs similar functions in other primates. Kondo, Saleem and Price (2003) showed that the temporal pole of monkeys is richly connected to orbital and medial prefrontal cortical nets, suggesting its involvement in the integration of affective, mnemonic and sensory information. Furthermore, the results of Croxson, Johansen-Berg, Behrens, Robson, Pinsk, Gross *et al.* (2005) revealed that the connectivity patterns between temporal and prefrontal cortices in humans and macaques are very similar.

Finally, it has been shown that the temporal pole plays an important role in object recognition. Lesions to this region especially impair the recognition and recall of specific entities, and familiar objects and faces (Nakamura & Kubota, 1996). Experimental results support the notion that this function might constitute a homology in monkeys and humans. Vogels (1999), for instance, demonstrated that neurons in the anterior temporal cortex of monkeys are involved in the processing of visual information related to objects, and that they are sensitive to the presentation of exemplars of a learned category. Likewise, other studies reviewed by Nakamura and Kubota (1996) suggest that lesions to the monkey temporal pole can lead to deficits in the recognition of the experimenter's gloves, food or live snakes, but not in the discrimination of unfamiliar objects or patterns.

Dorsolateral Prefrontal Cortex

The studies performed by Cela-Conde *et al.* (2004) and Jacobsen *et al.* (2006) revealed an increase in the lateral prefrontal cortex while participants rated the beauty of paintings, photographs and geometric designs. As mentioned above, this activity could be related to the actual decision about the beauty of visual stimuli presented in both studies.

Contrary to common assumptions about the prefrontal cortex, and specifically its lateral region, there is evidence that to a great extent its organization is functionally and cytoarchitectonically conserved in humans. Petrides and Pandya (1999) compared the connectivity patterns and cytoarchitectonic organization of the dorsolateral prefrontal cortex, which encompasses Brodmann Areas (BA) 8, 9, and 46, of macaques and humans. Their study did not reveal any new cytoarchitectonic areas in the human brain. In fact, these areas exhibited very similar features in human and macaque brains, whereby in both species the same traits could be used to distinguish the areas within this region. The degree of homology of neural architecture of this region is such that the same subdivisions of the areas can be found in both species (8Av, 8Ad, 8B, 9/46d and 9/46v). The comparative analysis of the ventrolateral prefrontal cortex of humans and macaques performed by Petrides and Pandya (2001) revealed a similar image of BA 47/12 and BA 45. The cytoarchitectonic criteria used to identify both areas in monkeys and humans were the same, as well as those used to differentiate them from dorsolateral cortex areas. Once again, even the finer subdivisions of area 45 (45A and 45B) were identifiable in both species.

Although the brain organization of monkeys has been studied mainly by means of lesion studies and that of humans by means of neuroimaging studies, these two species also present functional homologies related to the lateral prefrontal cortex. As noted by Petrides (2005), the lateral cortex of humans and monkeys is functionally organized along a caudal-rostral axis and a dorsal-ventral axis. Specifically, in both species the caudal region of the prefrontal cortex (BA 8) contributes to the flexibility of attentional shifts between stimuli and the selection of responses depending on learned conditional rules. At the rostral end of this axis, the mid-dorsolateral cortex (BA 46 and BA 9/46) is involved in working memory tasks that require monitoring the selection of stimuli or the occurrence of expected events. The contribution of the mid-ventrolateral prefrontal cortex seems to be important for other executive functions, including the selection and comparison of representations of stimuli stored in short-term and long-term memory, as well as judgments based on them (Petrides, 2005). Given that neuroimaging studies carried out on human participants and lesion studies on monkeys converge in their results, it seems adequate to consider that the functional and cytoarchitectonic of the human lateral cortex is a primitive trait.

However, it is true that certain aspects of the lateral prefrontal cortex distinguish humans from other primates. For instance, it is obvious that the involvement of these functions in certain human cognitive abilities, such as language, or even aesthetic preference, does not occur in non-human primates. Second, the kind of information on which these functions are performed also seems to differ. This was demonstrated by Denys, Vanduffel, Fize, Nelissen, Sawamura, Georgieva *et al.* (2004), who utilized fMRI to show that activity in the prefrontal cortex while monkey and human participants viewed visual objects was much stronger in the former than in the latter. The authors interpreted this finding as evidence of the multisensory nature of the information reaching the prefrontal cortex of humans, which contrasts with the primarily visual information that reaches the prefrontal cortex of monkeys.

Anterior Cingulate Cortex

The studies carried out by Vartanian and Goel (2004) and Jacobsen *et al.* (2006) identified activity in this region while participants rated visual stimuli as beautiful or expressed high preference for them. Above, we suggested that the involvement of this brain region in aesthetic preference could be related to the awareness of the affective state induced by aesthetically pleasant stimuli. Although the anterior cingulate cortex is cytologically distinguishable from the posterior cingulate cortex in both humans and monkeys, other conspicuous differences exist. The most obvious of these is the presence of two areas in humans (BA 33 and BA 32') that are not found in other primates (Amiez, Joseph, & Procyk, 2005; Vogt, Nimchinsky, Vogt, & Hof, 1995).

Additionally, Nimchinsky, Gilissen, Allman, Perl, Erwin and Hof (1999) have shown that the anterior cingulate cortex of great apes and humans contains a type of neurons, known as spindle cells, that have not been found in other mammals, including other primate species. In humans these neurons constitute 5.6% of pyramidal cells of transverse sections of layer Vb, and appear in clusters of between 3 and 6. Among other hominoids, a similar trend is observable in bonobos. Conversely, the relative abundance of these neurons is smaller in common chimpanzees, gorillas and orangutans, and exhibit little or no clustering. These results indicate that a new kind of neuron appeared in the anterior cingulate cortex at an early point in the hominoid lineage. The abundance of these neurons gradually increased throughout evolution, and they began clustering together. Nimchinsky and colleagues (1999) hypothesized that the main function of these neurons is to integrate affective information and transmit it to motor brain regions related with vocalization, facial expression or autonomic functions. Allman, Hakeem and Watson (2002) suggested that the increase in spindle cell proportion could be related to enhancements of emotional stability and self-control. Furthermore, they argued that together with the increase of the anterior frontal cortex, abundant spindle cells played a relevant role in the management of the requirements related to the growth of family size throughout human evolution.

Orbitofrontal Cortex

Kawabata and Zeki (2004) identified greater activity in the orbitofrontal cortex when their participants rated the stimuli as beautiful than when they rated them as ugly. A large number of studies have reported activity in this region when participants were delivered primary and abstract rewards, which suggests that its role in aesthetic preference could be to represent the reward value of visual stimuli.

The comparison of the orbitofrontal cortex of a considerable amount of macaques and humans revealed that the pattern of sulci and convolutions was similar in both species (Chiavaras & Petrides, 2000), though the human patterns were more intricate and variable than the monkey patterns. Despite this conservation in the sulcal pattern of the orbitofrontal cortex, and other general morphological and cytological similarities, which led Semendeferi, Armstrong, Schleicher, Zilles and Van Hoesen (1998) to consider that the state of BA 13 in humans was primitive, certain features of this region distinguish humans from other apes. For instance, in humans and bonobos BA 13 is relatively smaller than in other apes. Together with other aspects, this suggests that the number of orbitofrontal cytoarchitectonic subareas has increased throughout human evolution. Additionally, the cellular density of this area in humans is the lowest among all hominoids and, together with gibbons, they have the lowest grey matter index. This means that, relatively speaking, there is a greater space occupied by axons and dendrites (Semendeferi *et al.*, 1998).

Rolls' (2004) revision of the functions of the orbitofrontal cortex of primates revealed that this region is functionally conserved in humans, in the sense that, as is the case with monkeys, it includes representations of smell, taste, food texture, visual information received from the ventral stream, as well as necessary information for facial recognition. In primates this information is used to identify the stimuli that are being processed and to establish their reward value. Furthermore, both in humans and monkeys the orbitofrontal cortex is a crucial element in learning associations between stimuli and rewards, and modifying them when contingencies vary.

Frontal Pole

Jacobsen and colleagues' (2006) results showed that while participants rated the beauty of geometric visual stimuli, activity in the frontal pole was greater than when they rated the symmetry of the same stimuli. Previous studies have found activity in this brain region during the performance of evaluative judgments of a broad range of visual materials.

Petrides and Pandya (1999) compared the macaque and human BA 10, located in the frontal pole. This study revealed that the architectonic features that distinguish this area from the surrounding ones are the same in both species. This means that the kinds of neurons in this region, as well as their distribution in cortical layers, have varied little throughout the hominoid and human lineages.

Semendeferi, Armstrong, Schleicher, Zilles and Van Hoesen (2001) performed a qualitative and quantitative study of BA 10, comparing data taken from macaque, gibbon, orangutan, gorilla, chimpanzee, bonobo and human brains. Their results confirm that, with the exception of gorillas, BA 10 is easily identified in the frontal pole of all hominoid species. Yet, there are certain traits that distinguish BA 10 in human beings from other hominoids. First, its size is larger, both in relative and absolute terms, Second, although humans exhibit the largest absolute amount of neurons, neural density in this region is the lowest among all hominoids. This affords a greater space for connections with cells from the same and different areas, especially, as observed by Semendeferi *et al.* (2001), other association areas.

Summary

Our review has revealed that some areas shown by neuroimaging studies to be involved in aesthetic preference are relatively conserved in humans, while others exhibit a number of derived features. Occipital visual areas, whose activity during aesthetic preference has been interpreted as the correlate of emotional or attentional engagement, show a mosaic of novel and primitive features. Whereas areas supporting early visual processing seem to be largely conserved, those involved in later stages seem to have undergone modifications to a larger extent. Specifically, the processing of spatial information and the organization of visual information to guide action seems to have been enhanced during human evolution, rather than object-centered visual analyses.

Activity within the temporal pole has been related to the creation of a mnemonic and emotional context for aesthetic preference. Our review has revealed that this region performs very similar functions in monkeys and humans. These functions are essential for the categorization and recognition of familiar objects and the integration of emotion, memory, and sensory information.

The orbitofrontal cortex presumably supports the representation of reward value of visual stimuli during aesthetic judgment tasks. It seems that to a large extent its sulcal pattern, cytoarchitecture, and functions, are conserved in the human brain. The only derived features appear to be an enlargement of area 10 and a reduction in neural density. A similar picture emerges after reviewing the comparative literature on the frontal pole, involved in the decision-making stage of aesthetic preference: A relative enlargement and reduction of the density of neurons. There is also a great cytoarchitectonic similarity between humans and monkeys in the other regions shown to be involved in decisions about the beauty of visual stimuli, the mid-dorsolateral and mid-ventrolateral cortex. In addition to a considerable enlargement during the evolution of our species, the main difference between monkeys and humans is that the dorsolateral regions of the latter seem to receive multisensory information, rather than mainly visual ones. Overall, the complexity of prefrontal cortex connectivity patterns seems to have increased after our lineage split from chimpanzees some 7-8 million years ago. Finally, our review of comparative work on the anterior cingulate cortex, which probably plays a role in the conscious awareness of the affective state during aesthetic preference, has revealed two major modifications in cytoarchitecture: The appearance of two novel cytoarchitectonic areas, and the clustering of a kind of neurons unique to great apes and humans.

Conclusions

The studies we have reviewed in the present work afford a series of considerations we feel should be taken into account by evolutionary approaches to aesthetic preference. First, we have argued that psychological research, summarized and accounted for by two recent models (Chatterjee, 2003; Leder *et al.*, 2004), and neuroimaging experiments suggest that it is misleading to conceive aesthetic preference as a single cognitive process or mechanism. These studies have shown that aesthetic preference is supported by the performance of multiple cognitive processes, related to perception, recognition, meaning, attention and decision-making, as well as affective processes that interact with the former. Second, an enormous body of literature shows that each of these processes plays an important role in the performance of several other tasks. None of the cognitive or affective processes that psychologists have found to be constituent elements of aesthetic preference seem to be exclusively involved in this capacity, and the same can be said of their neural correlates.

Third, our review of the comparative literature has shown that brain regions whose activity underlies aesthetic preference tasks do not share a common evolutionary pattern. In fact, the functional and cytoarchitectonic organization of some of these regions seem to have undergone little variation since the human and chimpanzee lineages split, whereas others exhibit conspicuous modifications. We believe that evolutionary approaches to aesthetic preference should be able to account for the evolutionary history of the neural correlates of such capacity. Specifically, these approaches should take into account that during the evolution of our lineage the following aspects seem to have varied little: (i) the substrates of early visual processes; (ii) the ventral visual processing stream and its involvement in the recognition and classification of objects; (iii) the cytoarchitectonic organization and connectivity of the temporal pole and its involvement in the generation of an affective and mnemonic context; (iv) the organization of the lateral, orbital and anterior prefrontal cortex, both at an anatomical and functional level. Additionally, evolutionary accounts of aesthetic preference should account for the relation between the appearance of this capacity and the following cerebral changes that occurred throughout human evolution: (i) The development of the dorsal processing stream, related to an enhanced analysis of spatial relations; (ii) the increase in the variety of sensory information reaching and processed by the prefrontal cortex; (iii) the increase in the connectivity among prefrontal cortical regions and between these and other brain regions; (iv) the appearance of novel cytoarchitectonic areas and a new kind of neurons in the anterior cingulate cortex, related to the awareness of one's own affective states.

These considerations suggest that some of the mechanisms that support the human capacity to appreciate beauty, those related to the visual representation of objects, the representation of their reward value and certain executive functions, were already present in our primate ancestors millions of years before the emergence of our species. Our appreciation of beauty, however, also required certain brain modifications that occurred throughout human evolution, presumably related to a greater integration of sensory information, an enrichment of spatial analyses, and an enhancement of the awareness of affective states. Aesthetic preference emerged in human beings, thus, by virtue of a kind of mosaic evolution, resulting from the integration of these derived processes and structures with the pre-existing primitive ones. In contrast to Tooby and Cosmides' (2001) position, we believe that it is not necessary to argue that the changes that afforded human beings the capacity to appreciate beauty were selected precisely for that motive. In fact, given that the brain regions identified by neuroimaging studies, as well as the cognitive and affective processes they support, do not participate exclusively in aesthetic preference, we believe they are the result of multiple selective pressures and that they must have conferred adaptive advantages in diverse spheres of human experience.

This view of the evolution of aesthetic preference, compatible with what is known about the cognitive and affective processes involved in aesthetic preference, their neural correlates and the comparative study of the human brain, is difficult to reconcile with two of the features that characterize approaches framed within narrow evolutionary psychology, as used in Bechtel's (2002) work: The view of aesthetic preference as a separate module or cognitive programme, and the notion that it was selected for in our Pleistocene hunter-gatherer ancestors. As mentioned above, evolutionary psychology in a narrow sense usually considers that aesthetic preference is grounded on a sort of beauty detector, a kind of module or autonomous programme, which performs a specific function to solve a specific adaptive problem. However, as noted by Shapiro and Epstein (1998), evolutionary approaches to human cognition do not necessarily require the adoption of this specific modular conception of the mind. They argued that it is misleading to identify cognitive processes with the tasks or objectives they are used for, as evolutionary psychologists do. Natural selection does not necessarily select different cognitive processes to resolve different adaptive problems. It is much more probable that diverse cognitive processes are involved in the solution of each of the adaptive problems pointed out by evolutionary psychologists, and that each of those processes would be useful to solve various adaptive problems. Bechtel (2002) and Atkinson and Wheeler (2004) argued that neuroscientifically sound modules do not refer to tasks such as cheater detection, mate selection, or the detection of beauty, but to finer grain sized information processing operations. These processes can be involved in a multitude of adaptively relevant tasks. In relation to aesthetic preference, there is currently little doubt among psychologists and neuroscientists that the best way to conceive aesthetic preference is as the result of a varied set of processes, from those related to the perception of the most basic features to those underlying deliberation and decision making.

Finally, it is usually considered that in order to explain the origin and evolution of human cognitive processes it is sufficient to understand the problems faced by our Pleistocene hunter-gatherer ancestors (Tooby & Cosmides, 1992). This assumption faces at least two main problems. First, during the Pleistocene period our ancestors lived in a variety of environments, not merely savannas (Foley, 1995). In fact, the Pleistocene period saw the expansion of humans in many parts of the world, from the African plains to the woodlands of Southeast Asia and the sub-glacial valleys of northern Asia and Europe. It is difficult to see that these environments share a stable common set of selective pressures, and that these were not present, say for instance, in the environments of our Pliocene ancestors. Furthermore, the view that the origin of human cognition is to be found in Pleistocene savannas ignores that rarely in evolution do completely novel systems or mechanisms appear. This is the essence of Darwin's (1859/1991) notion of descent with modification. Just as the origin of bipedalism and human erect posture cannot be understood without reference to previous forms of locomotion and skeletal designs, human mental and neural mechanisms did not appear in a vacuum. They are surely the result of modifications to mechanisms inherited from our ancestors, including Pliocene hominids, and even earlier primates. In the words of Bechtel (2002):

> "[...] evolution begins with existing complex entities and modifies them. This is very different than a picture of developing new modules de novo. Moreover, it imposes an important constraint on evolutionary models – that they be grounded on information about phylogeny and the ancestral condition before the acquisition of a new ability" (Bechtel, 2002; p. 223).

There is no question that Pleistocene environments exerted strong selective pressures on our ancestors in relation to the appreciation of beauty, among many other cognitive and physical traits. The literature provides evidence of changes in cognitive and neural systems throughout this period of human evolution. But the fact is, that these pressures were acting on brain structures and processes that had been inherited from earlier ancestors. Our review of the literature has shown that part of the cognitive and neural underpinnings of aesthetic preference were already in place millions of years ago in our primate ancestors, which we share with some current primate species. Hence, evolutionary approaches to aesthetic preference cannot limit the study of the selective pressures and environments to those affecting our savanna dwelling hunter-gatherer ancestors. We believe that knowledge of the selective pressures acting on early hominids, hominoids, and maybe even earlier primates, as well as their cognitive adaptations, will provide a rich and necessary complement to understanding the evolution of aesthetic preference.

Acknowledgements

This research was made possible by a research grant HUM2007-64086/FISO awarded by the *Dirección General de Investigación del Ministerio de Educación y Ciencia*.

References

Allen, G. (1880). Aesthetic evolution in man. *Mind, 5*, (p.445-464).

Allman, J., Hakeem, A. & Watson, K. (2002). Two phylogenetic specializations in the human brain. *The Neuroscientist, 8*, (p.335-346).

Amiez, C., Joseph, J.-P. & Procyk, E. (2005). Primate Anterior Cingulate Cortex and Adaptation of Behaviour. In: S. Dehaene, J.-R. Duhamel, M. D. Hauser & G. Rizzolatti (eds.), *From Monkey Brain to Human Brain* (p. 315-336). Cambridge, MA: MIT Press.

Atkinson, A. P. & Wheeler, M. (2004). The grain of domains: The evolutionary-psychological case against domain-general cognition. *Mind & Language, 19*, (p.147-176).

Barrett, H. C. & Kurzban, R. (2006). Modularity in Cognition: Framing the Debate. *Psychological Review, 113*, (p.628-647).

Barton, R. A. (2006). Primate brain evolution: Integrating comparative, neurophysiological, and ethological data. *Evolutionary Anthropology, 15*, (p.224-236).

Bechtel, W. (2002). Modules, Brain Parts, and Evolutionary Psychology. In: S. J. Scher & F. Rauscher (eds.), *Evolutionary Psychology: Alternative Approaches* (p. 211-227). Dordrecht: Kluwer.

Cela-Conde, C. J., Marty, G., Maestú, F., Ortiz, T., Munar, E., Fernández, A., Roca, M., Rosselló, J. & Quesney, F. (2004). Activation of the prefrontal cortex in the human visual aesthetic perception. *Proceedings of the National Academy of Sciences USA, 101*, (p.6321–6325).

Chatterjee, A. (2003). Prospects for a Cognitive Neuroscience of Visual Aesthetics. *Bulletin of Psychology of the Arts, 4*, (p.55-60).

Chiavaras, M. M. & Petrides, M. (2000). Orbitofrontal sulci of the human and macaque monkey brain. *The Journal of Comparative Neurology, 422*, (p.35-54).

Clay, F. (1908). The origin of the aesthetic emotion. *Sammelbände der Internationalen Musikgesellschaft, 9*, (p.282-290).

Croxson, P. L., Johansen-Berg, H., Behrens, T. E. J., Robson, M. D., Pinsk, M. A., Gross, C. G., Richter, W., Richter, M. C., Kastner, S. & Rushworth, M. F. S. (2005). Quantitative investigation of connections of the prefrontal cortex in the human and macaque using probabilistic diffusion tractography. *The Journal of Neuroscience, 25*, (p.8854–8866).

Darwin, C. (1859/1991). *The Origin of Species*. Amherst NY: Prometheus Books.

—. (1874/1998). *The Descent of Man*. Amherst, NY: Prometheus Books.

Denys, K., Vanduffel, W., Fize, D., Nelissen, K., Sawamura, H., Georgieva, S., Vogels, R., Van Essen, D. & Orban, G. A. (2004). Visual activation in prefrontal cortex is stronger in monkeys than in humans. *Journal of Cognitive Neuroscience, 16*, (p1505-1516).

Eibl-Eibesfeldt, I. (1988). The biological foundations of aesthetics. In: I. Rentscheler, B. Herzberger & D. Epstein (eds.), *Beauty and the Brain. Biological Aspects of Aesthetics*. Basel: Birkhäuser.

—. (1989). *Human Ethology. Foundations of Human Behaviour*. Hawthorne, NY: Aldine de Gruyter.

Flack, J. C. & de Waal, F. B. M. (2000). 'Any Animal Whatever'. Darwinian Building Blocks of Morality in Monkeys and Apes. *Journal of Consciousness Studies 7*, (p.1–29).

Foley, R. (1995). The adaptive legacy of human evolution: A search for the environment of evolutionary adaptedness. *Evolutionary Anthropology, 4*, (p.194-203).

Gil-da-Costa, R. & Hauser, M. D. (2006). Vervet monkeys and humans show brain asymmetries for processing conspecific vocalizations, but with opposite patterns of laterality. *Proceedings of the Royal Society, B, 22*, (p.2313–2318).

Jacobsen, T., Schubotz, R. I., Höfel, L. & von Cramon, D. Y. (2006). Brain correlates of aesthetic judgment of beauty. *Neuroimage, 29*, (p.276 – 285).

Kaplan, S. (1987). Aesthetics, affect, and cognition. *Environment and Behaviour, 19*, (p. 3-32).

—. (1992). Environmental preference in a knowledge-seeking, knowledge-using organism. In: J. H. Barkow, L. Cosmides & J. Tooby (eds.), *The Adapted Mind: Evolutionary Psychology and the Generation of Culture* (p. 581-598). New York, NY: Oxford University Press.

Kawabata, H. & Zeki, S. (2004). Neural Correlates of Beauty. *Journal of Neurophysiology, 91*, (p.1699-1705).

Kogan, N. (1994). On aesthetics and its origins: Some psychobiological and evolutionary considerations. *Social Research, 61*, (p.139-165).

Kondo, H., Saleem, K. S. & Price, J. L. (2003). Differential Connections of the Temporal Pole with the Orbital and Medial Prefrontal Networks in Macaque Monkeys. *The Journal of Comparative Neurology, 465*, (p.499-523).

Leder, H., Belke, B., Oeberst, A. & Augustin, D. (2004). A model of aesthetic appreciation and aesthetic judgments. *British Journal of Psychology, 95*, (p.489-508).

Lenain, T. (1997). *Monkey Painting*. London: Reaktion Books.

Marcus, G. (2004). The Birth of the Mind. New York: Basic Books.

Miller, G. F. (2001). Aesthetic fitness: How sexual selection shaped artistic virtuosity as a fitness indicator and aesthetic preferences as mate choice criteria. *Bulletin of Psychology and the Arts, 2*, (p.20-25).

Munakata, Y., Santos, L. S., Spelke, E. S., Hauser, M. D. & O'Reilly, R. C. (2001). Visual representation in the wild: How Rhesus Monkeys parse objects. *Journal of Cognitive Neuroscience, 13*, (p.44-58).

Nadal, M., Capó, M. A., Munar, E., Marty, G. & Cela-Conde, C. J. (in press). Constraining hypotheses on the evolution of art and aesthetic appreciation. In: M. Skov & O. Vartaninan (eds.), *Neuroaesthetics*: Baywood.

Nadal, M., Munar, E., Capó, M. A., Rosselló, J. & Cela-Conde, C. J. (2008). Towards a framework for the study of the neural correlates of aesthetic preference. *Spatial Vision, 21,* (p.379-396).

Nakamura, K. & Kubota, K. (1996). The primate temporal pole: Its putative role in object recognition and memory. *Behavioural Brain Research, 77*, (p.53-77).

Nimchinsky, E. A., Gilissen, E., Allman, J. M., Perl, D. L., Erwin, J. M. & Hof, P. R. (1999). A neuronal morphologic type unique to humans and great apes. *Proceedings of the National Academy of Sciences USA, 96*, (p.5268-5273).

Orban, G. A., Van Essen, D. & Vanduffel, W. (2004). Comparative mapping of higher visual areas in monkeys and humans. *Trends in Cognitive Sciences, 8*, (p.315-324).

Orians, G. H. (2001). An evolutionary perspective on aesthetics. *Bulletin of Psychology and the Arts, 2*, (p.25-229).

Orians, G. H. & Heerwagen, J. H. (1992). Evolved responses to landscapes. J. H. Barkow, L. Cosmides & J. Tooby (eds.), *The Adapted Mind: Evolutionary Psychology and the Generation of Culture* (p. 556-579). New York, NY: Oxford University Press.

Petrides, M. (2005). Lateral prefrontal cortex: architectonic and functional organization. *Philosophical Transactions of the Royal Society B, 360*, (p.781-795).

Petrides, M. & Pandya, D. N. (1999). Dorsolateral prefrontal cortex: Comparative cytoarchitectonic analysis in the human and macaque brain and corticocortical connection patterns. *European Journal of Neuroscience, 11*, (p.1011-1036).

Petrides, M. & Pandya, D. N. (2001). Comparative cytoarchitectonic analysis of the human and the macaque ventrolateral prefrontal cortex and corticocortical connection patterns in the monkey. *European Journal of Neuroscience, 16*, (p.291-310).

Plotkin, H. (2004). *Evolutionary Thought in Psychology. A Brief History.* Oxford: Blackwell.

Preuss, T. M. & Coleman, G. Q. (2002). Human-specific organization of primary visual cortex: Alternating compartments of dense Cat-301 and Calbindin immunoreactivity in layer 4A. *Cerebral Cortex, 12*, (p.671-691).

Rilling, J. K. (2006). Human and nonhuman primate brains: Are they allometrically scaled versions of the same design? *Evolutionary Anthropology, 25*, (p.65-77).

Rilling, J. K. & Seligman, R. A. (2002). A quantitative morphometric comparative analysis of the primate temporal lobe. *Journal of Human Evolution, 42*, (p.505–533).

Rolls, E. T. (2004). Convergence of sensory systems in the orbitofrontal cortex in primates and brain design for emotion. *The Anatomical Record Part A, 281A*, (p.1212-1225).

Sánchez, J. C. (2005). Estética y constructivismo: filogenia, historia y vida humana. *Estudios de Psicología, 26*, (p.173-193).

Sasaki, Y., Vanduffel, W., Knutsen, T., Tyler, C. & Tootell, R. (2005). Symmetry activates extrastriate visual cortex in human and nonhuman primates. *Proceedings of the National Academy of Sciences USA, 102*, (p.3159-3163).

Schenker, N. M., Desgouttes, A.-M. & Semendeferi, K. (2005). Neural connectivity and cortical substrates of cognition in hominoids. *Journal of Human Evolution, 49*, (p.547-569).

Schoenemann, P. T. (2006). Evolution of the size and functional areas of the human brain. *Annual Review of Anthropology, 35*, (p.379-406).

Sejnowski, T. J. & Churchland, P. S. (1989). Brain and cognition. In: M. I. Posner (ed.), *Foundations of Cognitive Science* (p. 301-356). Cambridge, MA: MIT Press.

Semendeferi, K., Armstrong, E., Schleicher, A., Zilles, K. & Van Hoesen, G. W. (1998). Limbic frontal cortex in hominoids: A comparative study of area 13. *American Journal of Physical Anthropology, 106*, (p.129-155).

Semendeferi, K., Armstrong, E., Schleicher, A., Zilles, K. & Van Hoesen, G. W. (2001). Prefrontal cortex in humans and apes: A comparative study of area 10. *American Journal of Physical Anthropology, 114*, (p.224-241).

Shapiro, L. & Epstein, W. (1998). Evolutionary theory meets cognitive psychology: A more selective perspective. *Mind & Language, 13*, (p.171-194).

Sigala, N., Gabbiani, F. & Logothetis, N. K. (2002). Visual categorization and object representation in monkeys and humans. *Journal of Cognitive Neuroscience, 14*(2), (p.187-198).

Smith, C. U. M. (2005). Evolutionary neurobiology and aesthetics. *Perspectives in Biology and Medicine, 48*, (p.17-30).

Tincoff, R. & Hauser, M. D. (2005). Cognitive Basis for Language Evolution in Nonhuman Primates. In: K. Brown (ed.), *Encyclopedia of Language and Linguistics* (p. 533-538): Elsevier.

Tooby, J. & Cosmides, L. (1992). The psychological foundations of culture. In: J. H. Barkow, L. Cosmides & J. Tooby (eds.), *The Adapted Mind*. New York, NY: Oxford University Press.

Tooby, J. & Cosmides, L. (1995). Mapping the evolved functional organization of mind and brain. In: M. S. Gazzaniga (ed.), *The Cognitive Neurosciences* (p. 1185-1197). Cambridge, MA: MIT Press.

Tooby, J. & Cosmides, L. (2001). Does beauty build adapted minds? Toward and evolutionary theory of aesthetics, fiction and the arts. *SubStance, 30 (Issue 94/95)*, (p.6-27).

Vartanian, O. & Goel, V. (2004). Neuroanatomical correlates of aesthetic preference for paintings. *Neuroreport, 15*, (p.893-897).

Vartanian, O. & Nadal, M. (2007). A biological approach to a model of aesthetic experience. In: L. Dorfman, C. Martindale & V. Petrov (eds.), *Aesthetics and Innovation* (p. 429-444). Newcastle: Cambridge Scholars Publishing.

Vogels, R. (1999). Categorization of complex visual images by rhesus monkeys. Part 2: single-cell study. *European Journal of Neuroscience, 11*, (p.1239-1255).

Vogt, B. A., Nimchinsky, N. A., Vogt, L. J. & Hof, P. R. (1995). Human cingulate cortex: Surface features, flat maps, and cytoarchitecture. *The Journal of Comparative Neurology, 359*, (p.490-506).

Weiss, D. J. & Newport, E. L. (2006). Mechanisms underlying language acquisition: Benefits from a comparative approach. *Infancy, 9*, (p.241-257).

Wilson, E. O. (1975). *Sociobiology: The New Synthesis*. Cambridge, MA: Harvard University Press.

CHAPTER THREE

IN SEARCH OF INNER WORLDS: ARE HUMANS ALONE IN THE MENTAL WORLD OF POSSIBLE FUTURES?

MATHIAS OSVATH

Abstract: The experience of an inner mental world constitutes much of what it feels like to be human. The inner world is an astounding result of biological evolution. It is not unlikely that other species than Homo sapiens would benefit equally from possessing an inner world. Nevertheless, this possibility has been more or less neglected in scientific studies seemingly due to philosophical ideas stemming from Descartes. One way to approach the question of the existence of inner worlds in animals is by investigating their abilities to travel mentally in time, as such recollection and planning abilities strongly imply an experienced inner world. We have conducted a series of experiments on chimpanzees and orangutans that robustly suggest the ability to travel mentally forward in time, to plan for future states. These findings considerably raise the probability that humans have company in the inner world.

Introduction

Humans are very much inner world dwellers, often just roaming but also navigating it with great accuracy and purposefulness. The world of humans does not disappear when the eyes are shut. Instead we experience life-like episodes from the past, things to come or scenarios that will never even occur. The inner world is an extraordinary place where just about anything can happen with only modest deliberate effort from the inhabitant. On some occasions we have limited control over what takes place in front of our mental senses, for example, when we dream in our sleep. As a matter of fact, the inner world nearly constantly accompanies us. We even suffer from great difficulties of evading it – the number of preachings on how to seize the moment is a telltale sign of this odd human disability. Actually, neurological data reveals that when our brain is at wakeful rest it cannot help but make us ponder on the future or the past, or to produce inner speech or simulate behaviours (Christoff et al, 2004; Andreasen et al, 1995; Ingvar, 1979).

The life-like mental world with its first person perspective attracts much philosophical attention, as it constitutes fundamental aspects of a self-consciousness. This inner world concept is a source of disputes, and raises intriguing and nearly insoluble questions about ontology and phenomenology. What is the nature of this world? Who is the one experiencing it and where does that one reside? What does it actually mean that it is "like being" a perceiver of the mental episodes? And so on.

However, such questions are not the topic of the upcoming text. Here the premises are put rather plainly: There exist cognitive processes that, in an every day sense, are interpreted as an inner world by us, and such processes probably have a biological fitness value, and should hence be viewed as evolved cognitive mechanisms. The overarching question asked is instead if the human animal is the only one who has evolved such a world? Empirical data from studies on prospective cognition in great apes will form the core of the answer. But first a possible evolutionary context for a mental world will be outlined. It will be followed by a brief discussion about some common views and ideas on the possibility of experienced inner worlds in animals – views that might create obstacles for the scientific study of animal cognition.

Evolving an Inner World

Many of us would argue that acting in our inner world is more or less synonymous with thinking. Manipulating the inner world is seemingly what we do when we think. We mentally design various things and scenarios that we at times transfer to the physical world we share with others. Despite the intuitiveness of this folk psychological concept, it does not hold for a closer look. Most of our cognition is as anonymous as the blood flow or the pumping in our cells (it is not until it fails to function that we might indirectly become aware of the processes). We are certainly not conscious of the bulk of our own cognition. The illusion of the inner world as the main stuffing of our heads is due to the inwardly directed beam of awareness that only seems to capture this inner world and not very much else (perhaps this "beam" even *is* the inner world).

So, in the dawn of cognition, lifelike inner worlds probably did not exist. Though the animal kingdom was devoid of this mental environment, many species soon started to carry cognitive models of different ecologically important aspects of the outer environment. Such models could be described as cognitive simulators or emulators (Gärdenfors, 1995; Grush, 1997; Hesslow, 2002). To put it simply, this kind of mental models test actions before the organism executes them. The simulators considerably improve the swiftness of actions, as the nervous system is relieved from the constant adjustments to the environmental feedback in order to inhibit or exhibit the signals governing the muscular tensions. An action, including the environmental feedback, is instead simulated within the organism before it is initiated and thus, a pre-fabricated action can be utilized at once. The fidelity of these simulators differs and depends on training and development in ontogeny and on phyologenetical requirements. Most animals entrust their lives to simulators though. If a chimpanzee or a squirrel does not accurately enough simulate the jump to the next branch, the consequences might be fatal. Trial and error does not suffice in a long range of ecological niches. When an animal develops a simulating system of some sort it can occupy a niche which otherwise would be blocked. Karl Popper, with famous elegance, illustrates the utility of the internal models by expressing it as they let our hypothesis die in our place.

Primordial simulators were only connected to immediate actions with a minimum of delay between incoming stimuli, the simulation and the execution of an action. Further on, simulators could, so to speak, run off-line (Grush 1997) and handle more complicated action possibilities (as in case of the jumping ape or squirrel). Later in evolution, or perhaps simultaneously, cross-modal simulators would have had notable fitness value, as they allow the coordination of the simulators of the different senses in an action, for example, eye-limb coordination.

It is not necessary for the owner of such cognitive simulators to be aware of them in order for them to work. Some are probably more or less innate and others must be filled with information from the environment, for example, during childhood. The environment in which they are used must have high degrees of regularity in the aspects relevant for the different simulators, otherwise it will not be possible to fill the simulators with generalizable information, either phylogenetically or ontogenetically.

In niches with less regularity due to more complex causal relationships or an often changing environment, or when novel solutions increase fitness, another simulating system is needed – a system that is not only dependent on the faithful replication of the well-established information stored in the simulators. Such a multifarious environment might, for example, be a social one, where the behaviours of other cognitively advanced creatures must be taken into account before acting. The need for simulating something novel or to foresee different possible infrequent or complex scenarios seems to have been met by the evolution of an experienced inner world. An experienced inner world gives the owner the great benefit of being able to compare and relate different experiences to each other. It is possible to compare a currently experienced state in the physical world to a possible future, one that is simulated in the inner world. Or, different inner world experiences could be compared to each other. On the basis on these experiences the individual can make choices and form decisions. This system could, in other words, be a provider of different emotional information to be used in a decision process. That is one important reason why the inner world is actually experienced, as experiences yield emotions. It should also be added that a mental life-like inner world are almost necessarily experienced, because it would clearly not be of much utility to construct such an environment in an organism that is blind to it. This argument must not be confused with sophism, rather it states that if a life-like inner world really exists, it is experienced by the owner – the *existence* of this world is yet another question.

Pessimistic Views on the Possibility of Non-Human Inner Worlds

There is a widespread tendency among empirically oriented scientists and philosophers to avoid serious considerations of the possibility of experienced inner worlds in non-humans. Such awareness is readily and seemingly by default excluded from most interpretations of experimental results or observational data from non-humans. Instead, great effort is put into formulating explanations lacking such awareness – resulting in filling animal heads with elaborate learning mechanisms from one end and draining off any experienced inner worlds from the other.

Part of the explanation for this reluctance is the so-called Morgan's canon, which has turned into the golden rule that any respected comparative psychologist must cherish and practice by. The canon was formulated by the pioneering comparative psychologist Lloyd Morgan (1894) and spells: "In no case may we interpret an action as the outcome of the exercise of a higher psychical faculty, if it can be interpreted as the outcome of the exercise of one which stands lower in the psychological scale". Strangely enough, this is an obvious expression of a belief in a *scalae nature,* which comparative psychologists on the other hand are taught to shun, as such a scale is incompatible with the idea of evolution. There can be no "higher" or "lower" in the teleological sense in the future blind process of evolution. For some reason the canon is regarded as a version of the principle of parsimony. This was clearly not Morgan's intention, who even defended the lack of parsimony in the canon. He expressed that a leaner explanation is not necessarily the correct one. The implicit and peculiar conclusion drawn by later researchers is that Morgan actually misunderstood what he himself meant – the canon is indeed about scientific parsimony (Radner & Radner, 1989).

Nevertheless, Morgan, misconstrued or not, is not to blame for the underlying assumption that awareness of the inner world is "higher" or more complex than many of the cognitive processes that substitute this awareness in the explanations of non-human behaviours. Rather, Descartes is usually held accountable for digging this trench between humans and the rest of the tens of thousands of species possessing a nervous system. Without much controlled empirical data and with a great deal of introspection, Descartes assumed that humans alone have an aware experience, and that animals are more or less complicated automats. The Cartesian view on the matter has had a substantial impact and has survived for almost four centuries in Western thinking. Despite that, this fact in itself is rather awkward, given the amount of research carried out since then, there are also other peculiar implications hidden in this approach. Cartesian adherents explicitly agree that this awareness is uniquely human, but what is less obvious is that this for them also implies that the ability is higher than non-aware abilities that could perhaps explain the same behavioural results. Simply because the ability is allegedly human, it appears – human equals cognitive height, and lower means non-human. Hence, and in purported accordance to Morgan's canon, the behavioural data from non-humans are interpreted in lower terms, that is, excluding an inner awareness. The consequences of this belief approaches circularity. Frequently used escape routes in explaining animal cognition are elaborate models of associative learning (Griffin & Speck, 2004; Byrne & Bates, 2006; Call, 2006). Theories of conditioned learning mechanisms undoubtedly have great merits in their explanatory and predictive powers in an array of circumstances, and often they do indeed offer the most parsimonious approach. However, when accounting for behaviours of animals that when displayed by humans clearly rely on other mechanisms than associative learning, for example, problem solving in novel situations, then associationistic explanations tend to become extremely convoluted. The associative learning models lose most of their predictive strength in such cases; they rather become *post hoc,* bringing the models close to *ad hoc*. In other words, it does not seem possible to foresee some animal behaviours by applying associationistic accounts; it is only after the exhibition of a certain behaviour that it is possible to come up with an explanation of associative learning.

Serious scientific endeavours to explore animal cognition must be cleared from ancient ideological debris, without discarding the principle of parsimony or picking up new heavy ideological baggage. Evolutionary theory should be an appropriate starting block. For example, the hypothesis that complex traits have common origin if shared by species in phylogenetic proximity, could most reasonably serve as the null hypothesis also when it comes to comparing human cognition to that of others. Most comparative psychologists would agree upon this, the problem is rather what criteria should be used to be able to say that the "same" cognitive traits are exhibited in two or more different species. Such disagreements do, however, not diminish the validity of the point I am trying to advocate: An experienced inner world would doubtlessly carry a biological fitness value for individuals from a range of species. This special cognitive simulation of an environment must be treated as an evolved biological trait and not as something outside the realm of nature and only by definition unique to humans.

Empirical Studies

There are different ways to probe the question about the existence of an inner world. In language competent creatures like humans one could listen to what is expressed and quite soon figure out whether there seem to be an inner world within the speaker, or we could of course just ask. When such communicative abilities are lacking or when we have problems interpreting them, one must rely on other behavioural criteria than language expressions. One of these has received much attention in recent years and is the display of the ability to travel mentally in time.

Mental time travel is closely related to the concept of episodic memories, coined by Tulving (1972). Episodic memories enable recollection of specific events as opposed to just knowledge about something. Tulving illustrates this by pointing out the difference between knowing the rules of chess and remembering ever having taking part in a chess game. This recall is the backward mental time travel. The forward part, projecting into possible futures, is believed to rely on the same cognitive system as episodic memories do (e.g. Atance & O'Neill 2001, 2005; Suddendorf & Busby 2005; Suddendorf & Corballis 1997, 2007; Tulving 2005). There is a growing wealth of neurological data to support that, in humans, the same brain structures are used when engaging in tasks that involves either episodic memories, planning or even the use of a theory of mind (for review see Buckner & Carroll 2007). The defining character of mental time travel is the experience of a first person view in the inner world. Accordingly, such travels involve, as mentioned above, a certain form of self-consciousness, called autonoetic in this context (e.g. Tulving 2005, Gardiner 2002).

The ability to plan for future drive states or to travel forward in mental time seems to be the greatest evolutionary gain with an inner world or an episodic system, as it gives a high flexibility to the individual that can regulate actions far in advance in an optimal way. As discussed, the inner world is assumed to be a human unicum, and the assertion of human exclusiveness in future-oriented cognition specifically is dubbed the Bischof-Köhler-hypothesis (Suddendorf & Corballis, 1997). It spells that non-humans cannot differentiate future states from current ones. This means that they are unable to plan for something that is different from their current drive state. Many animals are seemingly proficient planners when it comes to immediate drives, for example, if they are hungry they figure out sequences of different actions in a specific order to get to food. However, they are supposed to lack the ability to plan for a state that they are currently not experiencing. Such planning is not to be confused with behaviours governed by innate releasing mechanisms, like migration, hibernation or hoarding. The assumed animal disability does not, however, imply that humans should be able to plan without any mental reference to the future in the current situation. Rather, it does mean that animals lack the ability in one way or another to suppress a predominant current drive state in favour of a future one – to give priority to an imagined state before the currently experienced one. As mentioned above, this would be much of the point of experiencing an inner world in a fashion similar to the real world. Two experiences can be weighted together and compared. If it were shown that non-humans engage in mental time travel, and perhaps in planning specifically, then it would constitute firm evidence that humans are not alone in the world within.

The most extensive and cohesive research program on mental time travel in animals has been conducted on scrub-jays by a team in Cambridge (e.g. Clayton & Dickinson 1998, 1999; Correia et al 2007; Raby et al 2007). These corvid birds exhibit impressive behaviours related to caching, suggesting not only that they have episodic memories, but also that they plan for future drive states and even have a well developed intersubjectivity. These abilities appear to be analogous to the ones managed by a given core brain structure in humans, which is interesting considering the differences between avian and mammalian brains. Despite the fact that these experimental data appear convincing in establishing mental time travel in non-humans, there are several who do not find them conclusive. Mainly because it is argued that associative learning or innate mechanisms are not sufficiently excluded (Gilbert & Wilson 2007; Premack 2007; Suddendorf & Corballis 2007; Tulving 2005; Roberts 2002, 2006). There have also been studies on great ape ability to remember episodically or to use prospective cognition (Schwartz et al 2005; Mulcahy & Call 2006). However, they have received even more severe critique than the corvid-studies for not eliminating the possibilities of associative learning, and for lacking controls for the animal's mental state in the testing situation (Suddendorf & Corballis, 2007; Tulving 2005; Roberts 2002, 2006).

As it seems, the proponents for the existence of mental time travel in non-humans are gaining ground, however the case is still not settled.

Great Apes Use Self-Control and Envisioning in the Face of Future States

In a series of four experiments we wanted to find out whether great apes are able to out compete strong current drive states in favour of highly delayed future rewards, and if they use processes that could be described as envisioning when acting towards the future (Osvath & Osvath, 2008). It is essential to answer these questions as such abilities form the basis of true future planning and forward mental time travel, and hence are signs of a complex inner world of possible futures. It has not been shown before this study that great apes are able to compete with their immediate drives while taking actions for the future. The only previous study on great ape prospective cognition (Mulcahy & Call, 2006) did not control for the drive states. It has been argued that the subjects in that study might have been governed by the same drive, craving for grapes, throughout the experiment, and that their key behaviours were not an expression of actions made towards a deviating mental state in the future (Suddendorf 2006, Suddendorf & Corballis 2007). Equally important to measuring drive states is the exclusion of associative learning as an explanation of the results. Associatively learned behaviours are, as described above, often viewed as the opposite of behaviours originating from a conscious process; hence it must be convincingly shown that great apes are not influenced by such learning in the experiment. The Mulcahy and Call study was also criticized for not excluding this possibility. The main argument was that the same tool was rewarding throughout the experiment, and thereby it is not impossible that the relationship between the tool and the reward was purely associatively learned (Suddendorf & Corballis, 2007).

We engaged two female chimpanzees (*Pan troglodytes*) and one male orangutan (*Pongo abelii*) in our study. Before the experiments started, the subject received brief training in using a thin plastic hose as a straw to obtain a large reward of sweet and satiating fruit syrup from an apparatus. A keeper demonstrated the function of the tool by using it in front of the subject and then handing it over. All three subjects immediately and spontaneously reinserted the hose and used it appropriately by sucking up the sweet liquid. This was the only occasion during the whole experiment series where there was only a short delay between obtaining the tool and accessing the apparatus. Then the subjects were shown a tray containing four objects out of which one, the hose, was functional for acquiring the reward. The other three objects were previously used in different enrichment activities in the daily lives of the apes, and thus probably familiar. The subject was only permitted to select one of the items. Prior to this, a bias control was conducted on three apparatus naïve apes not participating in the study – there were no significant biases found in the selection of the objects. The selection was made when the apparatus was visible but unattainable in the reward room. Subjects were not allowed into the reward room until one hour had passed after their selection. All subjects selected the functional tool in the first training trial. After this training session, the apes were exposed to the reoccurring nature of the reward in order to create possible incentives for possessing a functional tool for the future. This was done by granting subjects without tools access to the reward room containing the baited apparatus. This frustrating procedure was repeated twice daily.

The baseline experiment. The following day the actual testing began and the apes were respectively called into a selection room. From this room it was not possible to see the reward room. When inside, the subject was offered a choice of the kind described above: Among three enrichment objects and the functional hose. After their selection they were let out to their everyday enclosure shared with the rest of their social group. Then a 70-minute delay followed until the baited apparatus was installed in the reward room and the subjects were granted access. In other words, to obtain the reward the ape had to select the functional tool in a location with no visual cuing to the reward site, save and protect the tool for more than an hour in a social environment and then use it when let into the reward room. All three subjects performed significantly above chance. Out of the 14 trials conducted on each subject, one orangutan and one chimpanzee selected the functional tool in 100 % of their choices, and the other chimpanzee selected it in 13 trials. The subjects obtained the reward in approximately 80 % of the trials when they had selected the appropriate tool. The tool losses in the chimpanzee trials were most often caused by a playing infant that misplaced the hose. The orangutan actually brought the tool in two (of three) cases that were recorded as losses, however, in a non-functional state due to being carried in the mouth (in relation to the teeth of a male orangutan the tool was quite fragile). The results of this first experiment constructed a baseline for further experiments in the series. And, perhaps more importantly, it did not only confirm the results from the previous prospection study on great apes (Mulcahy & Call, 2006), but it also showed that apes are able to take actions towards the future in a location that has not been associated with the forthcoming event. Furthermore, this experiment shows that the apes in a prospection task can cope with the cognitive load added by sharing their waiting time with the rest of their social group.

The self-control experiment. The first experiment did not control for competing drives when the subjects carried out the key behaviour. This control was addressed in the second experiment, where it was tested whether subjects could suppress the selection of an immediate reward in favour of the tool needed to obtain the larger future reward. The inhibitory ability investigated in this experiment is often described as self-control – the ability to select a delayed larger reward instead of an immediate smaller one (e.g. Tobin & Logue 1994). Self-control in humans is allegedly unparalleled by any other species. The purported impulsivity of non-human animals is taken as evidence for the hypothesis that they are mentally stuck in the present moment (Köhler, 1921; Roberts, 2002, 2006). The ability of self-control has been widely studied in a range of different research paradigms from psychiatry and psychology to animal cognition and economics. Self-control appears to be a demanding ability that develops late in ontogeny and is profoundly integrated in human cognition. The levels of self-control in humans predict the individual success in domains as different as interpersonal skills, academic achievement and psychopathology (Mischel et al 1989; Tangney et al 2004; Carlson & Moses 2001; Duckworth & Seligman 2005).

While human self-control might span over long delays, minutes and hours or even months and years, the ability of most of the tested animals appears restricted to seconds. Interestingly, it has predominantly been pigeons and white rats that have represented the non-human animals. Therefore it is not surprising that recent self-control studies on great apes, which are as closely related to pigeons and rats as we are, show impressive results (Beran et al, 1999; Beran & Evans, 2006; Evans & Beran, 2007; Rosati et al, 2007). Some results even indicate that chimpanzees in some contexts are more competent than humans in self-control tasks involving food rewards (Rosati et al, 2007). This makes great apes good candidates for being fellow inner world dwellers. Self-control is likely the most potent experimental control for the drive states in a planning context. The impulse to experience immediate satisfaction must be fought in favour of making the future slightly brighter than the present.

In this experiment all relevant aspects of the set-up in the previous experiment were kept, with one crucial exception: A much-favoured fruit, a grape, was added to the selectable items. If animals are unable to differentiate future states from current ones, they are expected to select the immediate reward. Out of the 14 trials conducted on each subject, one chimpanzee selected the functional tool 8 times, the other chimpanzee chose the tool 11 times and the orangutan selected it 9 times. The rest of the choices were the grape. These results are significant. The level of performance is comparable to that of humans (Forzano et al, 1992), which confirms the results of Rosati et al (2007). To date, these results actually show the highest level of self-control demonstrated in non-human animals.

The association control experiment. If unable to anticipate a future reward, it would be irrational to select something that is useful in the future instead of something that is immediately rewarding – there would not be an obvious beneficial point such behaviour. However, if the selection was not due to intentional cognition, but associative learning then it would not have required prospection. The tool would be selected more often than the fruit only because it would be loaded with a high reinforced value. More than a century of research in conditioned learning has not revealed any processes that could credibly explain the impressive levels of inhibition described above. Even so, it is worthwhile to control for any not yet discovered associative learning mechanism, as such mechanisms are readily and creatively suggested by proponents of non-cognitive explanations of animal behaviour.

The third of our experiments controlled for such associations. If there was an associatively learned link between the tool and the reward, then the tool would only carry an intrinsic value. Its value would not be cognitively related to its future function in the mind of the ape, it would rather be a reinforced arbitrary connection stored in the associative system of the animal. The results of the self-control experiment would then be an expression of the strength of this associative link and not of self-control at all. In other words, the animal would in fact have faced two immediate rewards, where the plastic hose in itself constituted a higher reward than the grape. To investigate this possibility we let the subject select between objects identical to those of the baseline experiment. When the subject had selected the hose as expected (from the results of the first experiment) and thus had it in its immediate possession, it was instantly offered a new choice, including a grape, making this set-up identical to the one in the second experiment. If the functional tool only carries its value in itself, the second selection should mirror the results in the self-control experiment. If, however, the tool had its value related to its future function, then the grapes would be the expected choice in the second pairing.

In the 14 trials conducted on each ape, no one selected a second functional tool when already possessing one, but maximized the rewards by selecting a grape. It is worth pointing out that this does not necessarily mean that the tool did not evoke positive associations in the apes, it probably did just as precious tools might evoke positive feeling in humans, rather the results show that the value of the tool is not associatively *learned*.

The envisioning experiment. It might be intuitively hard to understand the complex cognition about future states revealed in the described experiments without assuming that the apes actually envision those states. Nevertheless, this is a theoretical possibility and thus, more insights are needed into which processes are used when the ape is selecting with the future in mind. One way to control for whether a possible future event is simulated by envisioning, is through measuring the ability mentally to include a novel object in a possible future scenario. To face an object that has never before been encountered and then, based on its functional properties that are only inferred, integrate it in a plan would be a feat implying envisioning. Such complex simulations of the states in the environment are prime examples of what an inner world would be suitable for.

From this assumption we conducted the fourth and last experiment in this series. This time the apes were exposed to novel tools that were highly dissimilar to the original functional tool used in previous experiments. Two of the other selectable objects were also exchanged for novel objects. These were included on two grounds, one was their attention grabbing effect on humans and the other that one of them should bear some similar features to the functional hose but obviously not its function. The fourth object included in all the trials was the familiar bamboo stick, associated with honey extraction in enrichment activities. It could also be used to extract the fruit soup in a highly inefficient way (similar to eating soup with a fork). The functional tools were either pipes or hoses of different lengths (between 40 and 58 cm), materials (plastic, rubber, wood, metal, fabric), shapes (round, square and triangular) and colours (white, black, orange, beige, grey, multicolour, metal and wood) (see figure 1). Twelve trials with unique objects were conducted on each subject. Prior to these trials a bias control was conducted with the three naïve apes that took part in the control in the baseline experiment. In this control no functional object was selected significantly above chance. However, the three control subjects in three trials chose the same distracter.

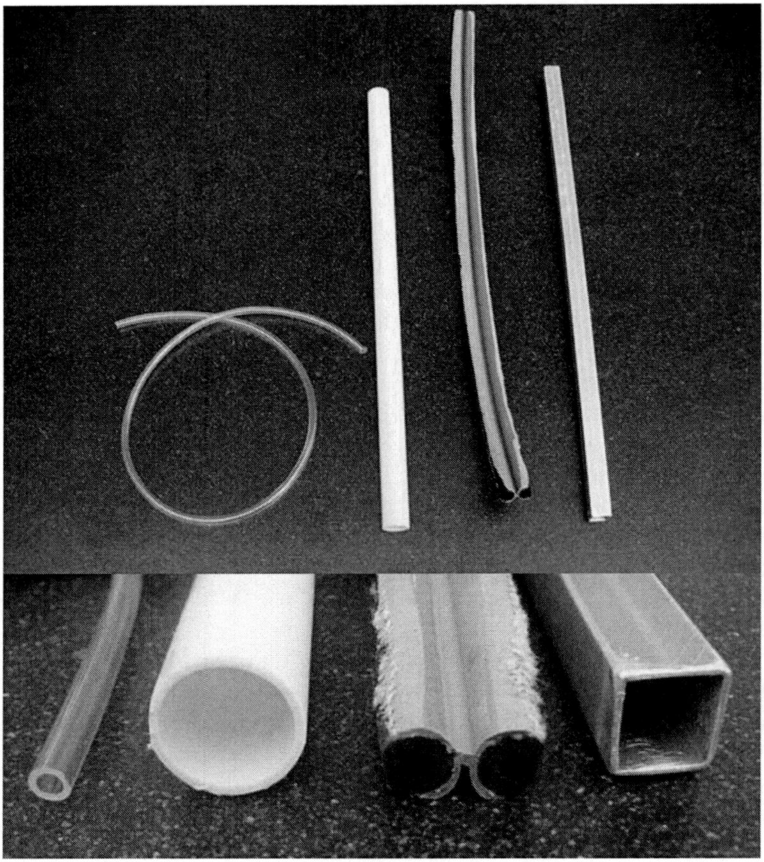

Figure 1: A comparison between the functional tools used in the study. The hose to the left was used in the first three experiments, and the other three constituted the novel and unfamiliar tools in the envisioning experiment (the plastic pipe was presented in the first trial, and the aluminium frame in the last). In the picture above their different appearance on the selection tray is shown, and in the in the picture below the different openings are compared.

Each object was rotated in front of the subject by an experimenter in a strict and controlled fashion in two directions with equal amount of time for each direction and object. The functional tool was never demonstrated in first or last position to avoid potential biases. To preclude gaze cuing the experimenter never looked directly at the objects or at the ape but at a point above and beyond the head of the subject (it should be added that this procedure was also applied in the other experiments).

This envisioning experiment differed from the rest in another important aspect. The chimpanzee participants were not sharing the waiting area with each other (though with the rest of the non-participating group), to avoid potential social learning.

The orangutan selected the functional object in 11 out of the 12 trials. One of the chimpanzees, Linda, chose the functional object in 10 trials, whereas the other chimpanzee succeeded in obtaining the correct tool in 9 trials. These results are significant. It is interesting to note that in the successful trials, the apes used the tool on the apparatus with no observable hesitation, even though some of the tools required a different body posture from the one in the previous experiments. Furthermore, when the subjects failed to select the appropriate tool, they did not try to use the non-functional one on the apparatus.

This series of experiments strongly suggests that great apes can use prospective cognition to act towards future events that are not currently experienced. And, the results were verified in another pilot study we conducted on another aspect of planning for the future.

The Other Side of the Coin – When Currently Satiated by the Reoccurring Reward

Competition with drive states can be performed in many ways. Self-control is perhaps one of the most potent examples of this ability. In particularly because it is fairly easy to control this process experimentally. However, it seems to be of equal ecological importance to be able to cope with the other side of the coin – to act towards the future when currently experiencing what could be thought of as a negative need. In other words, to suppress or disregard the current satisfaction to an extent that makes it possible yet again to be satisfied when the drive state reoccurs. Examples of cognitively overcoming physiological homeostasis in order to repeat the state of equilibrium could be filling up a water bottle when quenched, cooking or shopping for food when thoroughly satiated or preparing a night's rest in the morning.

In a pilot study we tested this ability on the male orangutan that participated in the previously described experiments. This experiment was based on tool behaviour in relation to a large food reward. The question to be answered was whether or not the orangutan would save the tool leading to the large reward immediately after the consumption of it. The first step in this experiment was to get an approximation and hence a control of when satiation was reached by the subject when offered this kind of reward. The food reward in this experiment was of the same kind as in the previous one: A thick and sweet fruit soup (rose hip berry soup). To a human, this soup is highly satiating due to its syrupy and sweet character. It has a high energy content with about 60 kcal per decilitre. To control for satiation, the orangutan was offered to drink from a bottle filled with 1.5 litres of soup. The subject was given this bottle of drink on three different occasions months apart. Two keepers with more than 30 years of experience of working with great apes, independently of each other made a judgment, based on the behaviour of the ape, when it started to become satiated. They agreed that drinking pace and general eagerness seemed to fade after about 1 litre was consumed. Based on these observations the reward size was set to 1 litre in the experiment.

In the experiment a paper box containing the reward was placed outside one of the two interconnected enclosure, in a counterbalanced fashion, three times a day with pseudo-randomized time intervals between. A thick and rather stiff rubber hose was the only tool that could be used to obtain the reward. If other tools were used, like sticks, then either the paper apparatus would break or the acquired amount of the reward would be extremely limited.

No training on the tool or on the apparatus was given before the experiment. The orangutan was not informed about the reoccurring nature of the reward prior to the experiment. The experiment started with the tool being planted within one of the two interconnected enclosures from where the subject was currently barred. The subject could not see the tool being planted. Then the ape was let into the enclosure and it was recorded when the tool was found. In this first trial it took 5 minutes before the tool was discovered. When 1 hour and 30 minutes had past, the subject was let into the other enclosure outside where the baited apparatus was installed. The orangutan had by now shredded the tool into tiny pieces and therefore could not obtain the reward. He tried using sticks and cloths, though with meagre results. This reward reoccurred three times during the day in and outside the different enclosures with time intervals no less than one hour and no more than three.

The next day the tool planting procedure was repeated. This time it took three minutes before the orangutan found it. After this followed an impressive sequence of three whole days when the orangutan kept track of and brought the tool with him when moving between the enclosures and going to rest at night. During these three days the orangutan successfully obtained the reward three times per day. And the key behaviour, saving the tool by bringing it with him after finishing up the reward, was recorded 10 times. The experiment had to be aborted after three days because of health considerations relating to the extremely high energy and sugar content in the reward.

Figure 2: The male orangutan Naong, who participated in both of the planning studies. Currently he is negotiating the problem of getting the honey trapped inside the small holes drilled into the log.

A control was also conducted by planting a non-functional tool following the same set-up as in the previous experiment. A blue plastic string with the same length as the hose was planted. The subject found it after 1 minute. It was not destroyed, but it was not saved either or brought to the next enclosure. The reward reoccurred three times daily, but the orangutan ignored the string as a tool (and instead returned to using sticks) and did not exhibit any saving behaviour related to it. Again, this confirms that tools are saved in accordance to their future function.

This pilot study indicates that apes are able to act towards a reoccurring opportunity to satiate a drive that is currently satiated. Firmer results on this ability would be achieved if the trials were extended with the same subject, as well as including other individuals. Furthermore, it should be noted that it probably does not require much effort for the ape to save the tool – this fact could be modified to reach further understanding in the planning skills of great apes. Nevertheless, these findings give additional verification to the hypothesis that great apes are able to plan for future states.

Discussion

The results from these planning studies on great apes are convincing evidence for the skill for planning for a currently non-experienced state. In humans, such tasks are solved by consulting the inner world, and comparing experiences from the first person perspective. The results are not readily explained by associative learning mechanisms. An associative account would require new, *post hoc* and unverified associative models. When humans display such abilities, it is viewed as a result of mental time travel, planning for the future and the possession of an inner perspective. Human children who solve far less complex tasks than the apes in these studies are viewed as mental time travellers (Atance & O'Neill 2005; Suddendorf & Busby 2005). Great apes are closely related to us and when they exhibit the same traits it should, according to the evolutionary null hypothesis, be regarded as an expression of the same trait derived from the latest common ancestor.

Interestingly, recent neurological data also confirms the results of these studies. It shows that chimpanzees share the high neural activity with humans at wake rest, which is well correlated with the experienced inner world (Rilling et al, 2007). It is indicated that some of the regions recruited to a greater extent in the chimpanzee brain than in the ones of the human subjects are related to emotional processing. This fits the idea of the experienced inner world as primarily a model for emotional information, which constitutes the reason for it to be experienced.

Experienced inner worlds might have evolved in other lineages than that of the great apes. As suggested by the studies on the scrub-jay abilities in mental time travel, it might be that corvids carry avian versions of inner worlds. Perhaps dolphins, or dogs and cats have their own worlds. If other inner worlds exist, naturally they should be species-specific, including the modal and other aspects important in the physical world of the animal. The common ground is that it is experienced from a first person perspective. The simple reason for this is because that kind of experienced simulator would have a high fitness value in some environments.

Obviously, there is the problem of the, in a philosophical sense, impossibility of knowing anything about the phenomenology of other's experiences. However, that is not really the issue here. The question is not about what it is like to experience something like an ape. I would know just as little about what it is like experiencing something being you (I presume the reader is human). The point is, that the phenomenon of experiencing an inner world is a biologically derived one, and should be viewed as a cognitive mechanism. "What it is like" is a superfluous question in such a context. It is, for example, possible to study biological locomotion, including its motor cognition, without asking what it is like habitually to walk on two legs if you are a human, or on four if you are a cat, or to swim with fins if you are a dolphin.

The problem of the phenomenology in others is intimately related to the problem of solipsism. If nobody else exists with a phenomenology like your own, then others would be rather like rational automats, and then you would be alone. Most scientists and philosophers seem to agree that solipsism is a fruitless view of the world. It is just more comfortable and less dull to assume that there are others out there with minds like your own, sharing roughly the same reality. Having agreed upon that, we use behavioural cues in one of our favourite sports: To figure out what is inside the world of our fellow humans. Perhaps it is time to include more players in the sport?

References

Andreasen N.C., O'Leary D.S., Cizadlo T., Arndt S., Rezai K., Watkins G.L., Ponto L.L., & Hichwa R.D. (1995). Remembering the past: Two facets of episodic memory explored with positron emission tomography. *American Journal of Psychiatry* 152, (1576-1585).

Atance, C.M., & O'Neill, D.K. (2001). Episodic future thinking. *Trends in Cognitive Science, 5,* (533-539).

Atance, C.M.,& O'Neill D.K. (2005). The emergence of episodic future thinking in humans. *Learning and Motivation, 36*, (126-144).

Beran, M.J., & Evans, T.A. (2006). Maintenance of delay of gratification by four chimpanzees (*Pan troglodytes*). *Behavioural Processes, 73*, (315-324).

Beran, M.J., Savage-Rumbaugh, E.S., Pate, J.L., & Rumbaugh, D.M. (1999). Delay of gratification in chimpanzees (*Pan troglodytes*). *Developmental Psychobiology, 34,* (119-127).

Buckner R.L., & Carroll D.C. (2007). Self-projection and the brain. *Trends in Cognitive Science, 11*, (49-57).

Byrne, R. W., & Bates L. A. (2006). Why are animals cognitive? *Current Biology, 16,* (445-448).

Call, J. (2006). Descartes' two errors: Reason and reflection in the great apes. In Hurley S., Nudds M. (eds.) *Rational Animals?* (219-234). Oxford: Oxford University Press.

Carlson, S.M., & Moses, L.J. (2001). Individual differences in inhibitory control and children's theory of mind. *Child Development, 72,* (1032-1053).

Christoff K., Ream J.M., & Gabrieli J.D.E. (2004). Neural basis of spontaneous thought processes. *Cortex* 40, (623-630).

Clayton N.S., & Dickinson A. (1998). Episodic-like memory during cache recovery by scrub jays. *Nature* 395, (272-274).

Clayton N.S., & Dickinson A. (1999). Scrub jays (Aphelocoma coerulescens) remember the relative time of caching as well as the location and content of their caches. *Journal of Comparative Psychology* 113, (403-416).

Correia S.P.C., Dickinson A., & Clayton N.S. (2007). Western scrub-jays anticipate future needs independently of their current motivational state. *Current Biology* ,17, (856-861).

Duckworth, A.L., & Seligman, M.E.P. (2005). Self-discipline outdoes IQ in predicting academic performance of adolescents. *Psychological Science, 16*, (939-944).

Evans, T.A., & Beran, M.J. (2007). Chimpanzees use self-distraction to cope with impulsivity. *Biological Letters*, doi: 10.1098/rsbl.2007.0399.

Forzano, L.B., & Logue, A.W. (1992). Predictors of adult humans' self-control and impulsiveness for food reinforcers. *Appetite, 19,* (33-47).

Gardiner, J.M. (2002). Episodic memory and autonoetic consciousness: A first person approach. In: Baddeley, A., Conway, M., & Aggleton, J. (eds.) *Episodic Memory – New Directions in Research.* (11-30). Oxford: Oxford University Press.

Gärdenfors, P. (1995). Cued and detached representations in animal cognition. *Behavioural Processes 35*, (263-273).

Gilbert, D.T., & Wilson, T.D. (2007). Prospection: Experiencing the future. *Science, 317*, (1351-1354).

Griffin, D.R., & Speck, G.B. (2004). New evidence of animal consciousness. *Animal Cognition, 7,*(5-18).

Grush, R. (1997). The architecture of representation. *Philosophical Psychology, 10,* (5-23).

Hesslow, G. (2002). Conscious thought as simulation of behaviour and perception. *Trends in Cognitive Science, 6,* (242-247).

Ingvar, D.H. (1979). "Hyperfrontal" distribution of the cerebral grey matter flow in resting wakefulness: On the functional anatomy of the conscious state. *Acta Neurologica Scandinavica, 60*, (12-25).

Köhler, W. (1921). Zur psychologie des schimpansen. *Psychologische Forschung, 1*, (2-46).

Mischel, W., Shoda, Y., & Rodriguez, M. (1989). Delay of gratification in children. *Science, 244*, (933- 938).

Morgan, C. L. (1894). *An Introduction to Comparative Psychology*. London: Walter Scott.

Mulcahy, N.J., & Call, J. (2006). Apes save tool for future use. *Science, 312*, (1038-1040).

Osvath, M., & Osvath, H. (2008). Chimpanzee (*Pan troglodytes*) and orangutan (*Pongo abelii)* forethought: Self-control and pre-experiencing in the face of future tool use.

Premack, D. (2007). Humans and animal cognition: Continuity and discontinuity. *Proceedings of National Academy of Sciences USA, 104*, (13861-13867).

Raby, C.R., Alexis, D.M., Dickinson, A., & Clayton, N.S. (2007). Planning for the future by western scrub-jays. *Nature, 445*, (919-921).

Radner, D., & Radner, M. (1989). *Animal Consciousness.* New York: Prometheus Books.

Rilling, J.K, Barks S.K., Parr, L.A., Preuss, T.M., Faber, T.L., Pagnoni, G., Bremner, J.D., & Votaw, J.R. (2007). A comparison of resting-state brain activity in humans and chimpanzees. *Proceedings of National Academy of Sciences USA, 104*, (17146-17151).

Roberts, W.A. (2002). Are animals stuck in time? *Psychological Bulletin, 128*, (473-489).

—. (2006). The questions of temporal and spatial displacement in animal cognition. In: Wasserman, E.A. & Zentall, T.R. (eds.) *Comparative Cognition: Experimental Explorations of Animal Intelligence.* (45-63). New York: Oxford University Press.

Rosati, A.G., Stevens, J.R., Hare, B., & Hauser, M.D. (2007). The evolutionary origins of human patience: Temporal preferences in chimpanzees, bonobos, and human adults. *Current Biology, 17,* (1663-1668).

Suddendorf, T., & Busby, J. (2005). Making decisions with the future in mind. *Learning and Motivation, 36*, (110-125).

Suddendorf, T., & Corballis, M.C. (1997). Mental time travel and the evolution of the human mind. *Genetic, Social, and General Psychology Monographs, 123,* (133-167).

Suddendorf, T., & Corballis, M.C. (2007). The evolution of foresight: What is mental time travel and is it unique to humans? *Behavioral and Brain Sciences, 30*, (299-351).

Schwartz, B.L., Hoffman, M.L., & Evans, S. (2005). Episodic-like memory in a gorilla: a review and new findings. *Learning and Motivation, 36*, (226-244).

Tangney, J.P., Baumeister, R.F., & Boone, A.L. (2004). High self-control predicts good adjustment, less pathology, better grades and interpersonal success. *Journal of Personality, 72,* (271-324).

Tobin H., Logue A. (1994). Self-control across species. *Journal of Comparative Psychology, 108*, (126-133).

Tulving, E. (1972). Episodic and semantic memory. In: Tulving, E. & Donaldson, W. (eds.) *Organization of Memory*. (382-403). New York: Academic.

—. (2005). Episodic memory and autonoesis: Uniquely human? IN: Terrace, H. & Metcalfe, J. (eds.) *The Missing Link in Cognition: Evolution of Self-Knowing Consciousness*. (3-56). New York: Oxford University Press.

CHAPTER FOUR

OBJECT FUNCTION AS A BASIC REQUIREMENT FOR LANGUAGE USE AND TOOL USE[1]

OSMAN S. KINGO & PETER KRØJGAARD

Abstract: Human beings excel in several domains in comparison to other species. Considering the domains of language use and tool use, human beings appear to be far superior even to our closest relatives. In this paper we shall attempt to argue that the ability to extract and understand the functional characteristics of physical objects may be a central skill that is manifest in human beings, and a skill that serves as a basic prerequisite in order to use language and tools at a substantially more advanced level than is the case for other living creatures.

What are the uniquely human features or characteristics that make human beings stand out in comparison to other living creatures? Although a range of domains beyond any doubt deserves attention when considering this question, two features, however, appear to be especially prominent: *Language use* and *tool use*.

Language, in a broad sense, is a truly magnificent achievement. Language makes communication in the present moment substantially more sophisticated, elaborated and with more nuances than non-verbal communication. Furthermore, language allows us to move beyond the present by providing the means to communicate about the past, the future, distant locations, persons, objects, issues and events that do not have to be present – or even exist (e.g., Nelson, 1996; Spelke, 2003). Language also provides the concepts by which we order the world into manageable and conventionalized units. Concept formation is not only a cognitive economical tool for the specific individual (Rosch, 1978); it also serves as the common and shared ground that makes true social communication

[1] This paper was supported by two grants from the Danish Research Council for the Humanities (Nos 273-05-0182 and 273-07-0009) to the second author.

possible (Mandler, 2004). If someone should attempt to list the most prominent candidates for the characteristics that set human beings apart from other species, even non-linguists would have difficulties not to include language on the list.

We believe that tool use is another central feature by which human beings excel. Tool use may not stand out to the same degree as language use, since other species to some extent also use tools. However, as we shall attempt to argue that tool use in human beings differs from tool use in other species in both quality and quantity. Contrary to other creatures of the earth, human beings *produce* tools, not only to ourselves but also and mainly to distant, potential users whom we do not know and are unlikely to ever encounter.

In this paper we shall attempt to argue that the ability to understand and extract the *functional characteristics* (e.g., Nelson, 1974) of objects may serve the purpose of being a basic requirement or prerequisite in order to excel subsequently in the two above-mentioned and apparently somewhat distant characteristics, namely language use and tool use.

The paper is structured as follows: In the first two sections we will briefly unfold how language use and tool use are truly special in human beings as opposed to other species. We do not claim to present proper reviews of the issues at stake, merely to provide sufficient evidence to substantiate the claim that human beings do indeed excel in these two areas. Given the fact that language seems more complicated and relates to a range of adjacent issues we have devoted more space to language than to tool use. Note further, that when referring to 'other species' we have taken the liberty to refer predominantly to the great apes. The third section offers a closer look at how infants and small children learn to extract the functional characteristics of objects. This area is closely related to the development of concept formation and has recently gained considerable renewed interest among researchers in the field after having been almost neglected for two decades. In the fourth section we will for the sake of comparison briefly attend to the role of functional aspects in the tool use of non-human primates. Finally follows a general discussion.

Language Use in Human Beings and Non-Human Primates

As already mentioned, the ability to use language in the broadest sense may very well be the single most salient difference in manifest achievements between human beings and non-human primates. However, in order to make a fair comparison of language between human beings and non-human primates, we will exclude spoken language as such. While spoken language is indeed an important part of language use in human beings, certain physiological features (motor control of tongue and supra-laryngeal tract, Byrne, 1995, p. 165) constrain the abilities regarding sound production in non-human primates, whereby general language abilities and physiological requirements, at least to some extent, are confounded. Thus, it is no coincidence that with our specific purpose in mind we employ the term language *use*.

Given the physiological deficits present in great apes regarding sound production, researchers interested in language use seem to have concentrated on *sign language* and *concept formation*. In the following argument these two domains will be treated in turn.

Sign Language or Vocabulary

The development of the human vocabulary is truly impressive. The average normal infant produces his or her first words around their first birthday. By the age of six the number of known words has increased to approximately 10,000 (Bloom, 1998), and depending on the method of measuring applied, the vocabulary of a typical college student has been estimated to consist of up to 150,000 words (Miller, 1977, in Siegler, DeLoache & Eisenberg, 2006).

In non-human primates, the picture is radically different. For almost a century, researchers have explored the possibilities of teaching some kind of sign language to non-human primates (Byrne 1995; Tomasello & Call, 1997). To date the most impressive case of word learning in non-human primates seems to be the accomplishments of the captive-born pygmy chimpanzee or bonobo (*Pan paniscus*) Kanzi (e.g., Savage-Rumbaugh, Murphy, Sevcik, Brakke, Williams & Rumbaugh, 1993). By the age of 7 Kanzi knew the meaning of about 150 words as estimated in blind testing. One interesting aspect of the achievements of Kanzi is the way he learned words. Earlier attempts to teach signs or words to non-human primates were based on explicit and direct teaching from a human being to a great ape. One of the subjects of such endeavours was Matata – Kanzi's mother. In the first 2½ years of Kanzi's life he received no training but observed the teaching that was directed towards his mother who generally faired rather poorly. Quite surprisingly, though, it appeared that Kanzi had profited substantially more from these instructions than his mother. At the age of 7 when he was given verbal instructions (e.g., "Pour the Perrier water in the milk"), Kanzi achieved approximately about the same level of correct responses as Alia, a 2-year old human child (59% correct responses for Kanzi vs. 54 % for Alia; Savage-Rumbaugh et al., 1993). While this is impressive in itself, one should note that a 2-year-old human child might not be the most obvious choice as an exponent for achievements in mature human beings.

One of the central demarcation lines when considering and comparing language abilities in human beings and non-human primates has been the ability actively to combine symbols in a way that reveals understanding of syntactic structure (e.g., Pinker, 1994) – which is often noted as a defining feature of language (Tomasello, 1994). For many years syntactic competence has been thought of as a domain simply beyond reach for non-human primates (e.g., Byrne 1995). However, very recent evidence with monkeys may have challenged this traditional view: Male putty-nosed monkeys (*Cercopithecus nictitans*) are, among other monkeys, known to produce acoustically distinct and loud calls presumably in order to inform conspecifics of dangers, such as approaching predators (Arnold & Zuberbühler, 2006). Arnold and Zuberbühler (2006) found that apart from such predator-specific calls, male putty-nosed monkeys regularly employed *combinations* of these distinct calls, often causing the group of monkeys to move. In subsequent experiments using recorded calls, combined calls were contrasted with distinct calls while monitoring the monkeys' group behaviour. The results revealed that combined calls did seem to have a specific semantic meaning to the group as they caused the monkeys to move over significantly larger distances than non-combined calls (ibid). Analogous combinations of calls with apparently specific meaning for conspecifics have also recently been found to be present in wild gibbons (Clarke, Reichard & Zuberbühler, 2006).

Such evidence obviously gives rise to the thought that some simple aspects of syntactic ability might be present in non-human primates. Hence, syntax may therefore not exclusively be human territory. However, the results could also be interpreted in a more cautious manner in which syntax may not be implied or needed. Note, that evidence indicating that two-component calls appear to have a specific semantic meaning for conspecifics may not, in our opinion, necessarily warrant syntactic *understanding* by neither producer nor receiver. The same level of competence may have been accomplished by simple trial-error learning. Whether syntax is truly beyond reach for non-human primates seems therefore to remain an empirical question to be explored further.

However, while evidence exists that non-human primates understand symbols, as well as reference, the jury is still out there regarding syntax. Obviously, this should not be taken as evidence that non-human primates are close to human beings in overall language use. Actually, it seems fair to say that the differences in magnitude regarding language use between human beings and non-human primates are no less than enormous. Furthermore, according to Berk (2006), there is no evidence to date that shows non-human primates are capable of producing complicated *and* novel sentences – a feat demonstrated daily even by pre-school children.

Concept Formation

What about concept formation then? A concept can be defined as a summary representation that constitutes the basic meaning or *intention* of a given notion (Mandler, 1997). According to Mandler (1997, p. 163) a concept answers the core question: "What kind of a thing is it?" In contrast, the *extensional* side of concepts deals with the question regarding "which things are tigers?" For many years the extension of a given concept was considered to be defined by some necessary and sufficient features that clearly and exclusively demarcated the given concept from other concepts (e.g. cars from bicycles or tables from chairs). This approach to concept formation, formalized by Frege (1952), was earlier called the Aristotelian or Classical view and is now at times called 'the defining attributes view' (Eysenck & Keane, 2005). One important feature of this Classical view is that all members of a given category are considered equivalent, that is, no car is more 'car' than other cars.

However, the Classical approach was severely criticized, first from philosophy and subsequently from psychology. In his *Philosophical Investigations* Wittgenstein (1958) argued convincingly - using 'a game' as an example - that it was simply impossible to specify the defining attributes of everyday concepts. Furthermore, from psychology empirical research on everyday concepts revealed that the typical user did not consider different exemplars of a given concept equivalent. Rather, there seemed to be an internal structure within a given concept indicating that some members of a given category (e.g., birds) were more *prototypical* (e.g., a robin) than others (e.g., a penguin). According to prototype theory, as this approach was called (e.g., Rosch, 1975; 1978), this aspect of concepts was called the *horizontal* dimension.

Prototype theory (Rosch, 1978) proposed and provided empirical evidence for a *vertical* dimension inherent in concepts, as well. Concepts were supposed to exist at three levels: Superordinate (e.g., furniture), basic level (e.g., chair), and subordinate level (e.g., bar chair). Another important criticism of the classical view from prototype theory was that even abstract, logical concepts (e.g., natural numbers) that actually *could* be defined in accordance with the criteria proposed by the Classical view were shown to possess prototypes (Lakoff, 1987).

Therefore, while the Classical view promised a beautiful, clear-cut and rule- based approach to concepts; the view was flawed by simply not corresponding to the concepts by which human beings live. However, it also should be noted that prototype theory is not the final word on concepts of adult human beings. Today, many researchers would insist that prototype theory would have to be accompanied by the *exemplar* approach (prototypes are usually specific exemplars, not just means of incidents) as well as, the *explanation* or *theory-based* approach (some conceptual distinctions are based on theory, not by resembling physical features; e.g., Wisniewski & Medin, 1994). Let us briefly look at concept development:

Empirical evidence from prototype theory indicated that the first words children learnt belonged to the basic level (Rosch, Mervis, Gray, Johnson & Boyes-Bream, 1976). However, prototype theory has not been successful in explaining how these concepts actually develop (Rosch, 1978). Besides, developmental psychologists have shown that children clearly establish concepts *before* they are able to provide a verbal account of these. As Mandler (1998) puts it: 'Infants think before they speak.' It appears that whether such preverbal concepts belong to the basic level depends, at least to some extent, on the method used to explore these concepts: For instance, if 3 to 4-month-old infants are shown visual stimuli, they typically dishabituate when watching an exemplar from another basic level category (e.g., a dog) as opposed to an exemplar from the basic level category to which the infants were habituated (Eimas & Quinn, 1994; Quinn, Eimas & Rosenkrantz, 1993). However, if older infants, say 7 to 11-month-old-infants, are allowed to manipulate the objects with their hands, they typically show prolonged investigation, when shown exemplars from another superordinate category (e.g., animals vs. vehicles), but not when presented to exemplars belonging to different basic level categories from the same superordinate level (e.g., horses vs. rabbits, or trains vs. busses) (Mandler & McDonough, 1993). To summarize, the concepts of human beings are not only highly efficient in ordering the world into suitable and cognitively economical concepts - they are also highly complex.

In non-human primates, the concepts are by all measures much more simple. Some studies have investigated the ability to sort abstract objects. For example, Garcha and Ettlinger (1979) studied whether chimpanzees, rhesus macaques and capuchin monkeys were able to sort wooden objects into three groups based on their shape, colour and size. The results were not impressive. None of the monkeys were able to arrange the objects above chance level within the first 100 trials. However, four of the five participating chimpanzees actually reached above chance level arranging – two within the first 100 trials and two within 650 trials (Garcha & Ettlinger, 1979).

In a more recent study on abstract categorization, Smith, Minda and Washburn (2004) investigated the categorization abilities of four male rhesus monkeys and 47 adult human students. The stimuli (presented on screens) could vary in the dimensions shape, colour and size with two options for each dimension. By means of feedback following each trial, the subjects were trained to categorize stimuli in accordance to six different types of pre-defined categories unknown to the subjects. Since the tasks were rather difficult, a substantial amount of trials were given. For each monkey 72,000 trials were carried out leading to a total of 288,000 trials. Each human subject was given 1,152 trials leading to a total of 54,144 trials. As expected, the results revealed that human beings learned the categories substantially faster than monkeys. However, what was probably more interesting were the results from the study in the second type of the six types of tasks instantiating a Boolian XOR problem to the subjects (e.g., if square, then white = A and black = B; if triangle, then black = A and white = B). When given this kind of tasks, human beings faired remarkably better than monkeys. According to Smith and colleagues, these results may indicate that monkeys build their categories exclusively by means of association and generalization. Human beings, in contrast, also employ *explicit rule learning* which appears to be a major advantage when dealing with XOR problems (Smith et al., 2004). In this respect, language use as such must be an effective tool when attempting to make rules explicit. Thus, explicit rule learning seems to be far more advanced in human beings than in monkeys.

While these results are far from convincing, it should be noted that such abstract objects are unlikely to have any functional significance to the participating subjects.

In a seminal study by Savage-Rumbaugh, Rumbaugh, Smith & Lawson (1980) objects with more direct significance for the subjects were used. Two language trained chimpanzees were taught to sort objects into groups of what human beings would consider as either 'food' or 'tools'. The two subjects learned fairly quickly to arrange the objects into these two distinct classes. Subsequently, the real objects were substituted with plastic tokens (lexigrams) corresponding to these categories. Furthermore, tokens referring to completely new objects were introduced to the categorization task. The results revealed that the two chimpanzees actually succeeded sorting even the novel tokens into the right categories as either food or tools. According to the authors, this indicated that the chimpanzees were able mentally to represent the referents of the lexigrams (Savage-Rumbaugh et al., 1980; for similar and more recent studies with baboons, see Bovet & Vauclair, 1998).

In more recent studies researchers have often used pictorial representations of the objects to be categorized. For instance, in a *same/different* design Wright, Rivera, Katz & Bachevalier (2003) investigated whether three capuchin monkeys could learn to indicate whether exemplars from categories (e.g., apples, buildings, cats, women's faces, flowers etc.) belonged to the same or to a different category. All three capuchin monkeys eventually reached 80 % correct scores, but only after 32 sessions of 100 trials each (Wright et al., 2003). Besides the substantial number of trials necessary to achieve above chance level in such studies, there are also often large individual differences amongst the participating animal subjects (e.g., Tanaka, 2007; Vonk & MacDonald, 2004).

The results from the seminal study by Savage-Rumbaugh and colleagues led to another important aspect of concept formation in non-human primates: When scoring in relation to categories is successful in non-human primates, it is almost always related to what Rosch (1978) would call the superordinate level (e.g., Murai, Kosugi, Tomonaga, Tanaka, Matsuzawa & Itakura, 2005) whereas basic level categorization is rarely seen. However, in a recent error study of mental representations of symbols with two bonobos (of which Kanzi was actually one) by Lyn (2007), the evidence revealed that the subjects made hierarchical categorizations. Thus, basic level categorization may not after all be a uniquely human characteristic.

To conclude, great apes do seem to have some ability to form concepts, although their competence is far from being equivalent to that of human beings. Furthermore, if the objects – or at least their referent – have no functional significance to the non-human primates, their ability to categorize appears enormously restricted in comparison to human beings.

Tool Use in Human Beings and Non-Human Primates

The amount and diversity of tool use in human beings is truly impressive. Some of the tools basically work as extensions of the human body. Although cranes and excavators are far more complicated to construct and use than basic tools like shovels, spears, hammers, knives and axes, they all share the idea that they basically work as extensions of the human body by which they make it possible to carry out work far easier than without tools.

However, within the last decades an entirely different kind of tools has been developed, that is, software-based tools, like cell phones, laptops and automotive navigation systems. Contrary to other tools, these computer-based tools typically have a graphical user interface in which the physical constraints (e.g., gravity, continuity) govern our macroscopic ecological niche are put aside. On such graphical user interfaces icons or symbols disappear and reappear fairly unnoticed. Besides, such advanced tools share the feature that their most powerful and intended affordances (cf. Gibson, 1979) are hidden to the naïve subject. Although there is reason to believe that we ought to keep the physical constraints that rule in the 'real world' outside the graphical display in mind when designing such interfaces (Nørager, this volume), the ability to manipulate symbols abstracted from the physical basis seems mandatory in order to benefit from such advanced tools.

Tool use in non-human primates is markedly different. According to Byrne (1995), tool use in our closest relatives, the chimpanzees, is restricted as opposed to human beings in the following way – all being related to the range of methods when attempting to manufacture tools:

> "No tools have been recorded made by addition of one item to another, and no tool has been made in the wild whose sole function is to make another tool" (p. 96-97).

However, Byrne (1995, p. 97) acknowledges that individual examples of great apes using a sliver as a sort of a wedge in order to stabilize a loosely seated anvil stone before attempting to crack nuts with a hammering stone comes close.

While we endorse Byrne's conclusion, we would like to add that the difference in tool use between human beings and non-human primates should not be restricted to differences regarding the manufacturing processes, although these are definitely important. As outlined above, some of the *end products* themselves (i.e. software-based tools) manifest in modern societies are qualitative different from the tools used among non-human primates.

In summary, we believe it is fair to say that language use and tool use is substantially more advanced and sophisticated in human beings than in any other primate. Although these two areas may seem quite different at first glance, we shall argue that the superiority regarding these domains manifest in human beings may, at least partly, derive from a special ability that enables us to extract the functional properties of objects – an ability already present even in human infants and small children.

Extraction of Functional Aspects of Objects in Infants and Small Children

How do infants successfully organize the world into the meaningful categories and concepts underlying language? The answer to this question is, obviously, very complex, and several decades of studies of categorization and concept formation have reflected this complexity (Cohen & Cashon, 2006; Kingo, 2008). At present, one of the thriving areas of investigation regarding infant categorization and concept formation consists of studies on object-function. Object-functions are especially interesting since they are conceptually rich constructs that embody causal relations between objects properties, actions on objects physical outcomes and the goals of purposeful agents (Booth, 2006).

The investigation of object-function as a crucial factor in infant categorization and concept formation has seen many changes in the last three decades. In the mid-70's, Nelson (1974) pioneered by putting forth the "functional core hypothesis", proposing that 1-year-old children in their earliest language development would first form concepts about things in the world by experiencing them in interaction with other people, and then would later learn and generalize object-names on the basis of these concepts. The core of these concepts was proposed to be the *function* of the things, what they did and what could be done with them. Taking the concept of a ball or "the idea of ballness" as an example, Nelson proposed that over time the child would compare the various relations into which the ball enters and synthesize those relations or functions that are *invariant* across events (e.g. "rolls", "bounces", "is picked up", "is tossed", "is caught" etc.). This synthesis of "functions" would constitute a *functional core* that would enable the infant to represent the notion of a ball *between* different encounters instead of representing each new instance of a ball separately (Nelson, 1974). In retrospect such a perspective on concept formation was in accordance with the general *Zeitgeist* of the 70's in this area of research (Nelson & Ware, 2002). Since then, though, the focus on studies of categorization and concept formation has moved from functional features to other object features such as visual static attributes of the objects *per se* or simple perceptual discrimination of pictures of various objects (e.g. Eimas & Quinn, 1994; Quinn, Eimas, & Rosenkrantz, 1993).

In recent years, a renewed interest in object-function has emerged. One of the shortcomings of the research in object-function in the first decades was the inability properly to separate functional attributes from other object attributes (Nelson & Ware, 2002). Now, due to the tremendous development in infant research methodology, such more sophisticated studies have become feasible, and object-function is once again moving onto centre stage. The new multitudes of methodologies have led to an increased need for a clear definition of the term object-function. Nelson has suggested that at least four different variations of the notion of function emerge from the child's point of view:

1. Actions on things, for example, throwing balls. This aspect conforms to Rosch's emphasis on common motor movements of persons engaged with the objects.
2. Independents activity of the things themselves, for example, a barking dog. This aspect may help to define animates versus inanimates.
3. Reaction of a thing to an action on it, for example, the rolling of a ball after it is pushed, reflecting a cause-and-effect relation between two actions or between an action and an end state.
4. The use (idiosyncratic or conventional) of a thing for human purposes, for example, drinking from a cup. This aspect reflects the relation of an object to goal activities. Conventional uses is the definition often applied to concepts of artefacts in discussions of conceptual understanding and object naming, but may not be the function that a child will identify as significant. (Nelson, 1979 cf. Nelson & Ware, 2002, p. 164-165).

Keeping in mind that object-function comes in different guises, we now present a small but relevant selection of contemporary studies and theories targeting object-function. We do this in order to emphasize the important position the concept of object-function has (re)claimed in the field of infant categorization and concept formation.

Some of the earliest examples of the more contemporary approach to object-function were the studies by Kelly Madole and colleagues in the mid-90's. These authors were interested in the assumed correlation between form and function in infant categorization (Madole, Oakes, & Cohen, 1993; Madole & Cohen, 1995). In one study, Madole, Oakes and Cohen (1993) let infants manipulate different objects in an object-examination task. An infant would be familiarized with a single object and was subsequently given a modified object to examine. Form and function were studied as separate factors since the modified object differed from the first in either form, function or in form *and* function. It was then measured if and how much the infant increased its attention to the modified object relative to the first object (a typical "out of category"-response from infants) and relative to a completely novel object. The stimuli objects were constructed from Lego® building blocks (see figure 1 for examples) and crossed two different shapes with two different functions resulting in four different objects. One function was *rolling* (the wheels were fixated on half of the objects to disable this function) and the other was *shaking* (half of the objects made a rattling sound when shaken). The novel object was a "mooing box" that mooed when it was tilted.

Figure 1

Pictures from Madole, Oakes & Cohen (1993)

The experimenters found, that 10-month-olds responded to changes in form only, while 14-month-olds responded to changes in form *or* in function (Madole et al., 1993). In the second experiment of this study, 10-, 14- and 18-month-old infants were familiarized with two objects in which a consistent form-function relationship was maintained. Infants were then tested with (1) a correlated object that maintained the familiarized form-function relationship, (2) an uncorrelated object that violated this relationship and (3) a novel object. The 10- and 14- month-old infants increased attention only to the novel object, while the 18-month-olds increased attention to the uncorrelated object, as well as the novel object. From these two experiments, it was concluded that there seems to be a developmental progression from attending only to the form of objects, to attending to form and function as separate properties, and finally to attending to the *relationship* between form and function (Madole et al., 1993).

In later studies, Madole and Cohen (1995) investigated whether the perceived form-function correlations depended on specific parts of the objects. In the first study, 14- and 18-month-old infants were found to pay attention to the correlations of the object-function and a specific object part. In studies two and three, objects were presented in such a way that the form of one part correlated with the function of a different part (contrary to the correlations most often found in real objects where the function of a specific part tends to correlate with the form of that specific part). Only the *younger* infants attended to the form-function correlations in these studies. This result was seen to suggest that the older children had acquired constraints on the kinds of correlations they recognize – this properly to adapt to or "tune in to" the typical correlations of real objects (Madole & Cohen, 1995). This trend has been confirmed by other studies such as Träuble and Pauen (2007) where 15-month-olds were found to acquire more *specific* functional knowledge from novel objects compared to 12-month-olds who primarily acquired *general* functional knowledge.

The research described above from the labs of Oakes, Madole and colleagues has partly challenged the functional core hypothesis suggested by Nelson. While maintaining that functional aspects play an important role in the formation of object concepts, these authors did not find any evidence of infant attention towards functional aspects before the age of 14 months. The functional core hypothesis rests on the assumption that functional aspects are essential in the concept formation of the 1-year-old, which in turn presupposes that infants are able to pay attention to these aspects at least from this age. However, in spite of the age discrepancies, the above-mentioned work by Oakes, Madole and colleagues confirms the relative importance of functional information in infant cognition and represents a "microanalytic" and more detailed approach to the study of object-function (Oakes & Madole, 2003). These two studies, though, primarily target the *fourth* notion of function as suggested by Nelson (described earlier).

Evidence regarding the importance of function in the sense of Nelson's *first* notion ('action on things') comes from Linda B. Smith (2005). She found that making 2-year-olds move an object horizontally or vertically defined the horizontal or vertical axis as the main axis of elongation and systematically changed the range of shapes seen as familiar. In other words, the child's own movement of the object affected the child's categorical responses even with very simple movement patterns (horizontal or vertical lines). Interestingly, making the child observe an adult perform the same movements did *not* affect the child's subsequent categorical responses. In a somewhat similar experiment, symmetrical or asymmetrical movement of a symmetrically ambiguous object effected whether or not this object was judged to be more or less symmetrical (symmetrical movement resulted in "symmetrical" judgements and vice versa). Again, watching others perform the actions did not change the range of shapes children judged to be members of the same category (ibid.). All in all, this study points to a very direct connection between *actions* on objects and the *categorization* of these objects. This connection has also been pointed out in other kinds of studies such as studies on action planning and the learning of rules and regularities in infancy (e.g. von Hofsten, 2007).

The distinction between self-action and the actions of others relates to the distinction between Nelson's first and third notion of function versus the second notion. That is, the distinction between actions on objects and the object's reaction to that action, on the one hand, and the independent activity of objects, on the other. Infants seem to be able to make this latter distinction at a very early age. Markson and Spelke (2006) found that 7-month-old infants were able to differentiate between the self-propelled motion of an object and the movement of an object by the experimenter's hand. Furthermore, infants from the same age group were able to retain the specific movement characteristics (self-propelled or passively moved) of different objects for some time (15 min.), indicating that the learning is robust and infants' expectations about self-propelled motion are enduring (ibid.). This study provides indirect evidence that infants are able to extract and retain different kinds of functional information from a very early age – well within the first year of life as originally proposed by Nelson (1974).

Mareschal and Johnson (2003) have argued that information on object-function plays another crucial role even earlier in development. They argue that young infants (4 months of age) are limited in the kinds of information they are able to integrate simultaneously. Specifically, the authors describe how the visual information of objects can be processed by two different neural "streams" or pathways in the cortex: The *dorsal* stream which primarily processes information on the location and movement of objects, and the *ventral* stream which primarily processes information on other object features such as shape, colour, texture, etc. At four months infants are not fully able to integrate these two information streams and are therefore forced to act based on only one of the two available kinds of information. The relevant point here is that it seems to be the object-function that determines which one of the neural pathways will dominate the infant's behaviour or actions towards the object. Three-dimensional objects, for instance, tend predominately to activate the dorsal stream since they afford handling and because location and movement are essential in such acts. However, two-dimensional pictures tend predominately to activate the ventral stream since such pictures do not afford handling and may therefore primarily provide the infant with information on the non-functional object features such as surface pattern and texture (ibid.). This is yet another example of the central role of object-function in early human infancy.

How do these different aspects of object-function play a role in concept formation? As mentioned earlier, Mandler (1997) has proposed that a concept in effect answers the question: "What kind of a thing is it?" Throughout her work, Mandler has maintained that conceptual information is not directly about the perceptual properties of an object *per se* but has more to do with what an object *does* or what it can potentially do (a door swings, opens and closes; a dog barks, runs, bites, etc.) (Mandler, 2004). Being an influential researcher in this area, Mandler has had many disputes with other researchers, but this claim of the primacy of object function in concept formation has taken few beatings and has spurred the greatest agreement among developmental researchers (Carey, 2000; Gibson, 2000; Murphy, 2004; Nelson, 2000; 2004; Quinn & Eimas, 2000; Quinn, 2004; Reznick, 2000; Shutts & Spelke, 2004; see also Kingo, 2008). Function, it seems, is closely tied to concept formation and thereby language.

By now, we hope it is evident that the kind of information collectively termed object-function is a crucial factor in early human categorization and concept formation, or at least the object of intense interest from a range of developmental researchers at present. Infants very rapidly extract and use information on object-function for many different purposes, among these the formation of concepts and mastery of their environment. The general trend is that more and more functional information becomes available with development. At the same time, though, the pool of information that is actually processed will meet an increasing amount of constraints due to an increase in background knowledge (Madole & Cohen, 1995; Oakes & Madole, 2003).

Extraction of Functional Aspects of Objects in Non-Human Primates

The motivational aspect is obviously very important when considering tool use in non-human primates (Tomasello & Call, 1997). Contrary to the prominent curiosity in human infants regarding objects in general, objects for non-human primates are primarily interesting when they are found to serve a direct purpose (Byrne, 1995). In that respect 'Curious George' should have been a child, not an ape! Thus, when investigating tool use in non-human primates, the functional aspects of the objects present are highly relevant indeed. In the present context one may ask how non-human primates fare in comparison to human beings when attempting to extract functional aspects of objects?

In a recent paper by Spaulding and Hauser (2005), the authors summarize the last decade of studies investigating what non-human primates understand of the tools they use, by stating that while the animals are clearly sensitive to the perceptually relevant features of the tools, they simply appear more or less to *lack* understanding of the physical properties the tools might possess (for instance, by choosing a perceptually similar tool, but made of an inadequate material – e.g. rubber instead of plywood; Povinelli, 2000, cited in Spaulding & Hauser, 2005).

Thus, the importance of functional aspects of tool use in non-human primates in comparison to human beings seems to contain a paradox: On the one hand, it seems unavoidable to take functional aspects of objects into account when considering tool use in non-human primates, since objects with no obvious functional relevance will have a very limited interest to the subjects. On the other hand - and contrary to what is the case for human beings - non-human primates display severe difficulties in disentangling the functional *core* of a given object; for great apes the functional aspects appear to be bound to featural aspects while 'hidden' functional properties predominantly remain beyond reach.

Discussion

We have now contended that the ability to extract functional information rapidly from objects is a crucial and early developing ability of human infants. For humans, this ability is central in the formation of concepts and in the development of tool use. We find that contemporary research on object-function in developmental psychology makes it clear that functional information plays a crucial role in infant cognition on several levels, and that functional information is a factor that constrains *and* expands the pool of possible interpretations of the physical surroundings, thereby partly shaping the human approach to the world. Furthermore, the contemporary approach to object-function shows us that infants do indeed possess the necessary skills to build representations similar to the functional cores suggested by Nelson (1974).

In non-human primates, it seems, functional information does not have the same impact on cognition. Although, non-human primates display some knowledge of object-function, this knowledge seems to be very restricted compared to humans and very purpose-specific. Contrary to human beings, it does not seem meaningful to describe the information non-human primates possess on objects-functions as a *functional core*. Functional cores are per definition complex since they are summaries of *various* actions with the same kind of objects. This is a kind of representation that is very different from what seems to be better described as somewhat simpler object-action-goal associations in the non-human primate world. Functional cores are *abstractions* and as such they allow humans to apply functional information beyond the present moment or specific problem. The implications and consequences of such ability are by no means small, as we shall now argue.

One influential approach to the meaning of objects is the theory of affordances by Gibson (1979). *Affordances* are what the world offers the individual. Depending on the physical characteristics, the skills available and the task at hand, a given subject can extract a range of different affordances from a given object. In the present context, it may be argued that affordances are more or less equivalent to *functional* characteristics of objects. The Gibsonian approach seems fairly straightforward when considering functional characteristics – or affordances – that are prominent to a given subject (e.g., large trees may offer shelter from the rain for creatures searching for that). However, other functional characteristics or affordances may not be likewise salient to the naïve subject. The functional characteristics or affordances of many man-made artefacts are not obvious unless you *know* what they are made for. Consider for instance artefacts like washing machines, computer routers or cell phones. While such objects may have many different and valuable affordances they were each made with a *specific* purpose in mind (e.g. Tomasello, 1999), and due to the complexity of the objects at stake, these purposes are not easily specified and subsequently 'picked up' directly in the visual array. Nevertheless, contrary to non-human primates, human beings have no problems using such artefacts. Note, that while non-human primates had difficulties separating functional characteristics of tools from their physical appearance, this is learned in late infancy in human beings.

Obviously, there is likely to be many requirements for a given subject in order to succeed in extracting the functional properties of a cell phone. Our point is simply that the superiority of human beings in this regard may, at least partly, derive from the ability to abstract functional characteristics that may – or may not – be prominent in the visual array.

The reader may consider highly advanced software-based artefacts as cell phones and equivalent as unfair items to base the arguments upon, since such objects are not only characterized by having somewhat 'hidden' affordances but also by requiring symbol manipulation in order to be handled – which is, as we have argued, another area in which human beings excel in comparison to other species.

However, the best example underscoring the claim that human beings appear to be light years ahead of other species in this respect may be present in the symbolic handling of other and much simpler objects. Let us take a brief look at *pretend play*. The stereotypical example of pretend play is when the human child at 1½ years begins to *pretend* to use a given object for a completely different purpose – as for instance when he or she uses a banana as a telephone, or a piece of wood as a car or a gun (e.g., Leslie, 1987). Note, that such applications are *not* based on inherent functional characteristics of the objects on which the pretended implementation takes place, nor necessarily by physical resemblances between the objects. Pretend play is not only impressive by implying the fragile emergence of meta-representations, as argued by Leslie (1987); pretend play can also, in our opinion, be seen as a fascinating manifestation of the ability to abstract a functional core of a given object (e.g., a cell phone) and pretend the implementation of this functional characteristic in an entirely different object. If the infant were unable to abstract the functional core of a given object, we would seriously doubt that pretend play would or could take place at all. When considering the necessary cognitive requirements in order to be able to think about and make new tools, the imaginary abilities manifest in pretend play must be a very powerful 'mental tool' indeed. Our claim is supported by the concluding paragraph by Tomasello and Call (1997, p. 70) in a brief section on symbolic play: "Apparently other species of primates have not been given the opportunity to display these same "symbolic" behaviours." Thus, we believe that the ability to abstract functional cores of objects may be an indispensable prerequisite for the superior tool use manifest in human beings as opposed to other species.

Let us now turn to language. As stated previously, the cognitive semantics of language may be the "fairest" area of comparison between the language of human and non-human primates since the ability of language production *per se* is largely governed by the specific physiological constitution of each species. We contend that an important factor in the excellence of human semantics may be our superior ability, not only to extract, but also to *abstract* functional information from experiences with our surroundings. The abstracts or summary representations of objects and events provide us with virtually context-free concepts, for instance, in the form of functional cores (Nelson, 1974; 1985) or other results of function/meaning analysis (e.g., Mandler, 1997; 2004). As with tools, the richness of functional information in our concepts allows us to use them very flexibly. Again, since our concepts are constituted by a range of functions or utilizations, we are able to apply these concepts in many different situations with very little effort. This flexibility may be one of the communication highlights of language as it allows for infinite word-word relations and complex (conventionalized) meaning "packages" (syntax or grammar). Naturally, language is more than functional information in a specific format, but language, as we know it, seems impossible without our inherent aptitude for discovering functions, and as far as we know, we are the only species with this highly developed aptitude. Non-human primates do not seem to be able to produce complicated or novel sentences or to be able actively to combine symbols in this flexible way (fully in parallel with their somewhat inflexible tool use). As the picture unfolds, it becomes more and more evident how keeping track of a multitude of functional characteristics for each object experienced gives human beings a huge advantage in the formation of concepts. At present we are still uncovering the nature of this ability in humans and in human infants in particular, but at this point we do know that functionality shapes human cognition from a very early age. Our functional aptitude provides us with a kind of multifunctional "mental tools" which enables us to contain, shape, produce and communicate our impressions of the world in a unique way.

Language use and tool use are somewhat different domains in which human beings excel in comparison to other species. In this chapter we have argued that the ability to extract and abstract the functional characteristics of objects may serve as a means to facilitate the excellence in human beings as opposed to other species in the aforementioned areas. In this vein we therefore welcome the renewed interest in infants' understanding of functionality in objects – an area that has remained central for a century within the field of comparative psychology.

References

Arnold, K., & Zuberbühler, K. (2006). Semantic combinations in primate calls. *Nature, 441,* (303).

Berk, L. E. (2006). *Child Development. 7th. ed.* New York: Allyn and Bacon.

Bloom, L. (1998). Language acquisition in its developmental context. In: D. Kuhn & R.S. Siegler (eds.), *Handbook of Child Psychology: Vol. 2: Cognition, Perception and Language* (5th ed.),(309-370). New York: Wiley.

Bovet, D., & Vauclair, J. (1998). Functional categorization of objects and of their pictures in baboons (papio anubis). *Learning and Motivation, 29,* (309-322).

Booth, A. E. (2006). Object function and categorization in infancy: Two mechanisms of facilitation. *Infancy, 10*, (145-169).

Byrne, R. (1995). The Thinking Ape: Evolutionary Origins of Intelligence. Oxford: Oxford University Press.

Carey, S. (2000). The origins of concepts. *Journal of Cognition and Development, 1*, (37-41).

Clarke, E., Reichard, U.H., & Zuberbühler, K. (2006). The syntax and meaning of wild gibbon songs. *Plos One, 1,* (1-10).

Cohen, L. B., & Cashon, C. H. (2006). *Infant Cognition.* Hoboken, NJ, US: John Wiley & Sons Inc.

Eimas, P. D., & Quinn, P. C. (1994). Studies on the formation of perceptually based basic-level categories in young infants. *Child Development, 65*, (903-917).

Eysenck, M. W., & Keane, M. T. (2005). *Cognitive Psychology*. (5th. ed.). Hove, UK: Erlbaum.

Frege, G. (1952). On sense and reference. In: P. Geach & M. Black (eds.), *Translations from the Philosophical Writings of Gottlob Frege.* Oxford: Basic Blackwell.

Garcha, H. S. & Ettlinger, G. (1979). Object sorting by chimpanzees and monkeys. *Cortex, 15,* (213-224).

Gibson, E. J. (2000). Commentary on perceptual and conceptual processes in infancy. *Journal of Cognition and Development, 1*, (43-48).

Gibson, J. J. (1979). *The Ecological Approach to Visual Perception.* Boston: Houghton Mifflin.

Kingo, O. S. (2008). The Concept of Concepts: Perceptual and Conceptual Categorization under Scrutiny. *Journal of Anthropological Psychology, 19,* (1-21).

Lakoff, G. (1987). Women, Fire and Dangerous Things: What Categories Reveal about the Mind. Chicago: The University of Chicago Press.

Leslie, A.M. (1987). Pretense and representation: The origins of "theory of mind". *Psychological Review, 94*, (412-426).

Lyn, H. (2007). Mental representation of symbols as revealed by vocabulary errors in two bonobos (*pan paniscus*). *Animal Cognition, 10,* (461-475).

Madole, K. L., & Cohen, L. B. (1995). The role of object parts in infants' attention to form^function correlations. *Developmental Psychology, 31*, (637-648).

Madole, K. L., Oakes, L. M., & Cohen, L. B. (1993). Developmental changes in infants' attention to function and form function correlations. *Cognitive Development, 8*, (189-209).

Mandler, J. M. (1997). Development of categorization: Perceptual and conceptual categories. In: G. Bremner, A. Slater & G. Butterworth (eds.), *Infant Development: Recent Advances.* (163-189). Psychology Press/Erlbaum (UK) Taylor & Francis, Hove: England.

—. (1998). Babies think before they speak. *Human Development, 41,* (116-126).

—. (2004). *The Foundations of Mind: Origins of Conceptual Thought* (1st ed.). New York: Oxford University Press.

Mandler, J. M., & McDonough, L. (1993). Concept formation in infancy. *Cognitive Development, 8*, (291-318).

Mareschal, D., & Johnson, M. H. (2003). The "what" and "where" of object representations in infancy. *Cognition, 88*, (259-276).

Markson, L., & Spelke, E. S. (2006). Infants' rapid learning about self-propelled objects. Infancy. Special Issue: Presence and quality of touch influence coregulation in mother-infant dyads, 9, (45-71).

Murai, C., Kosugi, D., Tomonaga, M., Tanaka, M., Matsuzawa, T., & Itakura, S. (2005). Can chimpanzee infants (*pan troglodytes*) for categorical representations in the same manner as human infants (*homo sapiens*)? *Developmental Science, 8,* (240-254).

Murphy, G. L. (2004). On the conceptual-perceptual divide in early concepts. *Developmental Science, 7*, (513-515).

Nelson, K. (1974). Concept, word, and sentence: Interrelations in acquisition and development. *Psychological Review, 81*, (267-285).

—. (1985). Making Sense: The Acquisition of Shared Meaning. London: Academic Press.

—. (1996). *Language in Cognitive Development: Emergence of the Mediated Mind.* Cambridge University Press, New York, NY: US.

—. (2000). Global and functional: Mandler's perceptual and conceptual processes in infancy. *Journal of Cognition and Development, 1*, (49-54).

—. (2004). A welcome turn to meaning in infant development: Commentary on mandler's the foundations of mind: Origins of conceptual thought. *Developmental Science, 7*, (506-507).

Nelson, K., & Ware, A. (2002). The reemergence of function. In: N. L. Stein, P. J. Bauer & M. Rabinowitz (eds.), *Representation, Memory, and Development: Essays in Honor of Jean Mandler.* (161-184). Lawrence Erlbaum Associates, Publishers, Mahwah, NJ: US.

Nørager, R. (2008, this volume). Why anthropological psychology? The importance of non-human primates, human infants and adults for understanding modern computerised technology.

Oakes, L. M., & Madole, K. L. (2003). Principles of developmental changes in infants' category formation. In: D. H. Rakison, & L. M. Oakes (eds.), *Early Category and Concept Development: Making Sense of the Blooming, Buzzing Confusion.* (132-158). New York, NY, US: Oxford University Press.

Pinker, S. (1994). The Language Instinct: How the Mind Creates Language. New York: William Morrow & Co.

Quinn, P. C. (2004). Multiple sources of information and their integration, not dissociation, as an organizing framework for understanding infant concept formation. *Developmental Science, 7*, (511-513).

Quinn, P. C., & Eimas, P. D. (2000). The emergence of category representations during infancy: Are separate and conceptual processes required? *Journal of Cognition and Development, 1*, (55-61).

Quinn, P. C., Eimas, P. D., & Rosenkrantz, S. L. (1993). Evidence for representations of perceptually similar natural categories by 3-month-old and 4-month-old infants. *Perception, 22*, (463-475).

Reznick, J. S. (2000). Interpreting infant conceptual categorization. *Journal of Cognition and Development, 1*, (63-66).

Rosch, E. (1975). Cognitive reference points. *Cognitive Psychology, 7,* (532-547).

—. (1978). Principles of categorization. In: E. Rosch, & B. B. Lloyd (eds.), *Cognition and Categorization* (27-48). Hillsdale, N.J.: Lawrence Erlbaum.

Rosch, E., Mervis, C. B., Gray, W. D., Johnson, D. M., & Boyes-Braem, P. (1976). Basic objects in natural categories. *Cognitive Psychology, 8*(3), (382-439).

Savage-Rumbaugh, E. S., Murphy, J., Sevcik, R. A., Brakke, K. E. Williams, S. L., & Rumbaugh, D.M. (1993). Language comprehension in ape and child. *Monographs of the Society for Research in Child development, 58,* Nos. (3-4).

Savage-Rumbaugh, E. S., Rumbaugh, D. M., Smith, S. T., & Lawson, J. (1980). Reference: the linguistic essential. *Science, 210,* (922-925).

Siegler, R., DeLoache, J, & Eisenberg, N. (2006). *How Children Develop. 2nd ed.* New York: Worth Publishers.

Shutts, K., & Spelke, E. S. (2004). Straddling the perception-conception boundary. *Developmental Science, 7*, (507-511).

Smith, J.D., Minda, J.P., & Washburn, D.A. (2004). Category learning in rhesus monkeys: A study of the Shepard, Hovland, and Jenkins (1961) tasks. *Journal of Experimental Psychology: General, 133,* (398-414).

Smith, L. B. (2005). Action alters shape categories. *Cognitive Science, 29,* (665-679).

Spaulding, B. & Hauser, M. (2005). What experience is required for acquiring tool competence? Experiments with two callitrichids. *Animal Behaviour, 70*(3), (517-526).

Spelke, E. S. (2003). What Makes Us Smart? Core Knowledge and Natural Language. Cambridge, MA, US: MIT Press.

Tanaka, M. (2007). Recognition of pictorial representations by chimpanzees (*pan troglodytes*). *Animal Cognition, 10,* (169-179).

Tomasello, M. (1994). Can an ape understand a sentence? A review of *Language Comprehension in Ape and Child* by E.S. Savage-Rumbaugh et al. *Language and Communication, 14,* (377-390).

—. (1999). The cultural ecology of young children's interactions with objects and artifacts. In: E. Winograd, R. Fivush & W. Hirst (eds.), *Ecological Approaches to Cognition. Essays in Honour of Ulric Neisser* (153-170). Mahwah, NJ: Erlbaum.

Tomasello, M., & Call, J. (1997). *Primate Cognition.* New York, NY, US: Oxford University Press.

Träuble, B., & Pauen, S. (2007). The role of functional information for infant categorization. *Cognition, 105*, (362-379).

von Hofsten, C. (2007). Action in development. *Developmental Science, 10*, (54-60).

Vonk, J., & MacDonald, S.E. (2004). Levels of abstraction in orangutan (*pongo abelii*) categorization. *Journal of Comparative Psychology, 118,* (3-13).

Wisniewski, E. J., & Medin, D. (1994). On the interaction of theory and data in concept learning. *Cognitive Science, 18,* (221-281).

Wittgenstein, L. (1958). *Philosophical Investigations.* Oxford: Blackwell.

Wright, A.A., Rivera, J.J., Katz, J.S., & Bachevalier, J. (2003). Abstract-concept learning and list-memory processing by capuchin and rhesus monkeys. *Journal of Experimental Psychology: Animal Behaviour Processes, 29,* 184-198.

CHAPTER FIVE

THE HUMAN ABILITY TO SINGLE OUT AND TRACK SPECIFIC OBJECTS THROUGH SPACE AND TIME: ORIGIN AND APPLICATION[1]

PETER KRØJGAARD

Abstract: This paper pursues the origin and application of the human ability to single out and track specific objects through space and time. Evidence from object based visual attention in adults, as well as, infant research on object individuation is reviewed, indicating that spatiotemporal information is important when tracking objects. However, recent results from comparative research indicate that basic aspects of object individuation may also be prominent in some non-human primates. Thus, when adult human beings display elaborated understanding of cultural artefacts beyond the level manifest in infants and non-human primates, some additional skills far beyond object tracking here-and-now must be required. It is proposed that the ability to abstract spatiotemporal threads of specific objects may be the uniquely human skill that fulfils this requirement.

"You are beautiful, but you are empty," he went on. "One could not die for you. To be sure, an ordinary passerby would think that my rose looked just like you—the rose that belongs to me. But in herself alone she is more important than all the hundreds of you other roses: because it is she that I have watered; because it is she that I have put under the glass globe; because it is she that I have sheltered behind the screen; because it is for her that I have killed the caterpillars (except the two or three that we saved to become butterflies); because it is she that I have listened to, when she grumbled, or boasted, or even sometimes when she said nothing. Because she is *my* rose. […] It is only with the heart that one can see rightly; what is essential is invisible to the eye." (Saint-Exupéry, 1996, p. 68).

---

[1] This paper was supported by two grants from the Danish Research Council for the Humanities (Nos 273-05-0182 and 273-07-0009).

Introduction

Considering everyday objects like nails, pieces of chalk, plates or ashtrays, it may not make any difference whether the object located on the table here-and-now is exactly the *same one* (numerical identity, Strawson, 1964) as the one encountered on the table five minutes ago, *or* just another identical looking exemplar of the *same kind* (qualitative identity, Strawson, 1964). However, when it comes to heirlooms, original pieces of art, favourite toys, pets, friends, children, loved ones – or the precious rose in *The Little Prince* - the difference between the same object and a replica becomes of paramount importance. Thus, in their everyday lives adult human beings are indeed capable of singling out and tracking specific object through space and time.

So one obvious question is: What is the origin of such ability? Here I shall attempt to pursue this question. In anticipation, I will present evidence and arguments that, on the *one* hand, suggest that the ability to single out and track specific objects through space and time can be traced back to infancy, as well as, to some extent dates back to our non-human primate ancestors; on the *other* hand, adult human beings seem to exploit this ability far beyond the level manifest in infants and non-human primates. I shall propose that when adult human beings excel in comparison to infants and non-human primates in their use of such competence, it may derive from the ability to *abstract* spatiotemporal threads of objects from one domain to another.

The paper is structured as follows: The *first* section of the paper is devoted to a brief outline of the different kinds of information that is usually considered relevant for keeping track of objects. The *second* section presents evidence from studies on object based visual attention in adults. The *third* and most comprehensive section focuses on infant research on object individuation. The *fourth* section is a brief outline of some of the few studies on object individuation that have been conducted with non-human primates. The *fifth* and final section offers examples of how the ability to single out and track specific objects paves the way for important achievements in early childhood, as well as, in adulthood: Regarding early childhood, I shall argue that such an ability is a necessary prerequisite for concept formation and word learning. As for adulthood, I will attempt to argue that adult human beings exploit this ability far beyond other primates in our appreciation of man-made artefacts and culture in general. Finally, I will propose what might be responsible for this uniquely human feature.

Featural and Spatiotemporal Information of Objects

It is commonly agreed that in order to keep track of a given object over time (e.g. a pen or a bicycle) one has to pay attention to at least *two* different kinds of object information, that is, featural and spatiotemporal information (e.g., Krøjgaard, 2004; Mammen, 1996, 2002; Xu, 1999).

Featural information refers to information regarding an object's specific properties of which colour, size, shape and texture are usually the most salient (Wilcox, 1999) whereas *spatiotemporal* information refers to basic information regarding object location and object motion. Adults seem to possess rather sophisticated knowledge regarding what solid objects at a macroscopic level can and cannot do.[2] For instance, an object without support will fall downwards until it meets a solid surface; one distinct object cannot occupy two different locations at the same time; two distinct objects cannot occupy the same location at the same time; objects move continuously in space and time, etc. (Spelke, Breinlinger, Macomber, & Jacobson, 1992; Xu, 1999). However, although our spatiotemporal knowledge may seem obvious and trivial, it is rarely reflected upon in everyday conversation. Only when our tacit expectations are violated, for instance when we cannot find our keys where we thought we left them, or when a competent magician demonstrates his tricks to us, our firm belief in spatiotemporal information becomes salient.

Cognitive Psychology: Approaches to Object Tracking in Adults

For quite some years featural information of objects was given significantly more attention within the field of cognitive psychology than spatiotemporal information (Krøjgaard, 1999b; Mammen & Krøjgaard, submitted). This was, for instance, evident in the seminal categorization studies by Bruner, Goodnow and Austin (1956). This intermediate neglect of the importance of spatiotemporal information in the earliest years of cognitive psychology is likely to have been inspired by the then dawning view that the human being could, or even should, be conceptualized as an information processing system - i.e., a computer (Gardner, 1987). Given that computers are excellent tools for realising featural representations (e.g. implemented in semantic networks) whereas time and space are less

---

[2] With the notion 'macroscopic level', I refer to solid objects at everyday-size like tennis balls, cups, and cars. At the quantum physical level the rules constraining object location and object motion may be substantially different.

intuitively obvious for a non-locomotive, stationary system, this tendency is probably not surprising. Thus, for some years, featural object information was considered to play the primary role within the field of cognitive psychology – also when attempting to keep track of specific objects (Krøjgaard, 1999b; Mammen, 1996).

However, in more recent years the understanding that featural information should play the primary role when adults categorize and track objects has been challenged from at least three different sides: (a) Common sense arguments; (b) evidence from experimental studies on object based visual attention in adults; and (c) formal accounts of the necessary prerequisites in order to make categorical judgements. These three challenges will be treated in turn.

Common Sense Arguments

As already stated, category membership was for some time considered to be exclusively based on the features of objects. However, the specific case of demarcating a category with only *one* member seems to give rise to serious problems in this respect. According to Pylyshyn (2001), human beings are bound to make use of some additional pre-conceptual visual indexing mechanism in order to be able to establish such a category. For instance, he writes:

> "If we could only refer to things in terms of their category membership, our concepts would always be related only to other concepts (the concepts for categories) and would never be grounded in experience. [...]. If our visual representations encoded a scene solely in terms of concepts or categories, then we would have no way to pick out or refer to particular individuals in a scene except through concepts or descriptions involving other concepts, and so on" (Pylyshyn, 2001, p. 129).

Thus, given the fact that human beings are indeed capable of establishing categories consisting of just one specific exemplar (e.g., the category of blue pens given to me by my grandfather), some kind of additional mechanism that goes beyond object features is simply necessary.

Empirical Evidence from the Multiple Objects Tracking Paradigm

The results obtained by the so-called *Multiple Objects Tracking* (MOT) paradigm have provided *empirical* evidence to support the claim that human beings are clearly capable of tracking specific objects through space and time without relying on object features or properties. In the MOT paradigm adult subjects are usually shown from 8 to 24 two-dimensional simple and completely identical objects on a computer screen. A small subset (the target objects) of these objects is then highlighted for a few seconds, and the subjects are instructed to pay specific attention to these target objects. Subsequently, the highlighting is removed, leaving all objects identical after which a test follows: All objects begin to move around on (apparently) random paths. After a while all objects stop and the subjects are typically asked either (a) to point out the target objects, or (b) state whether a given object was a target object or not. Results obtained with the former of these tasks revealed that subjects were perfectly capable of keeping track of up to four or even five objects (e.g., Sears & Pylyshyn, 2000; Yantis, 1992). As for the latter task, subjects typically produced the correct answer in at least 85% of the trials (e.g., Green & Bavelier, 2006; Pylyshyn & Storm, 1988).

Thus, the results derived from the MOT studies convincingly demonstrate that human beings are perfectly capable of keeping track of specific objects while *not* relying on the features of the objects. The empirical evidence obtained in the MOT studies simply rules out the idea that object features should be sufficient, since all objects displayed are completely *identical.* Hence, in these studies subjects must - in order to fare successfully when keeping track of specific objects - make use of *other* sources of information than featural information.

One obvious question entailing the results derived from the MOT studies is what kind of mechanism could possibly perform such a task? Pylyshyn's (1989, 2001) own candidate is the 'fingers of instantiation' mechanism, or in short FINST. According to Pylyshyn, the FINST mechanism, or 'visual index' is equivalent to a pointer or 'mental finger' that can be locked on or assigned to objects in the visual field. Two specific properties of the FINST mechanism are central. The first is that although the mechanism is supposed to track objects, it does no more than that; no *description* of the tracked objects is provided. Consequently, the FINST mechanism works at a preconceptual level. The second important feature of the FINST mechanism is that these visual indexes are supposed to be 'sticky' – which means, that when an object moves, the visual index moves along (Pylyshyn, 1989, 2001).

To summarize, in addition to common sense arguments, convincing empirical evidence shows that adults are capable of tracking specific objects through space and time, and they must do so by relying on some kind of spatiotemporal information; featural information alone is simply insufficient given the nature of the MOT tasks.

Formal Accounts of the Necessary Prerequisites in Order to Make Categorical Judgements

In a now seminal study Bruner and his colleagues showed that adults made categorical decisions by exclusively relying on featural object information (Bruner et al., 1956). However, by means of logical algebra, an axiomatic system and general topology, Mammen (1996) has convincingly argued that the results obtained in the study by Bruner and his collaborators (Bruner et al., 1956) are only valid if the number of objects to categorize amongst is finite, which in practical terms means *countable*. However, in the everyday world outside the laboratory, the number of relevant objects is not at all finite. Under circumstances where the number of objects on which to base the categories is beyond practical countability, Mammen (1996) has shown that featural information is insufficient; we will have to rely on spatiotemporal information too.[3]

Can this really be true? The following fictitious example may serve to illustrate the typical neglect of spatiotemporal information in our everyday thinking: Imagine a student who leaves his bicycle just outside the main entrance of the Department of Psychology in Aarhus. Later the same day, the student returns to pick up his bicycle in order to go home. How does he manage to find his specific bicycle among the many other bicycles also placed outside the department? A fairly straightforward answer would be that he looks for a bicycle with the same shape, colour, size, and brand, etc. as the one he happens to own. Following this account, the student seems to rely predominately, if not exclusively, on featural information in order to find his bicycle. Such an answer is not entirely wrong, but it is not completely correct either. What is forgotten in this chain of thought is that the student has already made a crucial (although often unreflected) decision *before* he attempts to track down the relevant features of his bicycle: He has chosen to look for his bicycle at the right global location (that is, outside the main entrance). He is not looking for his bicycle in Kansas or Berlin, but at the specific location in Aarhus, where he left it, and this strategy reduces the number of bicycles to choose from drastically. This may seem trivial but it is not. The point is, the decision

[3] Mammen's (1996) axiomatic system is comprehensive and cannot be given here. However, a more condensed version of the theory is on its way (Mammen & Krøjgaard, 2008).

about choosing the right location for the search for the bicycle is not based on featural, but on spatiotemporal information. However, choosing the right location seems so obvious and trivial that we tend to neglect, or even forget, the importance of such competence.

To summarize, from the field of cognitive psychology that investigates adult cognition, converging evidence exists from at least three different approaches that - contrary to earlier theories - featural information is not sufficient information when human beings attempt to single out and track specific objects through space and time.

Evidence from the Cradle: Infant Research on Object Individuation

In order to fully disentangle a given phenomenon manifest in adult human beings, one is very likely to benefit from studying the phenomenon in question as it evolves through the ontogenesis (Krøjgaard, 2005). Can infant research shed light on the development of the general purpose tool of being able to single out and track specific objects through space and time? The answer to this question seems to be a clear "yes".

Setting the Stage: The Heritage from T.G.R. Bower

T.G.R. Bower was a pioneer in at least two respects. First, he was one of the first infant researchers to insist on using tasks where gross motor achievements were not the key dependent variable. Second, Bower argued and demonstrated convincingly that the issue of object identity was an important and worthwhile research domain – also for infant researchers (Krøjgaard, 2004). Inspired by Michotte's (1962) work with adults, Bower began to present events like the following to infants (see Fig.1).

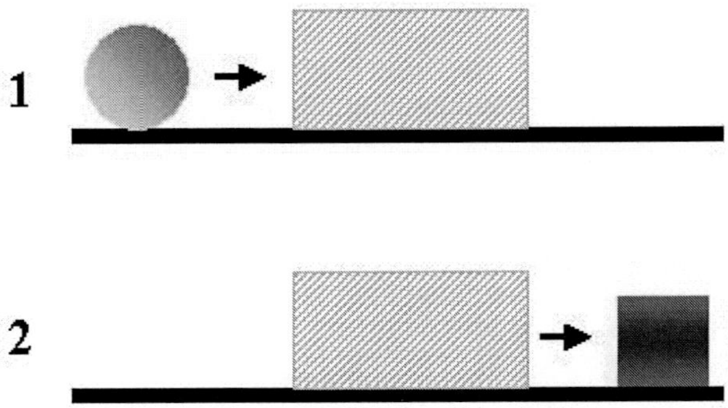

Figure 1. Schematic representation of a basic scenario in which one object moved behind an occluder followed by the appearance of a differently looking object on the other side of the occluder.

In principle the displayed scenario can be interpreted in at least two distinct ways. One might think that the ball remains behind the occluder while the cube moves into vicinity. Alternatively, one might parse the sequence as involving a ball turned into a cube – i.e. the so-called 'tunnel effect'. According to Bower, the key question in order to decide between these two interpretations is as follows: Does the event involve *one* or *two* distinct moving objects (Bower, Broughton & Moore, 1971)? Note how the simple surprise reactions to the scenario displayed in Fig.1 do not inform the experimenter about which of the two interpretations outlined above actually causes the surprise reaction. Thus, while Bower succeeded in making object tracking an important and worthwhile field of infant research and successfully specified the important question regarding the number of objects present, he ultimately failed to present a convincing design that provided "answers" from infants to this specific question. Fortunately, a quarter of a century later Xu and Carey (1996) actually managed to devise such a design.

Object Individuation in Infancy

As we will see in the following, Xu and Carey (1996) attempted to provide an answer to exactly this specific question put forth by Bower, namely, how many objects are present in a given scenario? Today, studies that address this specific question are usually referred to as studies on *object individuation.* Object individuation is defined as the ability to decide the number of distinct objects present in a given scenario

(Krøjgaard, 2004). In their now seminal study, Xu and Carey presented 10-month-old infants to a basic scenario in which two objects (e.g. a ball and a cup) moved in and out, one at a time, from behind a centre placed occluder (see Fig. 2).

Figure 2.

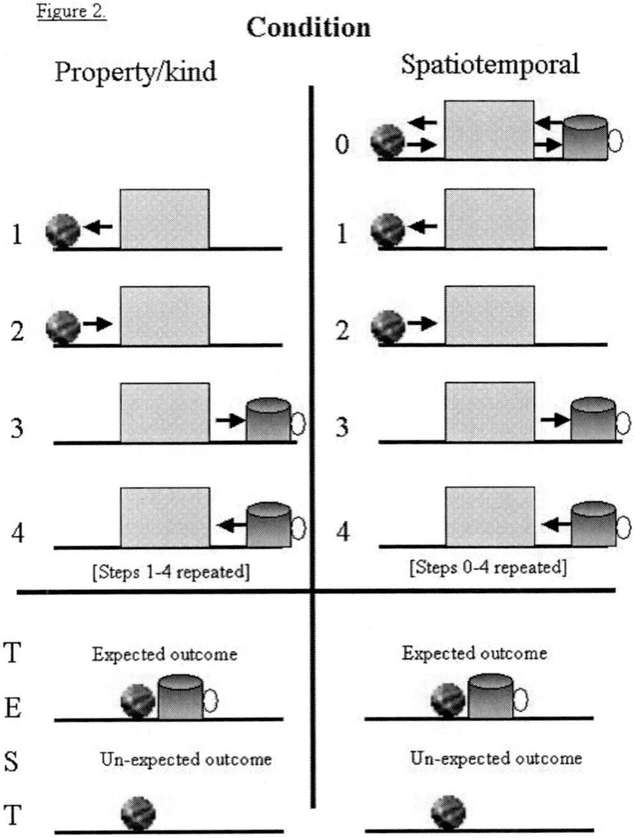

Figure 2. Schematic representation of the Property/kind and the Spatiotemporal condition of the Xu and Carey (1996) study.

Infants were assigned to one of two slightly different conditions of which the introduction sequences differed. In the property/kind condition the objects were never shown simultaneously but only *one* at a time. However, in the spatiotemporal condition *both* objects were initially presented at the same time. Subsequently followed a test shared by both conditions. The screen was removed revealing either an *expected outcome* (both objects present) or an *un-expected outcome* (only one object present) in balanced order. By means of the so-called violation-of-expectation (VOE) paradigm, the infants' looking time was measured and analysed. The results revealed that whereas the 10-month-old infants in the spatiotemporal condition reacted with surprise (looked relatively longer) when one of the objects was missing, their same-aged peers in the property/kind condition did not. Thus, when provided with unequivocal spatiotemporal information about the number of objects present, the infants succeeded in individuating the objects, whereas when spatiotemporal information was ambiguous and the infants had to rely on featural object information, they *failed* to individuate the objects. The results obtained by Xu and Carey (1996) have been replicated in a number of studies using similar designs that apply the VOE paradigm (Bonatti, Frot, Zangl, & Mehler, 2002, Experiment 3; Krøjgaard, 2000; 2003) as well as, in manual search tasks (Feigenson & Carey, 2003; Van de Walle, Carey, & Prevor, 2000; Xu & Baker, 2005; Xu, Cote & Baker, 2005).

Thus, on the surface, the results derived from Xu and Carey's studies (1996) and their replicators seem to indicate that spatiotemporal information is primary to featural information when infants attempt to individuate objects. However, Wilcox and Baillargeon (1998a) have argued that the immediate superiority of spatiotemporal information as opposed to featural information may be task dependent and only manifest when task demands are high. Note how the design employed by Xu and Carey (1996) forces the infants to *remember* what happens in the introduction in order to be able to come up with the correct 'answer' in the subsequent test. Removing a screen and revealing a single object behind the screen is only surprising provided that the subject is aware that a different number of objects (e.g. two objects) *should* be present behind the screen. Such a demanding design is by Wilcox and Baillargeon coined an *event mapping* design, because the subject will have to 'map' information obtained in a proceeding introduction onto the test in order to respond successfully (Wilcox & Baillargeon, 1998a).

This cognitively demanding event mapping kind of design is contrasted with a less demanding design, named an *event monitoring* design, in which everything takes place in a single on-going sequence. Thus, in the event monitoring design the infant will, at least in principle, be able to parse the relevant information just by 'monitoring' the sequence here and now; no 'mapping' from previous introduction sequences is needed (Wilcox & Baillargeon, 1998a). The clever wide-screen/narrow-screen design devised by Wilcox and Baillargeon (1998a) serves as an illustrative example of event monitoring designs:

Infants 9.5 months of age watched a basic scenario in which a ball moved consecutively from vicinity to a hideout behind a centre located screen from where 'it' (apparently) re-appeared as a box and *vice versa* in a single on-going sequence. The infants were allocated to one of two different conditions (see Fig. 3).

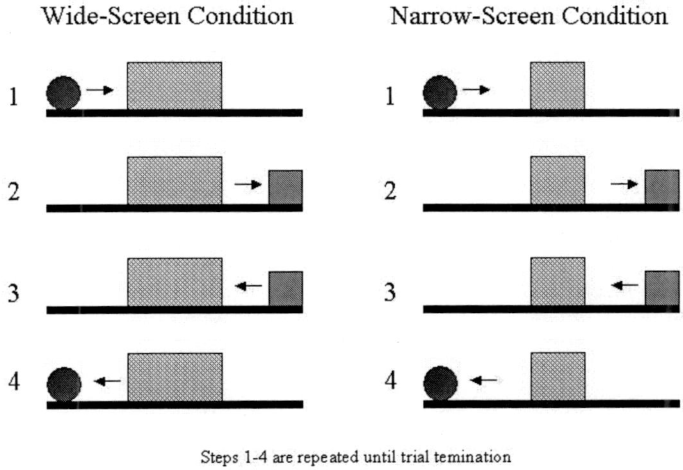

Figure 3. Schematic representation of the wide-screen/narrow-screen design employed by Wilcox and Baillargeon (1998a).

In the *wide-screen condition* the screen used was sufficiently wide to hide both objects at the same time. However, in the *narrow-screen condition* the screen was so narrow that both objects due to their combined width clearly could *not* be hidden at the same time behind the screen. Using the VOE paradigm, the reasoning was the following: If based on the sizes of the objects, the infants reasoned that both objects could be simultaneously hidden behind the wide screen, but *not* behind the narrow screen, then the infants in the narrow-screen condition should look reliably longer at this test event than their same-aged peers in the wide-screen condition. This was exactly what the results revealed, and similar results were subsequently obtained with 7.5-month-old infants (Wilcox & Baillargeon, 1998a, exp. 3 and 4) and 4.5-month old infants (Wilcox & Baillargeon, 1998b).

Thus, the results obtained by means of the wide-screen/narrow-screen design have revealed that when the cognitive demands are reduced, then infants below 10 months of age are indeed capable of individuating objects by relying to a large degree on featural means.

It should be noted that while Xu and Carey (1996) under what might be considered cognitively demanding conditions made a direct comparison between infants' use of featural and spatiotemporal information, Wilcox and Baillargeon (1998a, 1998b) - while reducing the cognitive load - only investigated infants' attempts to individuate objects by featural means. Thus, until recently no one had succeeded in conducting a direct comparison of infants' use of featural and spatiotemporal object information under test conditions where the cognitive load was reduced to a minimum – i.e. in an event monitoring design.

However, in a recent series of studies, Krøjgaard (2007) devised such a design. Infants 9.5, 8.0, and 6.0 months of age were presented to a 'rotating screen design' in which they witnessed a basic scenario where a cut-open cylinder rotated around stationary objects (see Fig. 4).

The infants were tested in one of two different conditions: An Object Chance Condition and an Object Disappearance Condition. Following habituation, the infants were presented to expected and un-expected test events. While the expected test events were the same for infants in both conditions, the un-expected test events differed: In the (un-expected) *Object Change Condition,* one of the objects had apparently changed appearance (from a Brio Clown to a Lego Man or *vice versa*) every time the cut-open cylinder revealed the contents. Thus, the un-expected Object Chance event involved a *featural* object change.

In the (un-expected) *Object Disappearance Condition,* the target object disappeared and reappeared on each two consecutive turns by the cut-open cylinder, that is, a *spatiotemporal* violation.

By means of the VOE paradigm the looking time of the infants was recorded and subsequently analysed. The results revealed that while the oldest (9.5-month-old) infants in both test conditions reacted with surprise when watching the un-expected test events as opposed to the expected test events, the results from the younger group of infants were different: Whereas the 8.0-, as well as, the 6.5-month-old infants in the Object

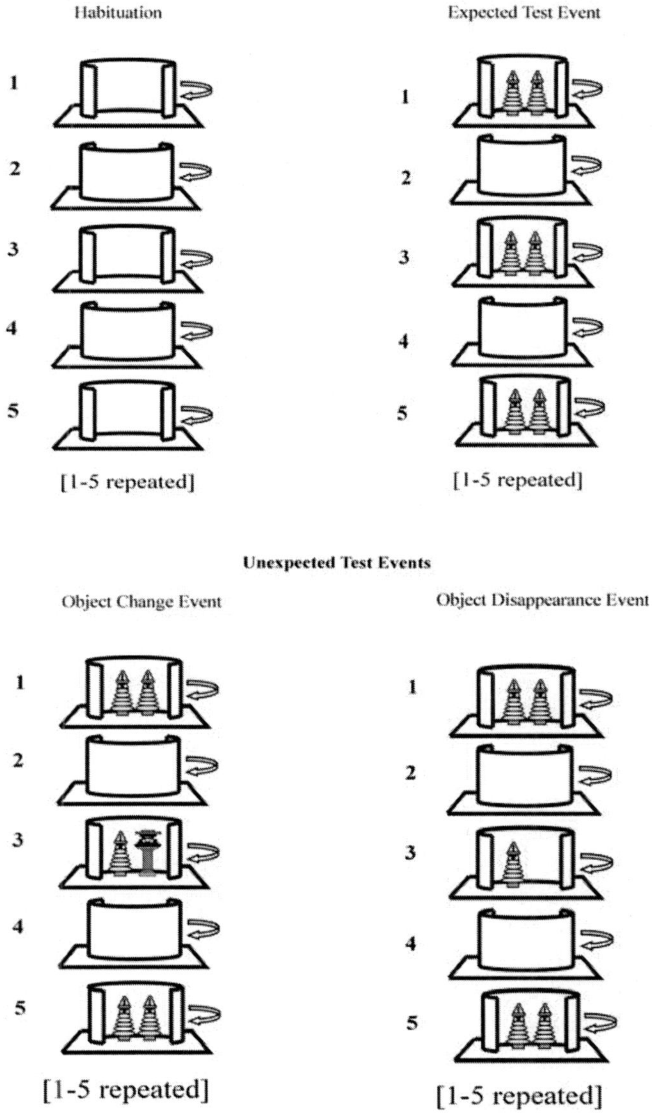

Figure 4. Schematic representation of the habituation and test events applied in Krøjgaard (2007).

Change Condition *failed* to individuate the objects, their same-aged peers in the Object Disappearance Condition *succeeded* (Krøjgaard, 2007).

Note, given the fact that all test events in the design employed by Krøjgaard (2007) unfolded in one on-going sequence, the design qualifies to the group of simpler and less cognitively demanding event monitoring designs (cf. Wilcox & Baillargeon, 1998a). Consequently, the immediate interpretation of the results derived from Krøjgaard's (2007) study may be that spatiotemporal information seems to be generally superior to featural information when infants attempt to individuate objects.

However, such a bold conclusion does not seem warranted. Even though the rotating screen design employed by Krøjgaard (2007) qualifies as the less demanding event monitoring kind of design, the negative results obtained by the younger infants in the Object Change Condition appear somewhat surprising, considering the evidence derived from the wide-screen/narrow-screen design by Wilcox and Baillargeon (1998a, 1998b) where infants down to 4.5 months of age succeeded in individuating objects.

How can this divergence in results be accounted for? At least two obvious differences between the rotating screen design and the wide-screen/narrow-screen design may explain why the infants in the rotating screen design were less successful in individuating objects than their peers who participated in the wide-screen/narrow-screen design: First, while the objects in the wide-screen/narrow-screen design moved, the objects in the rotating screen design remained stationary. Second, while the objects in the wide-screen/narrow-screen design were occluded for approximately 1 second out of each 5 seconds sequence, the objects in the rotating screen design were occluded for approximately 2.2 seconds out of each 4 seconds period. Thus, although both designs can be said to belong to the same category of relatively simple (event monitoring) designs, there is reason to believe that the rotating screen design, all things equal, is more cognitively demanding than the wide-screen/narrow-screen design (Krøjgaard, 2007).

In another very recent experiment Krøjgaard (in press) attempted to eliminate the potentially negative influence of using (a) stationary objects, and (b) long occlusion times (see Fig. 5).

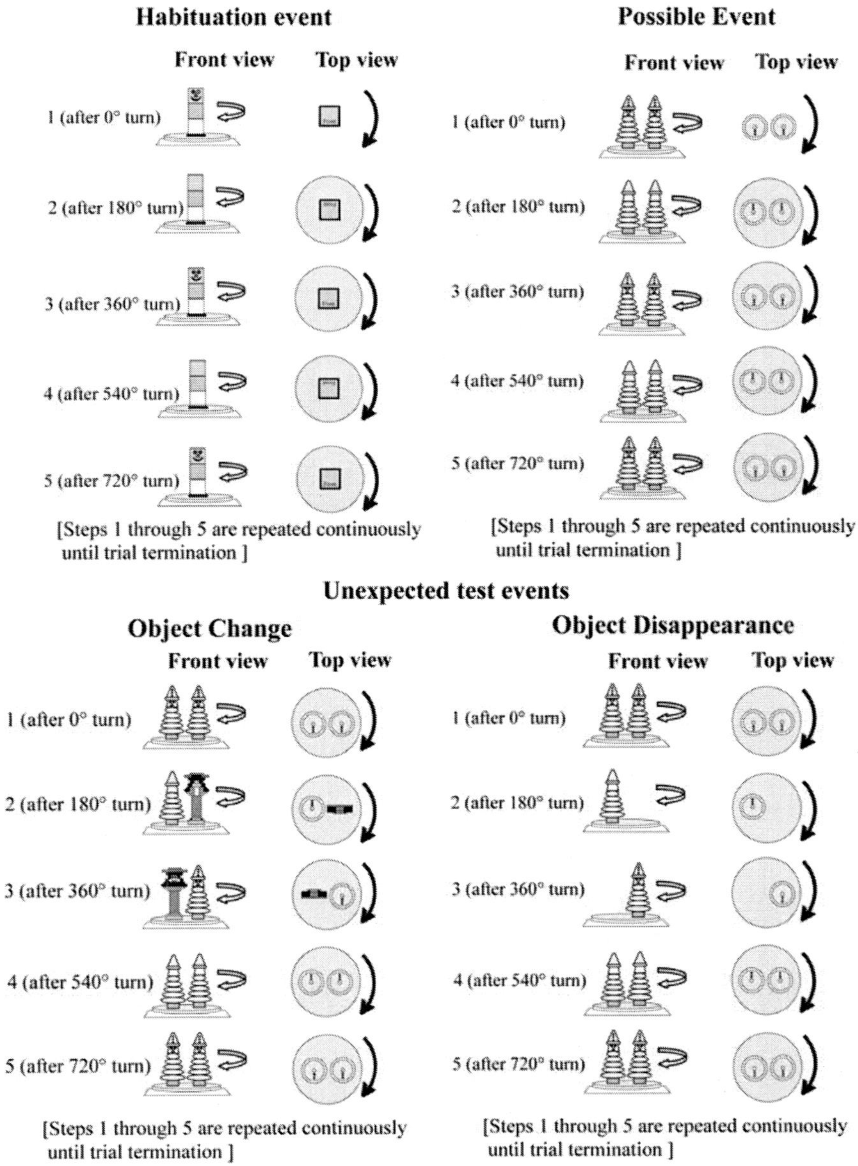

Figure 5. Schematic representation of the habituation and test trials applied in Krøjgaard (in press).

Using a similar setup, the rotating screen was replaced by a rotating platform on which the objects could be placed and hereby move. As displayed in Fig. 5 the un-expected events were carried out without having the objects occluded at all. This was made possible by using a mirror arrangement in which two synchronically rotating platforms were in play although everything from the point of view of the spectator appeared to take place straight ahead (see Fig. 6).

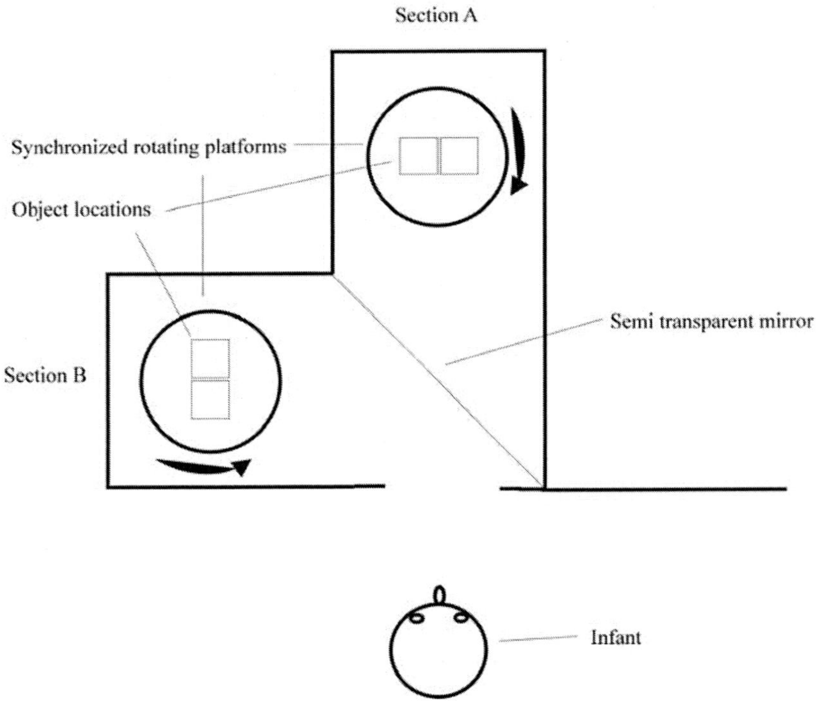

Figure 6. Schematic representation of the mirror arrangement. By instantaneously turning on and turning off the lights in Sections A and B respectively, the viewer's focus can be changed from A to B (and vice versa), although the viewer experiences the event as taking place in *one* location.

Thus, the mirror arrangement made it possible to reduce the occlusion time to nil, and hereby entirely eliminate the cognitive load induced by momentary concealment of the objects (Krøjgaard, in press).

Two structurally identical experiments with 8.0-, and 6.5-month-old infants were conducted by means of the 'rotating platform design'. The results revealed that the infants in both test conditions managed to individuate the objects regardless of which condition they were allocated. Thus, by using a more dynamic scenario with moving objects and by eliminating the eventual negative effect of occlusion, the infants were indeed able to individuate objects successfully regardless of whether they primarily had to rely on featural or spatiotemporal information (Krøjgaard, in press).

To summarize, there is still no definite conclusion regarding the question of whether spatiotemporal information is generally superior to featural information when infants attempt to individuate objects (for opposing views, see e.g., Xu, 1999, 2003 and Needham & Baillargeon, 2000). More research is simply needed. However, there seems to be no doubt that infants are capable of individuating objects, and that spatiotemporal information does play an important role when doing so.

Infant Research: Summary

What can be learnt from research on object individuation in infancy? First of all, the research outlined here provides compelling evidence that the ability to single out and track specific objects through space and time is not restricted to adults. On the contrary, some of the basic building blocks constituting such an ability seem to be quite evolved even among young infants – at least here-and-now. Thus, even early in their first year of life, infants pay attention to featural, as well as, spatiotemporal information when attempting to track objects in space and time – the two kinds of information that also play prominent roles when adults track objects. We have also seen evidence that when in doubt, spatiotemporal information seems to override featural information. Hence, spatiotemporal information may very well be even more basic than featural information. Accordingly, and returning to our fictitious young student trying to find his bicycle outside the department, it may not be trivial at all that our immediate (but insufficient) explanation is likely to be one stating that our student's search will be based solely on featural information. We may simply have forgotten the extent to which spatiotemporal information provides the ground on which featural information stands out. This is one of the reasons why it remains an important question to study the timing of infants' use of featural and spatiotemporal information respectively when attempting to individuate objects, because it helps us outline the ontogenetic emergence of these core abilities. One might speculate that if researchers of cognitive psychology had started out studying infants, and not adults, they may never have believed that one could safely neglect spatiotemporal information.

Object Individuation in Non-Human Primates

When researchers attempt to demarcate the specifically human, several important tools are available. In recent years evolutionary psychology has become one of the most rapidly growing approaches in this respect (e.g., Katzenelson, 2007). And studying our closest relatives is an obvious window in pursuing such a goal (Byrne, 1995; Tomasello & Call, 1997). Thus, one of the central questions entailing considerations regarding the human ability to single out specific objects in space and time is whether our closest relatives (e.g. non-human primates) possess competencies similar to the ones outlined above.

Until recently there seemed to be consensus that non-human primates were somewhat inferior when considering their ability to identify objects and make categorizations. For example, in their comprehensive review of the literature, Tomasello and Call (1997) wrote the following a decade ago:

> "Few animal species have any skills at sorting objects into groups the way that human infants and children do with such facility during the preschool years, and the ecological relevance of such skills for most species is not easy to determine. Primates have some relevant skills (combining their skills of categorizations and manipulation) but they seem to be extremely limited compared with humans" (183-184).

Although the abilities in non-human primates are indeed inferior to especially adult human beings (see below), the results from a few recent studies actually suggest, that some of the apes may seem capable of individuating objects:

Santos and her colleagues (Santos, Sulkowsky, Spaepen & Hauser, 2002; Flombaum, Kundey, Santos & Scholl, 2004) have recently pioneered in conducting studies on object individuation in rhesus macaques (*Macaca mulatta*). For example, Santos and her collaborators (Santos et al., 2002) used a modified version of the manual search paradigm originally devised for human infants (e.g., Van de Walle et al., 2000). Rhesus macaques were shown food objects (fruits) being placed in a closed box filled with leaves. Subsequently, they were allowed to search for the food items. However, due to a trick the rhesus macaques found their hidden food items to have undergone changes in either kind (i.e. featural violation) or number (i.e. spatiotemporal violation). Search time was recorded, and the results indicated that the rhesus macaques searched significantly longer when either the kind or the number of the appearing objects had been changed as opposed to controls (Santos et al., 2002). Thus, analogue to human infants, rhesus macaques seem capable of individuating objects by featural, as well as, by spatiotemporal means.

In another experiment Flombaum et al. (2004) showed that members of the rhesus macaques displayed the 'tunnel effect': When watching an event in which an object moved behind an occluder and subsequently reappeared on the other side of the occluder as a differently looking object, the apes reacted (based on their reaching performance) as if only a *single* object was involved in the event (cf. Fig. 1; see also Santos, 2004). This reaction from rhesus macaques resembles the interpretation reported from human adults presented to similar scenarios (Michotte, 1962; Xu, Carey & Quint, 2004).

Thus, resembling some of the findings from human infants, a limited set of studies have shown that rhesus macaques are able to individuate objects successfully by means of featural or spatiotemporal information. Provided that additional evidence will converge with the above recent findings, it seems to be the case that at least *some* of the basic building blocks required for being able to single out and track specific objects through space and time, here-and-now, are indeed manifest in some of our ancestors.

Tracking Objects: Uniquely Human Applications

As the evidence above shows, there is reason to believe that aspects of the ability to single out and track specific objects through space and time may be present in a few non-human primates too (i.e. rhesus macaques). Hence, as such, the ability to single out and track specific objects through space and limited durations of time cannot be said to be a uniquely human competence. Nevertheless, as I shall argue, human beings seem able to exploit this ability to a level far beyond what other species are capable of. In the following, I will direct attention to two specific areas where this seems to be the case. First, we will take a look at how the ability to single out and track specific objects through space and time appears to be a necessary precondition when infants and young children begin to form concepts and learn words. Second, I will argue that such ability is necessary to conceive of man-made artefacts and culture in general.

Concept Formation

Through the first years of the ontogenesis, the vocabulary of children expands with impressive speed. Children usually say their first word around 12-13 months of age, and by the age of 6 years they typically have a vocabulary consisting of approximately 10,000 words, implying that children between the ages of 1 and 6 on average have to learn more than 5 new words daily (Bloom, 1998). One obvious way of learning nouns is what has been coined 'the original word game' (Brown, 1958), in which an adult directs a child's attention to a specific object (e.g., a butterfly) for example by pointing while saying: "This is a butterfly!" In order actually to learn the new word in such a context, a range of necessary conditions will of course have to be fulfilled. I will not go through all these (e.g., being able to understand the intention of the adult, hearing the concept, remembering the concept, etc.), but would just like to draw attention to one necessary prerequisite that often seems to be overlooked or neglected when considering concept formation or word learning – often at the expense of 'language games' (Wittgenstein, 1958). As the reader may already have guessed, I am referring to the ability to single out and track specific objects. However, Spelke displays the relevance very clearly when she writes:

> "There is a core conception, *physical object*, of which sortal objects such as "table" and "horse" are specific examples, and this conception does the major work of singling out physical bodies and tracing them through time. [...] There must be some way of apprehending books and following them through time that does not itself depend on an already developed sortal concept. It is reasonable to suppose that a general conception of *physical object* fulfils this function" (Spelke, 1988, p. 231, italics by author).

Spelke believes further that such *core knowledge* regarding physical objects is hard-wired from birth (Spelke, 1994). Note, however, that the argument that word learning should rely on the ability to single out and track specific physical objects in space and time is not contingent upon the nativist claim endorsed by Spelke. Hence, it is by no means surprising that others have reached the same conclusion about the importance of being able to single out objects as a precondition for concept formation (Kingo, 2008; Mandler, 1997; Nelson, 1974).

Understanding Man-Made Artefacts

By introducing the term *affordance,* Gibson (1979) tried from an ecological perspective to show how meaning to a large extent already was 'out there' in the relation between the subject and the object. The theory of affordances may provide a satisfactory framework for understanding what each single object may offer a given subject. However, one might argue that some affordances are more salient than others and that some affordances ultimately may not be out there for simple information pick-up without *a priori* knowledge of the object in question. For example, while a hammer may easily afford hammering for a human being (among many other affordances, of course), it does not seem obvious, at least not to me, that some of the powerful affordance related to the computer, I am currently writing on - e.g. storage of information and sending emails - should be available and ready in the visual array just to be picked-up by a subject that never seen or heard of a computer before. Such 'affordances' seem to require earlier acquired knowledge in order to be perceived effectively.

Another potential criticism regarding the Gibsonian approach to the perception of meaning is the inability to distinguish between the meaning of natural objects, on the one hand, and man-made objects or cultural artefacts, on the other. While both natural objects (e.g. stones) and man-made objects (e.g. stone axes) offer a large number of different affordances, only the man-made objects are *intended* for *specific* purposes (Bærentsen & Trettvik, 2002; Mammen, 1986; Tomasello, 1999). This important difference between natural and man-made objects seems difficult to grasp within a Gibsonian framework.

Consider, for instance the following thought experiment:[4] Imagine we have two identically looking pieces of flint stone, both in the shape of an arrowhead. One of the stones (stone A) is a natural result of a storm in which a giant tree hit a larger piece of flint stone, causing the stone to break into two pieces of which one of the pieces by coincidence got the shape of a perfect arrowhead. The other stone (B) is the result of several hours of meticulous work by a competent Stone Age man. Since the two 'arrowheads' A and B are completely identical from a featural point of view, it seems impossible to distinguish between them armed with a Gibsonian framework. However, even if we cannot tell the two 'arrowheads' apart, human beings are perfectly capable of *understanding*

[4] The thought experiment is inspired by an equivalent one given by Jens Mammen (personal communication).

and *acknowledging* the difference between the one that was made by a fortunate event in nature (stone A) and the one resulting from intentional work by a skilful Stone Age man (stone B). If human beings were ignorant of such differences, there would be no point in exhibiting fragments of old Greek vases, ancient coins or original stone arrowheads in museums; no one would have reasons to be annoyed if original pieces of art were duplicated (cf. Mammen, 1993; Krøjgaard, 1999a) - and *The Little Prince* would have no reason to bother about his unique and precious rose.

However, adult human beings are indeed sensitive to such differences. One might ask what competencies allow human beings to understand clearly the differences between stones A and B or the importance of museum exhibits (e.g. fragments of an ancient vase) of which the present affordances are rather limited. Mammen (1996) has convincingly argued that the ability to single out and track specific objects through space and time is an important prerequisite for such ability. However, although necessary, such ability may not be *sufficient*. Given that rhesus macaques - just like human infants - may be endowed with the basic abilities necessary in order to individuate objects, as seen earlier in this paper, one might speculate what adult human beings *add* to this ability when they demonstrate such elaborated understanding of cultural artefacts, as mentioned above. I will tentatively suggest that one important competence that adult human beings may apply when grasping the ontology of man-made artefacts is their ability to *abstract* the spatiotemporal threads related to specific objects and implement these in new domains or contexts. Note how following the spatiotemporal path of a given object (e.g., an arrowhead) here-and-now is simply insufficient to understand the historical thread of the thousands-of-years-old piece of flint stone. Without the ability to abstract and envision the history of the arrowhead abstracted from its current appearance, the difference between stones A and B would be incomprehensible.

I believe that several pieces of evidence implicitly support the claim proposed above: Tool use in human beings is far less bound to specific contexts than it is the case for non-human primates (Byrne, 1995; Tomasello & Call, 1997). Intentional teaching to conspecifics is far more developed in human beings than in non-human primates (Byrne, 1995; Tomasello & Call, 1997). One-trial learning is much more common in human beings than in non-human primates (cf. Krøjgaard, 2006). Program-level imitation is far more prominent in human beings than in non-human primates (Byrne, 1995). While these well-established differences between human beings and non-human primates diverge strongly with regard to sociality, purpose, and application, they all *share* the feature of relying on the ability to *abstract* certain aspects of a given situation and subsequently apply the abstracted material to another domain or context. I do not claim that this is the only competence relevant for understanding man-made artefacts, but I do claim that such ability plays a significant role. In short, the argument is that if one was unable to single out and track specific objects through space and time and subsequently abstract this information from one domain into another, then it seems very difficult, if not impossible, to follow stone B or vase fragments through space and time on a larger scale. Furthermore, without such ability man-made objects and consequently culture simply seems beyond comprehension. Thus, while *sharing* may be a uniquely human feature as argued by Tomasello and colleagues (Tomasello, Carpenter, Call, Behne & Moll, 2005), I believe that specific *cognitive* skills (i.e. the ability to abstract from one domain or context into another) need to be added to the unique human repertoire. Social-cognitive prerequisites appear to be insufficient as the examples from the present paper should hopefully illustrate.

Concluding Remarks

Here I have attempted to pursue the origin of the ability to single out and track specific objects through space and time. The results from a substantial amount of experimental studies indicate that basic aspects of such ability (e.g. object individuation) are already present and operative, at least to some extent, in young infants. Meanwhile, a few recent studies have revealed that similar basic aspects of such ability may very well also be present in at least one species of non-human primates, that is, rhesus macaques. Thus, while adult human beings may indeed employ the ability to single out and track specific objects through space and time when understanding man-made artefacts and culture in general, they must do more than track objects here-and-now in order to produce the advanced level of elaborated understanding of culture that characterises adult human beings, but not non-human primates. I have tentatively proposed, that the added tool may be the ability to *abstract* the spatiotemporal threads that constitute the tracking of objects. Such claim cannot be proven, but only falsified. If, for instance, evidence that indicate that some non-human primate is capable of abstracting spatiotemporal threads related to objects from one domain or context into another is obtained, then the claim will be falsified. Provided that the 'abstraction hypothesis' proposed here is valid, we would also need to consider *when* and *how* it comes into play through the ontogenesis of human beings.

Another important area for further research is comparative psychology. Given that the amount of evidence on object individuation in non-human primates is limited, we simply need to conduct more comparative studies on these and related issues.

Attempting to produce relevant pieces to the large and complex jigsaw-puzzle of specifying the eternally prominent question regarding who we are as human beings, evolutionary psychology has beyond any doubt proven an important and worthwhile approach. Throughout this paper I have tried to demonstrate that investigating basic human competencies as they evolve early in the ontogenesis is another powerful tool. If we are to succeed in uncovering what is uniquely human we may need to a larger degree than hitherto has been the norm to *combine* these two approaches.

References

Bloom, L. (1998). Language acquisition in its developmental context. In: D. Kuhn & R.S. Siegler (eds.), *Handbook of Child Psychology: Vol. 2: Cognition, Perception and Language* (5th ed), (309-370). New York: Wiley.

Bonatti, L., Frot, E., Zangl, R., & Mehler, J. (2002). The human first hypothesis: Identification of conspecifics and individuation of objects in the young infant. *Cognitive Psychology, 44,* (388-426).

Bower, T.G.R., Broughton, J., & Moore, M.K. (1971). Development of the object concept as manifested in changes in the tracking behaviour of infants between 7 and 20 weeks of age. *Journal of Experimental Child Psychology, 11,* (182-193).

Brown, R. (1958). How shall a thing be called? *Psychological Review, 65,* (14-21).

Bruner, J.S., Goodnow, J.J., & Austin, G.A. (1956). *A Study of Thinking.* New York: Wiley.

Byrne, R. (1995). The Thinking Ape: Evolutionary Origins of Intelligence. Oxford: Oxford University Press.

Bærentsen, K.B., & Trettvik, J. (2002). An activity theory approach to affordance. In: *Proceedings of the Second Nordic Conference on Human-Computer Interaction* (51-60) NordiCHI 2002. New York: ACM Press.

Feigenson, L., & Carey, S. (2003). Tracking individuals via object files: Evidence from infants' manual search. *Developmental Science, 6,* (568-584).

Flombaum, J.I., Kundey, S., Santos, L.R., & Scholl, B.J. (2004). Dynamic object individuation in rhesus macaques. *Psychological Science, 15,* (795-800).

Gardner, H. (1987). The Mind's New Science: A History of the Cognitive Revolution. (2nd. ed.) New York: Basic Books.

Gibson, J.J. (1979). *The Ecological Approach to Visual Perception.* Boston: Houghton Mifflin.

Green, C.S., & Bavelier, D. (2006). Enumeration versus multiple object tracking: The case of action video game players. *Cognition, 101,* (217-245).

Katzenelson, B. (2007). Evolutionspsykologi. [Evolutionary psychology]. In: B. Katzenelson & B. Karpaschoff (eds.), *Klassisk og moderne psykologisk teori* (301-323). København: Hans Reitzels Forlag.

Kingo, O.S. (2008). The concept of concepts: Perceptual and conceptual categorization in infancy under scrutiny. *Journal of Anthropological Psychology, 19,* (1-21).

Krøjgaard, P. (1999*a*). Spædbarnets erkendelse af objekters numeriske identitet, In: Problemstillingens relevans og en gennemgang af nogle centrale *visuel tracking* studier [Infant's cognition of object's numerical identity. In: Relevance of the issue and a review of central visual tracking studies]. *Nordisk Psykologi, 51,* (1-15).

—. (1999b). Spædbarnets erkendelse af objekters numeriske identitet, II: E.S. Spelkes forskning, et uløst problem og perspektiver [Infant's cognition of objects' numerical identity II: The research by E.S. Spelke, an unsolved problem and perspectives]. *Nordisk Psykologi, 51,* (81-97).

—. (2000). Object individuation in 10-month-old infants: Do significant objects make a difference? *Cognitive Development, 15,* (169-184).

—. (2003). Object individuation in 10-month-old infants: Manipulating the amount of introduction. *British Journal of Developmental Psychology, 21,* (447-463).

—. (2004). A review of object individuation in infancy. *British Journal of Developmental Psychology, 22,* (159-183).

—. (2005). Spædbarnsforskningens relevans for almenpsykologien. [The relevance of infant research in relation to general psychology. (Target article with open peer commentaries)]. *Bulletin fra Antropologisk Psykologi, 15,* (6-51).

—. (2006). Comparative psychology, developmental psychology, and methodology. *Journal of Anthropological Psychology, 17*, (86-94).

—. (2007). Comparing infants' use of featural and spatiotemporal information in an object individuation task using a new event-monitoring design. *Developmental Science, 10,* (892-909).

—. (in press). A direct comparison of infants' use of featural and spatiotemporal information when individuating objects in a non-occlusion event monitoring design. *European Journal of Developmental Psychology.*

Mammen, J. (1986). Erkendelsen som objektrelation. [Cognition as object relation]. *Psyke & Logos, 7,* (178-202).

—. (1993). The elements of psychology. In: N. Engelsted, M. Hedegaard, B. Karpatschof & A. Mortensen (eds.). *The Societal Subject* (29-44). Aarhus: Aarhus University Press.

—. (1996). *Den menneskelige sans* [The Human Sense. 3rd edition, 1st edition published in 1983]. København: Dansk Psykologisk Forlag.

—. (2002). Mapping the subject: The renewal of scientific psychology. *Bulletin fra forum for Antropologisk Psykologi, 11,* (77-89).

Mammen, J., & Krøjgaard, P. (submitted). Being able to decide objects' numerical identity: A formal account of choice and sense.

Mandler, J.M. (1997). Development of categorisation: Perceptual and conceptual categories. In G. Bremner, A. Slater, & G. Butterworth (eds.), *Infant Development: Recent Advances* (163-189). Hove, England: Psychology Press.

Michotte, A. (1962). *Causalité, Permanence, et Réalité Phénoménalés* [Phenomenal Causality, Permanence and Reality]. Louvain: Publications Universitaires.

Needham, A., & Baillargeon, R. (2000). Infants' use of featural and experiental information in segregating and individuating objects: A reply to Xu, Carey and Welch (2000). *Cognition, 74,* (255-284).

Nelson, K. (1974). Concept, word and sentence: Interrelations in acquisition and development. *Psychological Review, 91,* (267-283).

Pylyshyn, Z.W. (1989). The role of location indexes in spatial perception: A sketch of the FINST spatial-index model. *Cognition. 32,* (65-97).

—. (2001). Visual index, preconceptual objects, and situated vision. *Cognition, 80,* (127-158).

Pylyshyn, Z.W., & Storm, R.W. (1988). Tracking multiple independent targets: Evidence for a parallel tracking mechanism. *Spatial Vision, 3,* (179-197).

Saint-Exupéry, A. (1996). *The Little Prince.* London: William Heinemann Ltd. (French version, 1945).

Santos, L.R. (2004). 'Core Knowledge': A dissociation between spatiotemporal knowledge and contact mechanics in a non-human primate? *Developmental Science, 7,* (167-174).

Santos, L.R., Sulkowsky, G.M., Spaepen, G.M., & Hauser, M.D. (2002). Object individuation using property/kind information in rhesus macaques (*Macaca mulatta*). *Cognition, 83,* (241-264).

Sears, C.R., & Pylyshyn, Z.W. (2000). Multiple object tracking and attentional processing. *Canadian Journal of Experimental Psychology, 54,* (1-14).

Spelke, E.S. (1988). Discussion, Section C: The ontogeny of perceptual and causal knowledge. In L. Weiskrantz (ed.), *Thought Without Language* (229-236). Oxford, England: Clarendon Press/ Oxford University Press.

—. (1994). Initial knowledge: Six suggestions. *Cognition, 50,* (431-445).

Spelke, E.S., Breinlinger, K., Macomber, J, & Jacobson, K. (1992). Origins of knowledge. *Psychological Review, 99,*(605-32).

Strawson, P.F. (1964). *Individuals.* London: Methuen.

Tomasello, M. (1999). The cultural ecology of young children's interactions with objects and artifacts. In E. Winograd, R. Fivush & W. Hirst (eds.), *Ecological Approaches to Cognition. Essays in Honour of Ulric Neisser* (153-170). Mahwah, NJ: Erlbaum.

Tomasello, M. & Call, J. (1997). *Primate Cognition.* New York and Oxford: Oxford University Press.

Tomasello, M., Carpenter, M., Call, J., Behne, T., & Moll, H. (2005). Understanding and sharing intentions: The origins of cultural cognition. *Behavioral and Brain Sciences, 28,* (675-735).

Van de Walle, G., Carey, S., & Prevor, M. (2000). Bases for objects individuation in infancy. Evidence from manual search. *Journal of Cognition and Development, 1,* (249-280).

Wilcox, T. (1999). Object individuation: Infants' use of shape, size, pattern, and color. *Cognition, 72,* (125-166).

Wilcox, T., & Baillargeon, R. (1998a). Object individuation in infancy: The use of featural information in reasoning about occlusion events. *Cognitive Psychology, 37,* (97-155).

Wilcox, T., & Baillargeon, R. (1998b). Object individuation in young infants: Further evidence with an event-monitoring task. *Developmental Science, 1,* (127- 142).

Wittgenstein, L. (1958). *Philosophical Investigations*. Oxford: Blackwell.

Xu, F. (1999). Object individuation and object identity in infancy: The role of spatiotemporal information, object property information, and language. *Acta Psychologica, 102,* (113-136).

—. (2003). The development of object individuation in infancy. In J. Fagan & H. Haynes (eds.), *Progress in Infancy Research. Vol. 3,* (159-192). Mahwah, NJ: Lawrence Erlbaum.

Xu, F., & Baker, A. (2005). Object individuation in 10-month-old infants using a simplified manual search method. *Journal of Cognition and Development, 6,* (307-323).

Xu, F., & Carey, S. (1996). Infants' metaphysics: The case of numerical identity. *Cognitive Psychology, 30,* (111-153).

Xu, F., Carey, S., & Quint, N. (2004). The emergence of kind-based object individuation in infancy. *Cognitive Psychology, 49*, (155-190).

Xu, F., Cote, M., & Baker, A. (2005). Labeling guides object individuation in 12-month-old infants. *Psychological Science, 16,* (372-377).

Yantis, S. (1992). Multiple visual tracking: Attention and perceptual organization. *Cognitive Psychology, 24*, (295-340).

CHAPTER SIX

WHY EVOLUTIONARY AND DEVELOPMENTAL COGNITIVE PSYCHOLOGY? THE IMPORTANCE OF NON-HUMAN PRIMATES & HUMAN INFANTS FOR UNDERSTANDING ADULT USE OF MODERN COMPUTERIZED TECHNOLOGY

RUNE NØRAGER

Abstract: State of the art human technologies like cell phones and computers have become indispensable to many in our daily lives. To just as many, however, the uses of such modern computerized technological devices also pose daily challenges. Rigorous research in primate tool use has provided numerous insights that can be put to use to understand how this forefront of human technology can be made easier and more intuitive to use and thereby assist in overcoming these challenges. Based on evolutionary and developmental cognitive psychology, it is identified that: 1) Modern computerized technology lacks sufficient support for the basic dynamics that characterise tool use in primates. 2) These basic dynamics are supported by corresponding cognitive resources, the nature of which is well studied in both animal cognitive research and infant developmental cognitive research.

Based upon this, it is hypothesized that these functional dynamics can be implemented back into computerized technology and hereby support basic cognitive abilities in humans. Preliminary results, which support this approach, are reported from experimental studies derived from this theoretical framework.

Background: Framing the Problem

As a scientist in cognitive psychology and a technology consultant, I often experience that people, in general and professionals like industrial designers and software engineers, are puzzled as to what role psychology can possible play in the development of technology. My argument is usually that with intricate knowledge about how humans perceive, think and act, we are able to design technology that matches these skills in the best possible way. Much like the ergonomics of a chair requires knowledge about the human biomechanical physiology, so too do cognitive ergonomics require knowledge about cognitive dynamics. In this article I shall pursue this argument in a broader scope, namely by following a direct line from cognitive research of non-human primates over infant cognitive development and adult cognition to the design and development of modern computerized technology. Paradoxically, it is the basic cognitive ergonomics of basic human tool use that lend themselves very well to be studied in its pure form in primates and infants that will provide some novel insights of how to design advanced human computerized tools.

It is important to stress that although much attention is devoted to technology in the article, and especially computerized technology, this is motivated by the unique possibilities captured hereby to investigate both theoretically and experimentally key aspects of what is particularly human in a novel way with human adults. More specifically, computerized technology allows us in experimental conditions to isolate and manipulate key variables of human cognition that are also in focus in non-human primate research and human infant developmental research. In summary, the applied field of human-computer interaction has the potential to provide research in animal cognition with an additional string to play on in the quest to isolate the particularly human aspect and how it unfolds under various circumstances.

Tools: Causalfunctional Objects, Purposeful Artefacts and Computerized Technology

The evolution of human tool use can be described in various ways depending on which characteristics of the tools and their use one focuses on. From a psychological cognitive perspective, the evolution of human tool use can be described as a function of the cognitive complexity reflected in the tools. Such a description starts with the transition from basic stone tools, that functioned as crude amplifications of natural capabilities, over composite tools like the axe and bow, where knowledge of mechanical forces make it possible to transform these forces and modulate them. The next step in this progression is the addition of externally powered tools that finally have cybernetic dynamics of automatic control and regulation incorporated, like the fly-ball governor in steam engines[1]. In this paper, however, a somewhat different distinction of technological evolution will be made that is motivated by how primate tool use can inform human computerized tool use. The distinction used could be considered a psychological cultural-historical equivalent to the pure cognitive psychological lineage outlined briefly above. In this psychological cultural-historical optic, three categories or distinctions about tools are motivated. These will briefly be outlined and then discussed in light of current research.

First, some tools are considered basic objects with purely causalfunctional properties in relation to other objects. They are not tools as such but merely temporary means to an end. All objects in this category of tools can be used as means to a specific end, as long as they share the relevant causalfunctional properties. A rock can be used for a number of purposes such as a weight, a crude hammer, a doorstopper, etc. Tools in this category are therefore only constrained by their inherent physical causalfunctional properties. Objects thus only momentarily become tools as a consequence of the immediate needs of the primate or human who picks them up.

The *second* category consists of purposeful tools, i.e., tools that in a social and cultural context have been developed to realise and attain certain goals. This class of tools can be argued to be distinctly human since it requires a social and culturally shared tool practice (Bærentsen, 1989; Klix, 1980). Furthermore, this aspect of tools is heavily integrated with the use of language that functions as a categorical marker for the intentionality imbedded in these tools (Jaswal, 2006; Kemler Nelson, Russell, Duke & Jones 2000). Although a tool from a purely causalfunctional perspective might serve various different uses, they are meant for very specific purposes that constitute a subset of the causalfunctional possibilities and sometimes only a singular use. This relation plays out at the dinner table when children are carefully instructed that the cup is used for drinking and not for any other purposes, although initially the child constantly challenges this narrow cultural use. The linguistic cultural aspect of tools thus narrows down the use of tools as a subcategory of their causalfunctional possible use. Although different objects may be used to achieve the same goal, they are typically not due to the constraints of cultural labels and practices.

The *third* category consists of computerized technology. It distinguishes itself from the other two categories by its lack of a functional material, physical substance (Bærentsen, 2000). The causalfunctional properties of a cell phone's physical form do not in any way correspond to its intentional functional design purpose (ibid.). The functionally intended design purpose of computerized technology can therefore only be grasped by the functional logic of the software code embedded in it. The screen of the cell phone and the buttons on the front give a hint at this functionality from a symbolic discourse but usually there is more than meets the eye. This is why we need elaborate manuals to inform us about how to operate and access the hidden functionality imbedded in the software. The functionality designed into software codes in programmed technology therefore distinguishes itself from the functionality imbedded in physical artefacts in that the realm of intended use and possible use are the same (Bærentsen & Trettvik, 2002). As most parents will know, the proper canonical use of a cup can be negotiated and relativised. i.e., although the cultural linguistic dimension imposes constraints on the use of such artefacts, these constraints can be broken. In computerized functionality, however, there is a one to one correspondence between the intended use and the possible scope of use (Bærentsen & Trettvik, 2002), i.e., there is no other meaningful use than the culturally intended use designed into the software[2].

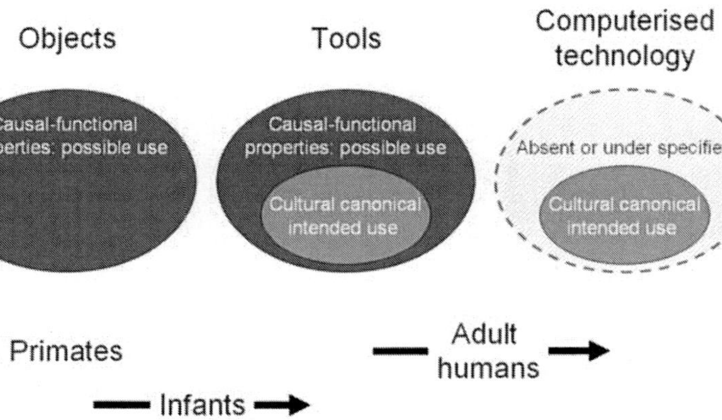

Figure 1. Cultural-historical distinction of tool use evolution

To qualify this triadic segmentation of tool use three lines of research will be outlined:

1. Primates' conceptual understanding of objects and their causal-functional properties.
2. Human infants´ understanding of tool use and rational agents' tool use.
3. Neuropsychological studies of tool use and understanding in brain injured patients.

The goal of this theoretical analysis of current research findings in comparative psychology, infant cognitive development and neuropsychology is tentatively to outline the dynamics of the interplay between the causal-functional and cultural dimension of tool use, i.e. the first and second tool category. Even though many questions are still unanswered and debated in these lines of research, there is enough established consensus to address computerized technology from this combined primate and infant developmental approach.

[ad. 1] For a more detailed review of this lineage I refer the reader to Bærentsen (1989) and Nørager (2004).

[ad. 2] Colleagues from computer science typically contest this sharp distinction by saying that software functions can also be negotiated. While this is true to some degree for skilled software professionals like hackers, this is still a far cry from qualitative new uses like those invented by children with the cup at the dinner table.

Primates' Understanding of Objects and Causal-Functional Properties

In a number of experiments it has been demonstrated that primates have a fundamental repertoire of representational systems that constrains the behaviour of objects in space and time based on spatio-temporal information (Weiss & Santos, 2006). Although there are limitations to these abilities, such as interactions between objects when they are occluded (Flombaum, 2004), it does provide primates with the basic ability to predict how objects behave in time and space, i.e., their causalfunctional properties. This is a fundamental basis for productive tool use and in studies by Hauser (1997), Santos et. al. (2003) and Evans & Westergaard (2004), it has been found that primates can directly perceive the relevant causalfunctional properties of tools and base their choice of tools on the most optimal relation between their end goal and the causalfunctional properties of the tool. In these studies *functionally* irrelevant properties like colour, texture and pattern were ignored by the primates (ibid.). Basic causalfunctional representational skills have also been documented in non-primates like the Caledonian crow (Weir & Kacelnik, 2006). The ability to represent objects over time and space and their causalfunctional relations therefore seems to be a basic ability in a range of animals that does not seem to require linguistic nor advanced intellectual reasoning skills reserved only for humans (Weiss & Santos, 2006).

Tools as Cultural Artefacts: The Design Stance

Human tools are special due to the lasting intentionality that they represent. This is what Daniel Dennett calls the design stance of tools (Dennett, 1987). Studies have demonstrated that when a human actor uses a tool to an end that involves surplus movements that have no direct causal function, like tapping on the side of a box with a stick, these are ignored by chimpanzees but closely mimicked by children of age 3 to 4 (Horner & Whiten, 2005). This difference can be taken to indicate that primates are only sensitive to the causalfunctional properties of the scene, whereas children are also sensitive to the properties of the cultural dimension of the scene (Horner & Whiten, 2005; Casler & Keleman, 2007). The reason for the primates' lack of sensitivity to cultural norms beyond pure causalfunctional aspects might be due to differences in human and primate cultural schemes. However, studies by Whiten et. al. (2005) suggest that primates are indeed capable of distinguishing between different cultural tool use approaches as long as these distinguish themselves with regard to purely causalfunctional properties. This can be taken to account for how primates have schemes similar to humans that allow them to represent qualitatively different approaches to the same problem.

It is important to realise that in such re-enactment experiments, the children are indeed sensitive to the *intentionality of the actor* and not merely copying the action sequences at the programme level. This has been demonstrated in studies by Meltzoff (1995) where infants as young as 18 months would not blindly reinact the actual (failed) behaviour of an actor but recognized and re-enacted the intended goal-directed behaviour. Young children are thus sensitive to the intended goal of the actor and not merely making a programme level imitation of the action sequences. Furthermore, studies by Whiten et. al. (2006) have demonstrated that three-year-old children appreciate the hierarchical structure of action sequences and how these can be flexibly interrelated.

This sensitivity to the cultural design stance of tools coincides with the early appropriation of linguistic skills. In a study by Gergely, Bekkering, and Király (2002) it was investigated whether 14 months old children would readily imitate an adult's goal directed activity under two different constraints. In one condition children would watch an adult actor wrapped in a blanket, with the hands inaccessible, turn on a lamp placed on the table with the head. In the other condition the actor would also wear a blanket but have the hands placed next to the lamp, while still using the head to turn on the lamp. The use of the head instead of the hands was thus either constrained by physical conditions or by choice of the actor alone.

Of the children who witnessed the scene where the actor's hands where occupied, a majority would instead use the hands to switch on the lamp. In the condition where the actor's hands were free, a majority of the infants would copy the actions of the actor and switch the lamp on with their head as well.

In the interpretation of the results by Gergely et al. (2002) children are on the one side subject to the same emulation as primates, opting for a more direct approach using the hands to achieve the same goal as the actor. On the other side, in the absence of any physical constraints, children seem able to evaluate the scene and appreciate the rationality of the actor and hereby go beyond mere emulation. At three to four years of age when language is well mastered, children are also sensitive to the excess cultural aspects of tool use that have no apparent causalfunctional meaning (Horner & Whiten, 2005).

The cultural canonical design stance can be seen as comparative to the causalfunctional constraints by further limiting what objects can and should be used for by introducing an "ought to" or "should" dimension nested inside the realm of possible "can" uses (Trettvik, 2006). Sensitivity to social information about the canonically intended use of artefacts is very outspoken in young children and helps them to quickly form enduring function based categories about canonical uses of tools (Casler & Keleman, 2005, 2007). The constraints of the cultural dimension are further documented in that the names children extend to objects are based on their shared causalfunctional use rather than their overall featural similarity (Casler & Keleman, 2007).

There are, however, limitations to the constraints of the cultural design dimension. Studies by Asher & Kemler Nelson (2007) with 3-4 year olds demonstrated that uses of tools that seemed to oppose their inherent causalfunctional constraints, i.e. implausible use, would generate more inquiring questions about such tools' functionality, as opposed to when use of the tools was causalfunctionally plausible. Such findings can be taken to account for that at the age of 3-4 years, excess cultural movements in the use of tools are considered acceptable as long as these fall inside the constraints of the causalfunctional properties. Cultural norms can thus narrow the causalfunctional uses of objects and tools but not to the extreme.

Although primates do not seem to be sensitive to cultural excess tool use dynamics, studies by Whiten, Flynn, Brown, and Lee (2005) have documented what could be described as a primordial sensitivity to cultural tool use norms. In their studies two strategies for freeing a nut from a contraption was introduced to two groups of chimpanzees by means of a high ranking female that had observed a human instructor perform the action. Both strategies were viable and functionally optimal in their own right from a causalfunctional perspective and did not involve excess cultural movements. Whiten et. al. (2005) reported a conformity bias in chimpanzees that had discovered both strategies towards using the predominant strategy of their companions.

Broad empirical findings thus support that infants and primates have advanced representational skills that allow them to appreciate causalfunctional relations between objects. Sensitivity to cultural norms of tool use that cannot be derived from the objective physical constraints alone seems however, to be beyond primates and preverbal infants. Only in young children with verbal abilities do we find sensitivity to unique human culturally constrained tool use dynamics.

From Objects to Cultural Tools in Infants and Adults

Human tool use is combined both of causalfunctional properties and a canonical design stance (Hauser, 1997) – then "can" and "should" dimension. The transition from handling and seeing artefacts not only as sensory-motor objects, but also starting to know of their intended cultural canonical use is profoundly documented in the scale error phenomenon, which has been formally investigated in recent years (Deloache et. al., 2004; Ware et. al., 2006). In scale error situations 18 to 30-months old children have the possibility to play with toys such as a car, a slide or a couch that are life-size, i.e., scaled to their bodily dimensions. Research assistants will encourage the children to slide down the slide, sit inside the car and on the couch. Later in the experiment small miniature toy replicas of the same objects are introduced into the play. The scale error phenomenon arises when the children try to use these miniature toys in the same fashion as the life-size toys. They will, for example, try to fit their feet inside a toy car and slide down a slide both of which are so small they fit on the palm of their hands. It is very clear that nothing is wrong with their sensory-motor functions since their object manipulation skills are precise and adequate when they grasp and manipulate the toy replicas. In contrast, their intentional goal is not adequately adjusted to the toy replica. They try to sit on the small couch and even try to enter the small car that does not even allow their foot to fit inside. The children will put considerable effort into realising their cultural knowledge driven canonical use of the objects with the small copies and consequently, look very puzzled and displeased when they fail to attain this goal. The toy replicas seem to activate cultural canonical use representations of the artefacts and force a knowledge driven activity programme onto an object that does not correspond causalfunctionally to it. Whereas these studies involved scale errors in young children and objects related to their own body, recent scale errors have also been investigated in relation to scale errors between two objects in the world such as a doll and a toy bed (Ware et. al., 2006). It thus seems that scale error is a general phenomenon that occurs in children's interaction with toys, i.e. cultural artefacts, during their first years of language use.

The scale error is explained as a neurological perception-action dissociation in early life that later in adults becomes fully integrated (DeLoache et. al., 2004, Ware et. al. 2006). Similar perception-action dissociation are also documented in adult neurological patients where brain injuries selectively can damage patients' conceptual understanding of what tools are for while they retain the ability to handle the tool skilfully. Comparatively, other patients are unable to handle tools in a meaningful way but are still able to identify the artefacts, name them, and describe their function (Daprati & Sirigu, 2006; Goodale & Milner, 2006; Johnson-Frey, 2004). These complimentary neurological systems have become known as the "what" versus "where" systems (ibid.). The "what" system supports conscious naming and identification of artefacts and is believed to be a more recent evolvement than the evolutionary older "where" system that supports the sensory-motor handling of objects (Goodale & Milner, 2006; Velichkovsky, 1990). This distinction resembles the causalfunctional "can" dimension and the cultural intentional "should" dimension.

Theoretically, this dual aspect of artefacts has been analysed inside the Activity Theory framework as an extension of James Gibson's concept of affordance that mainly focused on natural biological sensory-motor affordances (Bærentsen & Trettvik, 2002). This has traditionally invoked a number of problems since Gibson's theoretical work was hard to conceptualise in relation to cultural artefacts. Bærentsen & Trettvik (2002) thus posit the need for a dual understanding of affordances that both have a biological evolutionary background and a cultural historical background.

Biological evolutionary founded affordances are called adaptive operations since they are evolutionary adaptations to the physical world (ibid.). Affordances that are cultural-historical founded are called conscious operations since they are seated in the individual's conscious interaction and knowledge about cultural specific norms of the use of artefacts (ibid.). Such affordances are not directly perceivable in the objective physical properties but manifest themselves as extra material social dynamics that unfold themselves over time.

When scale errors are appreciated inside this cultural-historical tool use dimension, it follows that scale errors are not merely a dissociation between cognitive functions but between basic sensory-motor object handling skills and their cultural-historical meaning. The scale error dissociation between the causalfunctional properties of an object and the uniquely human cultural design stance can therefore be hypothesized not to occur in animals. Indeed from an evolutionary perspective, it would be highly maladaptive for animals to be subject to scale errors. Scale errors are instead to be understood as the epiphenomenon of infants' transition into a world not only of material objects but also of extra material cultural norms.

Human tool use can thus be considered as a unique cultural dimension that furtherconstrains the natural causalfunctional properties of objects. The unique properties of computerized technology can therefore be unfolded in a cultural historical perspective as a kind of tool that is only constrained by the cultural dimension but not by the physical causal-functional dimension. This viewpoint is further developed in a deeper analysis of the historical development of computerized technology.

Computerized Technology

It is the central argument of this article that modern computerized technology is functionally separated or abstracted from the material and physical basis that characterises material tools. Even though a computerized tool such as a cell phone or a video recorder are tangible physical objects, their functional meaning, i.e., what they essentially are and do as tools, is primarily constituted by the software code that controls and regulates the relations between the physical buttons and the graphical representations we see on the menu screens, i.e. how the tool ultimately behaves. This makes computerized technology, no matter what shape or form it comes in, a distinct kind of technology quantitatively different from all other technologies and tools that in their way of functioning are subject to the constraints of the physical world. The software codes used to create computerized technology are formalised symbolic languages that define logical procedures. Software is thus not constrained by the physical laws whereby it is possible to violate these laws as it happens when an object on a computer screen suddenly disappears (Bærentsen, 2000). In this way the discourse for how to understand and conceptualise disembodied software as a language is principally in the domain of structural semantics (Johnson, 1987). Computerized technology in a functional sense can thus be seen as a reflection of unique, complex human linguistic capabilities rather than basic tool making skills based on causalfunctional manipulations.

The paradox of computerized technology is that ever since its advent in the early fifties, as the epiphany of human technological development, there has been a gradual movement towards more and more support for more basic human cognitive abilities in the way we interact with such technology – for example by opening it up to out causalfunctional sensory-motor skills by making it more tangible (Dourish, 2004). It is this recapitulation that now calls for evolutionary and developmental psychology to provide novel insights into how this process might continue and existing achievements advance further.

Symbolic Tools: From Command Lines to Direct Manipulations

In the earliest days of computerized technology, interaction with computers was a painstaking procedure that required computer scientists who essentially had to manually construct a new computer for each new task. With the advent of more generic and automatic instruction and input routines, it became possible to execute simple programmes that the scientists did not have to create from scratch each time. Despite the complexity even at this level, it did however, facilitate the spread of computers and computerized technology to a broader scope of uses – though still heavily framed inside scientific and military purposes. While modern computerized technology still rests on the same logical foundation as the early computers, the ways in which we interact with these devices have progressed dramatically from the initial starting point.

Computers were quickly fitted with a keyboard in order to provide easy input to the computer in combination with screens and printers to generate output. Interaction with the computer proceeded by typing in commands in a text interface like the one depicted below. Such command line interfaces require extensive knowledge of the meaningful vocabulary that can be entered and executed by the computer.

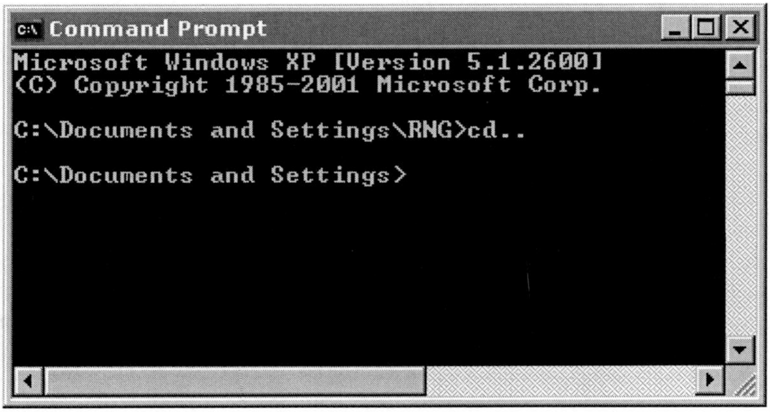

Figure 2. Command Line Interface (CLI)

Alternatives to the command line interfaces were investigated from a very early point since they were cumbersome and difficult to use even by trained specialists. The human mind is simply not naturally geared to remember specific command syntaxes and large amounts of abstract information. To overcome the disadvantages of the command line interface, the graphical user interface (GUI) was invented (Dourish, 2004; Tomitsch, 2003). In a GUI instead of remembering all sorts of commands and associated variables, the available and meaningful commands in a given context are made readily available directly as a graphical representation of buttons that can be clicked with a pointing device such as the mouse. It was also argued that the use of pictograms and icons instead of text provided a superior support for memory of the meaning and functions of the buttons (ibid.)

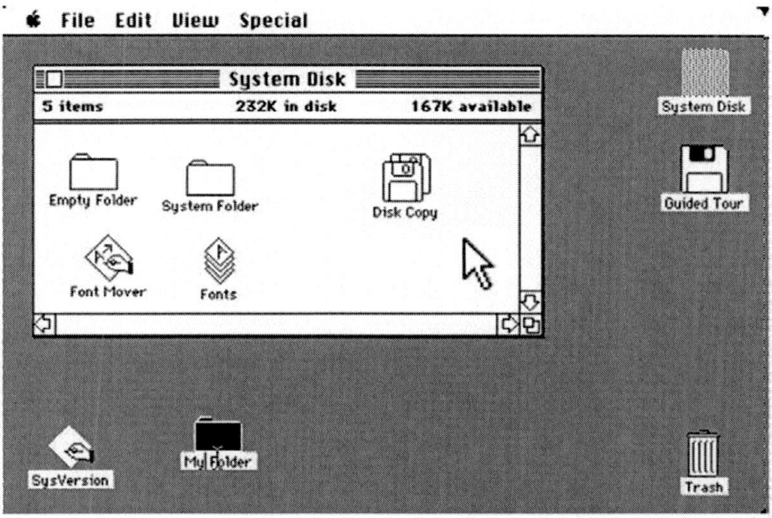

Figure 3: Apple Macintosh desktop graphical user interface (GUI) from 1984 and mouse

Modern computerized tools like the cell phone, the e-mail programme and video recorder draw heavily on the GUI symbolism in the form of text and icons in the way we operate and interact with them. The symbolic content of the graphical user interfaces are typically manipulated indirectly via buttons, such as "play" and "rewind" that are pushed directly with the finger or indirectly with a pointing device such as the computer mouse. The motor skills we use are thus highly stereotyped across many different types of computerized tools and functions. Consequently, computerized tools do not distinguish themselves in the way we interact with them from a sensory-motor causalfunctional perspective, but only by the meaning of the symbols we push and manipulate. A separate goal that grew out of graphical user interfaces was the concept of direct manipulation (Hutchins et. al., 1985). In direct manipulation the user acts directly with the mouse (or some other input device) on the graphical objects on the screen that bypasses indirect commands. For example, a picture in a text document can be resized directly by pulling its corners as an alternative to specifying its size as height and length in centimetres in a dialogue box. Direct manipulation has become even more direct with the advent of touch screens and applications to match them as seen in the Apple iPhone. In this way the pointing device has been altogether circumvented so the users, for example, can rotate and scale pictures up and down in size directly on the screen with pinching gestures.

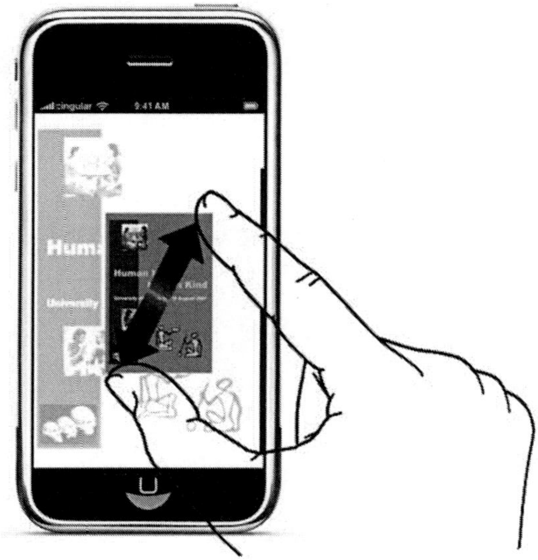

Figure 4. "pinch" gesture to resize a picture on the touch screen of the Apple iPhone

A final movement to be mentioned here is tangible user interfaces or simply tangibility, where the rich interaction from the direct manipulation framework is radicalised by letting all sorts of physical objects represent aspects of the immaterial elements from the graphical user interface (Dourish, 2004). One illustrative example is the conceptual marble phone answering machine where each new incoming voice message is represented by a marble and placed in a little tray on the machine that can be picked up and handled like an object. Where traditional immaterial voice messages are stored inside the answering machine only to be accessed and listen to by pushing buttons on a menu, the marble answering machine gives the messages a physical form by coupling them to the physical marble token (Wensveen, 2005). The goal is to exploit the physical constraints that material objects are per default subject to and hereby access our accustomed natural understanding of such objects and their behaviour (ibid).

Design of user-friendly technology draws on multiple disciplines, such as anthropology, sociology, psychology, etc. The approaches based specifically on cognitive theories, to make technology more user-friendly, from the GUI over direct manipulation to tangibility, all have underlying assumptions about what makes them more user-friendly. These assumptions typically fall into two categories:

The *first* is to reduce complexity and unburden our cognitive resources. For example, the transition from command line interfaces (CLI) to graphical user interfaces (GUI) reduces the need for the users to learn and remember all the viable commands and hereby unburdens the memory of the user. The direct manipulation approach similarly does away with various indirect commands that act on graphical representations of objects and allows the users to manipulate the object directly, with a number of limited and highly generic gestures. This further reduces the number of specific commands that have to be learned and memorised. The implicit notion behind these solutions is that we have a limited amount of cognitive resources that should be used sparingly.

The *second* approach has a radically different underlying assumption that is complimentary to the first. This approach acknowledges that we have a relatively limited amount of *conscious intellectual* resources but that these are only the tip of the iceberg (Bærentsen, 2000; Rasmussen, 1986). The vast majority of human cognitive resources are made up of highly automated subconscious resources (ibid.). Whereas intellectual resources are characterised as highly flexible and adaptable, the subconscious resources are in contrast rigid and optimised to solve very specific real world tasks, such as the manipulation of objects. The tangibility approach mentioned last reflects an effort to draw on these resources by letting immaterial data and logical functioning occupy tangible physical objects and so by proxy gain access to these basic cognitive resources.

A quick assessment leaves no doubt about the tremendous success of the classical graphical user interface (GUI). The widely used Microsoft Windows desktop GUI metaphor has become a standard on most computers in the world. The traditional GUI with menus and icons is also adopted in most cell phones and many other devices. There is thus no doubt that the approach to reduce complexity has been more viable than the approach to integrate and provide support for more basic cognitive resources.

If we compare the two approaches, GUI and tangibility, cognitive psychology has devoted much attention as to higher order intellectual skills as for example how adult memory works. Such rigorous research has helped pave the way for the GUI and similar solutions. In comparison, the tangibility approach mainly builds on an acknowledgement that we have resources devoted to handle objects naturally but only have a very crude understanding of the actual basic cognitive mechanisms involved. Without such a detailed knowledge of the basic cognitive dynamics, the only way to integrate support for these basic resources is to recast immaterial functional aspects of the technology directly in the shape of physical objects. Paul Dourish has framed the problem with the tangibility approach very well in his 2004 book "Where the Action is":

> Our understanding of the nature of these problems is, so far, quite limited, certainly in comparison to the more traditional international style that characterizes most interactive systems today [GUI based]. The theories that govern traditional interaction have only limited applicability to this new domain. At the same time, tangible computing has been explored, largely, as a practical exercise. Most prototypes have been developed opportunistically, driven as much by the availability of sensor technology and the emergence of new control devices by a reasoned understanding of the physicality in interaction. We have various clues and pointers, but there is no theory of tangible interaction. Why does tangible interaction work? Which features are important, which are merely convenient and which are simply wrong? How does tangible computing mediate between the environment and the activity that unfolds in it?
> —Dourish, P. (2004, p.52-53). My insertions in brackets

In this light it should potentially be possible to access the benefits of the vast resources by intimate knowledge of their structure and transform that into more flexible solutions that can be adopted by the widespread GUI solution. Recent years' research in primate and infant cognition has done exactly that and has made it possible to outline the structure of the basic cognitive resources.

Basic and Intellectual Cognitive Resources

As humans we are easily fooled to believe that the majority of our cognitive resources help us solve problems like interpretation of the time schedule at the bus stop and filling out an insurance policy. Both tasks are highly intellectual in nature and draws on our cultural cognitive resources. Such skills are part of what Jens Rasmussen (1986) calls the conscious processor, since they are tied to conscious processes and we have to put in a conscious effort to engage them. Contrary to our phenomenological experience, however, most of our cognitive resources are tied to basic sensory-motor skills which, for instance, make it possible for us effortlessly to catch a ball and navigate between people on a crowded sidewalk (Bærentsen, 2000; Rasmussen, 1986). These sensory-motor skills are part of what Rasmussen (1986) calls the subconscious processor. The broad distinction between higher order symbolic intellectual processes and lower order basic sensory-motor processes is general to most psychological theories of cognition although it is labelled under different names (Velichkovsky, 1990, 2002). In the following brief outline of the characteristics of the conscious and subconscious processor the terminologies used by Jens Rasmussen (1986) will be adopted.

According to Rasmussen (1986), the *conscious processor* is highly adaptable to novel situations due to its flexibility and rational analytical problem solving abilities. It operates with symbolic information and is thus closely tied to our linguistic abilities and our conscious attentional control. This in turn makes it relatively slow, sequential, effortful and limited by the capacity of the short term memory. This dependence on conscious control and willed effort also makes it very fragile and susceptible to breakdowns in stressful and multitasking situations. The conscious processor relies on learned data processing strategies and models. Human performance on tasks that rely heavily on the conscious processor will as a consequence show great variability that depends on the efficiency of the strategy used (Rasmussen, 1986). The conscious processor is also the high-level coordinator of the subconscious system.

The *subconscious processor* deals with data in terms of what Rasmussen calls time-space signals as opposed to symbolic information as it is naturally evolved to deal with functions of sensory-motor control. The major limitation of this subsystem is that it is only capable of dealing with familiar, frequently encountered situations like the ones the organism has adapted to. The high degree of specialisation of the subconscious processor makes it highly efficient and combined with massive capacity that stem from parallel processing, it is both fast and effortless. In contrast to the conscious processor this system is more robust against stress and breakdowns.

Conscious processor	*Subconscious* processor
Slow, sequential, effortful	Fast, parallel, effortless
Capacity limited by short term memory	Very high capacity for data processing
Flexible	Rigid
Symbolic based reasoning	Deals with time-space signals/data
Large repertoire of data-processing models and strategies	Responsible for perception, motor control, and intuitive judgement.
Rational – analytical thinking	Information is decoded from sensory input and higher-level features extracted
High level coordinator of subconscious subsystems	Main data processor
Off-line operation, achronic	Real-time operation, synchronic

Figure 5. Traits of the conscious and subconscious processor. Adopted from Vincente & Rasmussen, 1988

The distinction between basic sensory-motor skills (the subconscious processor) and higher order intellectual skills (the conscious processor) ties in very well with the duality of tools as both objects and cultural artefacts. To fully appreciate the cognitive premises of human tool use, and especially computerized tool use, we must therefore draw on theories that address both the dynamics of the conscious and subconscious processor.

Higher order intellectual skills have traditionally been well studied inside the information processing cognitive science framework (Velichkovsky, 1990, 2002). Indeed many theories of human cognition co-evolved in close relation with the field of human-computer interaction (Hoff, 2004). There is thus a historical correspondence between the logics of computerized technology and the theories of higher order cognitive skills. Theories and methodologies of human use of computerized technology are as a consequence naturally focused on higher order cognitive skills (ibid.). Methodologically and theoretically there is thus a large repertoire to draw on in the design and development of modern computerized technology in relation to cultural aspects and conscious higher order intellectual skills.

The theoretical and methodological corpuses that support how we understand the role of basic sensory-motor processes of the subconscious systems in relation to computerized technology are in contrast much less evolved (Dourish, 2004; Bærentsen & Trettvik, 2002). Several reasons can be mentioned that may account for this. *First* the historical aspects of the correspondence between computerized technology and higher order cognitive skills naturally evolved this area in contrast to basic skills that did not have a natural counterpart in the development of programmed technology (Hoff, 2004). *Second*, the phenomenological bias of higher order cognitive skills naturally drew attention to their role in behaviour. *Third*, computerized technology was initially developed in the realm of highly intellectual uses by skilled professionals operating large complex industrial installations with a focus on supervisory tasks (Bærentsen, 1996; Rasmussen, 1986). *Fourth*, owing to the legacy of developmental psychologist Jean Piaget, the implicit notion of humans born as cognitive blank slates, the subconscious processor was easy to conceive merely as a highly adaptable system of sensory-motor space time signals without any phylogenetic derived structure. In other words the subconscious processor was a structureless slave to higher order intellectual dynamics one needed not concern about (Nørager, 2004). *Finally*, the dynamics of the conscious and subconscious processor makes it difficult to isolate problems that occur when computerized technology does not support the subconscious processor. For example, when you move the mouse to a position where you expected something to be on the computer screen – such as the print icon in Microsoft PowerPoint - this rapid and precise movement is supported by basic orientation and positioning skills. You then suddenly realise that the print icon has disappeared or been relocated. In such a breakdown in our use of the computerized artefact the focus shift causes conscious attention to be drawn to the nature of the problem whereby, for example, search routines helps us realise that the print icon has been relocated higher in the menu list due to that how other seldom used menu items have been removed.

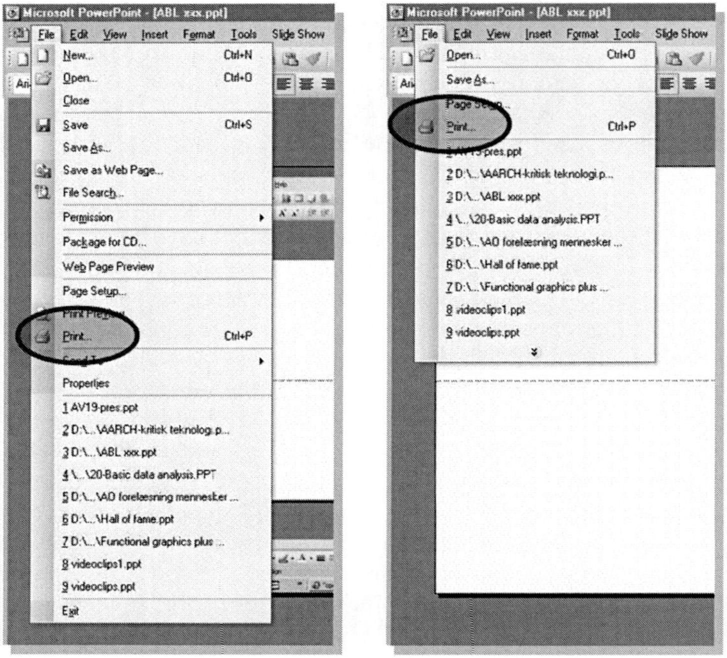

Figure 6. Dynamic pruning of seldom used menu items in Microsoft Office causes the position of the other menu items to shift around.

What poses a problem or incompatibility with basic cognitive skills will most likely not be so in relation to conscious cognitive intellectual skills since they exhibit flexibility, adaptability and rational analysis. We simply acknowledge that the icon is in a new position and consciously guide the mouse cursor to this new position and click it. In other words, the problem seems to dissolve the moment we consciously focus on it!

To counter these historical and methodological biases primate cognitive research and developmental psychology seem well suited to investigate these basic abilities and their dynamics in their pure form for several reasons. *First*, since much research has already been done, it is only sensible to try and apply it in a novel way. *Second*, higher human intellectual skills are very flexible and adaptable, which makes us so prominent in the animal kingdom. Unfortunately, in a research setting these very skills often make it difficult to investigate the limits and evaluate breakdowns of specialised basic cognitive skills. The reason being that when breakdowns occur higher cognitive abilities will simply shift in and take over. Hereby the incompatibility problem as such ceases to exist. *Third*, basic cognitive skills are often tacit in nature and not immediately accessible to human phenomenological introspection (Bærentsen & Trettvik, 2002). In experimental conditions we therefore cannot rely on verbal reports. Paradoxically, the cornerstone of methodologies to evaluate usability of computerized technology and one of the most widely used inspection protocols in human-computer interaction is the "think aloud protocol" where users are asked to verbalise their continuous stream of conscious thought when they use and evaluate a computerized product (Rubin, 1994). *Fourth*, the conscious experience we have of how we interact with the physical world has proven in some ways to be contradictory to its true nature and sometimes rather counterintuitive (Hermer-Vazquez, Spelke & Katsnelson, 1999), which further biases the conscious phenomenological approach.

The non-verbal premises of research with animals and human infant have generated novel and ingenious research designs that allow us directly to probe the dynamics of basic cognitive skills without having to rely on verbal reports. This has also positively influenced research on adult humans and provided us with optics that overcome challenges normally associated with experiments that probe basic subconscious abilities.

Primate Cognition In Computerized Technology

The attempts so far to tap into basic cognitive resources of the subconscious processor has primarily been to let the immaterial functional logic of computerized software code inhabit physical tangible objects. As Dourish, however, underlines we have no precise knowledge of what the functional effective aspects of this solution is, other than it allows the user to interact naturally with these physical computerized objects (Dourish, 2004). With intricate knowledge about what the functional properties that make tangible computing work, it can be argued that in addition to letting computerized functionality inhabit physical objects we can also let the physical functional constraints of the physical world populate and inhabit the computerized world. Hereby we can potentially profit both from the advantages of the widespread classical GUI approach and the benefits from the tangibility approach. The possible viability of this approach can be illustrated with the evolvement of animated cartoons.

In the early days of animated cartoons it was only possible to tell very basic and short stories. Longer animated stories with complex narratives did not work because the characters did not seem convincing to the audience and generally lacked a natural look and feel which made them ungainly and awkward (Lasseter, 1987). One of the great feats of Walt Disney was to overcome these limitations by making detailed kinematic studies of animate and non-animate object behaviour dynamics. These detailed studies where crystallised into eleven animated cartoon principles that among others capture the functional essence of how animate and non-animate bodies behave in the physical and social world (ibid.). By implementing these kinematic dynamics into the cartoons, Disney made cartoons look "right" and thereby opened the cartoon industry to longer and more complex narratives. Indeed Disney and his cartoonists with these principles had the skills to create arbitrary anthromorphic characters and endow even a sack of flour with animate properties. Profoundly enough, when the computer industry made it possible to generate animated characters that on the surface looked almost like the real world - something hand animated cartoons could never achieve - they failed because they had overlooked Disney's original intrinsic cartoon animation techniques. It took John Lasseter and his team at PIXAR studies to reintroduce the animation techniques to computer animated cartoons (Lasseter, 1987). The lesson learnt once again was that the functional aspects of the physical world in relation to agents and objects lie in the intrinsic dynamics and not in the phenomenologically "defining" surface characteristics. The ability to endow even triangles with mental states when the dynamics of the animation are right is beautifully illustrated in the studies by Abell, Happé & Frith (2000) with normal and abnormal autistic children's appreciation of abstract animated characters.

Such examples illustrate that with the proper knowledge of the intrinsic functional dynamics these can be transposed onto other objects, even though they may lack the surface characteristics of the natural objects. It was therefore natural that the work of Disney and Lasseter was adapted into graphical user interfaces that people in general also tend to perceive and conceptualise in categories of scenes, objects and agents (Bærentsen, 2000). In their article from 1993 Chang & Ungar (1993) outline how principles from animation can be applied to the user interface. However, they call for further research to determine and clarify the detailed dynamics and the effects of doing so (ibid). Like tangible computing although it works we do not know why.

To summarise, the approach to endow the graphical user interface with functional dynamics from the physical world, towards which our subconscious processor is naturally adapted, seems warranted. The approaches by Disney, Lasseter, Chang & Ungar, however, still suffer from the same bottom up experiential driven approaches which Dourish challenges with the tangibility approach. In addition, we need a coherent theoretical framework from which hypotheses can be derived and subjected to experimental investigations. The core knowledge framework most noticeably represented by the research of Elisabeth Spelke and her colleague is a good departure point towards such a framework.

The Core Knowledge Paradigm

To recapitulate on what we have addressed to far the following points are outlined:

• Computerized technology is functionally abstracted from the material basis that characterises material tools.

• The human subconscious processor is optimised to deal with certain functional dynamics of the material aspects of tools.

• Research and practical examples suggest that these functional dynamics from the
physical world can be implemented into immaterial graphical user interfaces.

The next challenge is thus to address systematically what these functional dynamics are, which the subconscious processor is optimised to deal with. One particular coherent framework that also builds on rigorous research has been advanced by Elisabeth Spelke and her colleagues. Based on an adoption of Fodor's contemplation of basic cognitive modules, they have devised the idea of core knowledge (Spelke, 1994, 2000; Spelke & Tsvikin, 2001; Spelke & Kinzler, 2007). The core knowledge research paradigm has so far experimentally outlined four core knowledge domains (Spelke, 1994; Spelke & Kinzler, 2007). These are representational systems that relate to:

1) Object dynamics
2) A small exact and large approximate number system
3) Spatial orientation based on geometrical relationships (ibid.).
4) Agents and rational goal directed behaviour

Although the object representational system has been studied most intensively all four system have received much experimental and theoretical attention.

Due to the massive research attention related to the core knowledge paradigm, the field has moved forward quickly and generated many well documented findings alongside with novel and ingenious methodologies. It is important to stress that despite general agreement on many research findings the specific core knowledge conceptualisation of these findings is still very much debated. For a critical discussion of core knowledge I refer the reader to Cheng & Newcombe (2005), Newcombe & Huttenlocher (2003) and Newcombe & Uttal (2006). Based on the large amount of empirical evidence, it seems, however, warranted to adopt some of the most well established and generally agreed upon findings into the applied field of computerized technology design. Two of these findings are core knowledge about objects and their behaviour and core knowledge about spatial navigation.

One of the special features of the core knowledge framework is the special role of language as a resource that integrates and overcomes the constraints of the basic core knowledge systems (Spelke, 1994, 2000; Spelke & Tsivkin, 2001). This flexible role of language is dynamic which means that the extension of basic cognitive competencies into the cultural domain must continuously be furnished by higher order linguistic cognitive skills. If the linguistic skills are occupied, as it has been experimentally investigated in dual task paradigms, results indicate that adult humans in some ways base their behaviour in relation to orientation and other agents on the more isolated and basic functions of the core knowledge modules (Hermer-Vasquez, Spelke & Katsnelson1999; Newton & de Villiers, 2007). Since computerized technology by nature is very demanding on our linguistic and symbolic cognitive skills, as outlined previously, this aspect accentuates the need for computerized technology to support basic core knowledge dynamics of human cognition more directly.

Core Knowledge of Space and Objects in Graphical User Interfaces

The final step is experimentally to investigate the value and effect of designing key core knowledge dynamics into graphical user interfaces. Based on the theoretical analysis we developed the hypothesis that graphical interfaces constrained with core knowledge dynamics should:

1) Be more intuitive and easy to use
2) Be faster and generate fewer errors
3) Be more pleasurable to use due to fewer breakdowns and less intellectual cognitive strain.

In order to test this hypothesis two typical graphical users interfaces were identified that most users of contemporary computer systems should be familiar with. Both user interfaces came in two versions, A and B. The A versions violated either spatial- geometric orientation relations or spatiotemporal object dynamics. The other B-versions of the two graphical user interfaces were manipulated in order to comply with the respective core knowledge dynamics of space and objects. Both graphical user interfaces thus came in two versions that were graphically similar in their surface characteristics but different in their intrinsic functional dynamics.

The *spatial* core knowledge paradigm interface consisted of a windows folder list that users had to navigate in. The paradigm conceptually resembled a situation where you have to go in and out of offices located on a long office hallway. The test conditions would be akin to A) always (magically) to return to the entrance of the office hallway when you exit an office no matter how long the hallway it located. Alternatively B) you always return to the precise position in the office hallway where you entered the office.

The *object* core knowledge paradigm consisted of a list with graphical icons of TV channels. The list had to be resorted rearranged by moving the channels one at a time from the position in the list to the right and then either up or down to a new position. The paradigm conceptually resembled a situation where you have to move books around on a bookshelf. The movement from left to right and vice versa (in and out of the bookshelf) could either be A) instant (magically) or B) animated as a continuous movement.

In two series of counterbalanced experiments 56 participants solved a number of structurally equivalent tasks in both versions, A and B, of the two types of interfaces. Participants were recruited to control for experience with computers, age and sex. After the experiment the subjects were interviewed to probe for their phenomenological subjective experience.

The preliminary results from these experiments are very encouraging (Nørager, 2006, 2008A, 2008B). In both experiments the B-versions that did not violate the core knowledge dynamics were significantly preferred over the other A-versions by the participants. There was also a small difference in the time the participants took to solve the tasks with the versions that did not violate core knowledge being the fastest to use. In our analysis of the data we also quantified how confused the participants were in either of the two B-versions based on the movement pattern of the mouse cursor and how many errors they made. We found that in both types of graphical user interfaces the users made significantly fewer errors and were significantly less confused in the B-versions that did not violate the core knowledge dynamics. Some of these differences related to how confused the participants were turned out to be not only significant but quite extreme.

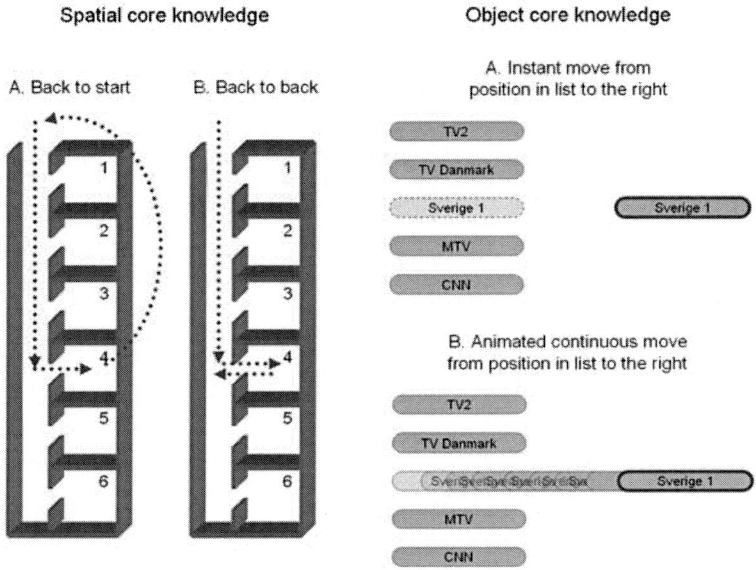

Figure 7. Spatial core knowledge was tested in a folder navigation paradigm akin to an office floor hallway. Object core knowledge was tested in a TV channel object move paradigm.

In our interpretation of the data we have focused on the relation between the relatively small functional time difference and the sometimes huge difference on more qualitative and phenomenological aspects. Here the theoretical framework provides us with explanatory insights in relation to the dynamics between higher order and lower order cognitive dynamics. Although the participants made substantial errors with the A-version graphical user interfaces that violated core knowledge dynamics and were highly confused by them, they were also able to recover from these breakdowns during the interaction. The recovery from a breakdown is, however, a demanding intellectual task, akin to problem solving, which moves focus from the current activity and causes cognitive strain. Although these breakdowns did not impact time consumption drastically in the A-version interfaces that violated core knowledge dynamics they did, however, irritate and annoy the users very much despite the quick recovery. This was also evident by some of the participants' overt exclamations of dissatisfaction and annoyance like deep sighs. This relation is probably what is reflected in the major preference difference in favour of the B-version user interfaces that did not violate the core knowledge dynamics. In summary, programmed technology that functionally constrains to the dynamics of the physical world is thus not only functionally superior but is also vastly more “aesthetically” pleasing to use.

Two other findings in the data are important in relation to the nature of core knowledge dynamics. *First*, both Hermer-Vasquez, Spelke & Katsnelson (1999) and Rasmussen (1986) note that persons’ performances based on intellectual competencies are more varied due to the differences in efficiency of various strategies adopted. Performance based on basic cognitive resources is much more similar due to the uniform subconscious nature and evolutionary background in contrast to individually and culturally appropriated skills. This contrast also seems to be reflected in our data. The variance on all measured dependent variables in the experiment was significantly larger in the data associated with A-version user interfaces that violated core knowledge dynamics. This can be explained by the greater need for the participants to rely on intellectual strategies to solve the tasks and overcome the breakdowns in the interaction associated herewith, i.e., some users adopted better intellectual compensatory solution strategies than others. *Second*, in the follow up interview to the experiments, the participants were asked for their preference regarding the two versions of the graphical user interfaces. As reported previously, the participants highly favoured the B-version graphical user interfaces. When asked why they preferred one over the other, almost half of the participants in the object interface paradigm were unable to account correctly for their preferences as related to the dynamical difference between the two. This fits very well with the subconscious nature of core knowledge dynamics. The participants are not able consciously account for the true nature of their dissatisfaction and pinpoint the origin of the breakdowns, simply because they do not have immediate conscious access to those cognitive layers. Unable to account correctly for the difference between the two versions as rooted in the functional dynamics, these participants would instead rationalise and provide *incorrect* accounts linked to difference in surface characteristics that was not actually there. Hermer-Vasques, Spelke & Katsnelson (1999) reported similar findings with adult subjects who were unable to account correctly for how they had reoriented in a dual task orientation experiment that forced orientation to rely on more basic core knowledge orientation skills.

To summarise, we have provided support for the hypothesis on different levels. The raw scores on performance provide support for the hypothesis that graphical user interfaces that do not violate core knowledge dynamics are easier and more intuitive to use. The users also experienced fewer breakdowns and therefore had to use less intellectual cognitive resources to overcome the breakdowns, which seem to make the graphical user interfaces more enjoyable to use, as reflected in the preference measure. The data structure, which closely matched each other across two very different tasks and user interfaces, also points to a common mechanism in play, which further supports the core knowledge dynamics as a common conceptual framework for understanding these data.

As a final note, it can be added that since the tasks deployed in these experiments were very easy it can be further hypothesized that if these experiments had taken place in a stressful situation or involved secondary tasks, the quantitative difference in time might have been more outspoken due to more severe breakdowns of the overall interaction with the non-core knowledge versions. This will be the focus in upcoming experiments.

Closing Remarks

Tool use from an animal and infant cognitive developmental perspective highlights the duality of human tool use as both causally, functionally and culturally constrained. On this background, computerized technology singles itself out by its lack of a causalfunctional basis. The importance of the causalfunctional basis was illustrated by the basic cognitive resources naturally dedicated to support activity in this respect. Animal cognitive psychology and infant developmental cognitive psychology have successfully studied the functional properties of these basic cognitive skills both in primates, human infants and adults. This research agenda has given rise to a coherent framework called core knowledge. Based on this approach, it seems possible to re-implement support for the vast basic cognitive resources in modern computerized technology. The deep nature of computerized technology as essentially disembodied from a causalfunctional basis is hereby possible to overcome. In a sense, the challenge is to re-establish the functional essence of the causal-functional properties in computerized technology.

The conception of the basic subconscious cognitive skills, which is reflected in the core knowledge theoretical framework, makes it very well suited to generate hypotheses about how human adults' use of computerized technology can be experimentally investigated. This was done in two experiments from which preliminary data on a number of levels provide support for the viability of this approach.

On a more general level, the animal and infant cognitive developmental framework adopted in this article dictates that the problems many users experience with computerized technology is not a generational problem that will disappear as younger generations become more accustomed to such technology. The problems with computerized technology lie at the very heart of our cognitive resources which have a long evolutionary history. Instead it might be said that younger generations are more accustomed to compensate intellectually for the shortcomings of computerized technology than older generations. Furthermore, due to the tacit nature of these subconscious skills we cannot expect either that technology will eventually evolve out of these problems on its own as many other technologies mature. Instead, it will require a focused scientific effort to find the functional dynamics of the core knowledge systems needed to help remedy the problem. Animal cognitive psychology and infant developmental cognitive psychology have proven to be two very promising directions to pursue in order to help guide the design of future computerized technology.

References

Abell, F., Happé, F. & Frith, U. (2000). Do triangles play tricks? Attribution of mental states to animated shapes in normal and abnormal development. *Cognitive Development*, 15, (p.1-16).

Asher, Y.M. & Kemler Nelson, D.G. (2007). Was it designed to do that? Children's focus on intended function in their conceptualization of artifacts. *Cognition*, in press.

Bærentsen, K. B. (1989). Mennesker og Maskine. In M. Hedegaard, V. Rabøl Hansen, & S. Thyssen (eds.). Et Virksomt Liv. Udofrdringen af Virskomhedsteoriens Praksis (p. 142-187). Aarhus Universitetsforlag: Aarhus.

—. (1996). Episodic knowledge in system control. In B. Holmqvist, P.B. Andersen, H. Klien, & R. Posner (eds.). *Sign of work. Semiosis and information processing in organisations* (p. 282-323). New York: Walter de Gruyter.

—. (2000). Intuitive user interfaces. *Scandinavian Journal of Information Systems*, 12, (p.29-60).

Bærentsen, K. B. & Trettvik, J. (2002). An activity theory approach to affordance. In: O. W. Bertelsen, S. Bødker, & K. Kuuti (eds.),

Proceedings of the Second Nordic Conference on Human-Computer Interaction (p. 51-60).
Casler, K. & Keleman, D. (2005). Young children's rapid learning about artifacts. Developmental Science, 8, 6, (p.472-480).
Casler, K. & Keleman, D. (2007). Reasoning about artifacts at 24 months: The developing teleo-functional stance. Cognition, 103, (p.120-130).
Chang, B. & Ungar, D. (1993). Animation: From cartoons to the user interface. Proceedings of the 6th annual ACM symposium on User interface software and technology, (p.45-55).
Cheng, K. & Newcombe, N.S. (2005). Is there a geometric module for spatial orientation? Squaring theory and evidence. Psychonomic Bulletin & Review, 12, 1, (p.1-23).
Daprati, E. & Sirigu, A. (2006). How we interact with objects: Learning from brain lesions. Trends in Cognitive Science, 10, 6, (p.265-270).
DeLoache, J.S., Uttal, D.H. & Rosengren, K.S. (2004). Scale errors offer evidence of perception-action dissociation early in life. Science, 304, (p.1027-1029)
Dennett, D. (1987). The Intentional Stance. Cambridge, MA: MIT Press
Dourish, P. (2004). Where the Action Is: The Foundation of Embodied Interaction. Cambridge: MIT Press.
Evans, T.A. & Westergaard, G.C. (2004). Discrimination of functionally appropriate and inappropriate throwing tools by captive tufted capuchins (Cebus paella). Animal Cognition, 7, (p.255-262).
Flombaum, J.I., Kundey, S.M., Santos, L.R. & Scholl, B.J. (2004). Dynamic object individuation in rhesus macaques. Psychological Science, 15, 12, (p.795-800).
Gergely, G., Bekkering, H. & Király (2002). Rational imitation in preverbal infants. Nature, 415, 755.
Goodale, M. & Milner, D. (2006). One brain – two visual systems. The Psychologist, 19, 11.
Hauser, M.D. (1997). Artifactual kinds and functional design features: What a primate understands without language. Cognition, 64, (p.285-308).
Hermer-Vazquez, L., Spelke, E. S., & Katsnelson, A. S. (1999). Sources of flexibility in human cognition: Dual task studies of space and language. Cognitive Psychology, 39, (p.3-36).
Hoff, T. (2004). Comments on the ecology of representations in computerised systems. Theoretical Issues in Ergonomics Science, 5, 5, (p.453-472).

Horner, V. & Whiten, A. (2005). Causal knowledge and imitation/emulation switching in chimpanzees (Pan troglodytes) and children (Homo sapiens). Animal Cognition, 8, (p.164-181).

Hutchings, E.L., Holland, J.D. & Norman, D.A. (1985). Direct manipulation interfaces. Human-Computer Interaction, 1, (p.311-338).

Jaswal, V.K. (2006). Preschoolers favour the creator's label when reasoning about artefact's function. Cognition, 99, B83-B92.

Johnson, M. (1987). The Body in the Mind. University of Chicago Press: Chicago.

Johnson-Frey, S. H. (2004). The neural basis of complex tool use in humans. Trends in Cognitive Science, 8, 2, (p.71-78).

Kemler Nelson, D.G., Russell, R., Duke, N. & Jones, K. (2000). Two-Year-Olds Will Name Artifacts by Their Functions. Cognitive Development, 71, 5, (p.1271-1288).

Klix, F. (1980). Erwachendes Denken. Berlin: VEB Deutscher Verlag der Wissenschaften .

Lasseter, J. (1987). Principles of Traditional Animation Applied to 3D Computer Animation. SIGGRAPH '87, Computer Graphics, Vol. 21, No. 4, (pp. 35-44).

Meltzoff, A. N. (1995). Understanding the intentions of others: Re-enactment of intended acts by 18-month-old children. Developmental Psychology, 31, (p.838-850).

Newcombe, N.S. & Huttenlocher, J. (2003). Making Space. The Development of Spatial Representation and Reasoning. Cambridge, MA: MIT Press.

Newcombe, N.S. & Uttal, D.H. (2006). Whorf versus Socrates, round 10. TRENDS in Cognitive Sciences, 10, 9, (p.394-396)

Newton, A.M. & de Villiers, J.G. (2007). Thinking while talking. Adults fail nonverbal false-belief reasoning. Psychological Science, 18, 7, (p.574-579)

Nørager, R. (2004) Ecological Cognitive Ergonomics. A theoretical and practical framework for constraining the immaterial functional logic of programmed technology in user interfaces with core knowledge dynamics. Psykologisk studieskriftserie, 7,8.

—. (2006) Users expect interfaces to behave like the physical world. In: O.W. Bertelsen, M. Brynskov, P. Dalsgaard, O.S. Iversen, M.G. Petersen & M. Wetterstrand (eds.). Sixth Danish Human-computer Interaction Research Symposium, Department of Computer Science, University of Aarhus, (p. 47-48).

—. (2008A). Basic spatial and orientation constraints in graphical user interfaces. Unpublished manuscript.

—. (2008B). Basic object dynamic constraints in graphical user interfaces. Unpublished manuscript.

Rasmussen, J. (1986). Information Processing and Human-Machine Interaction: An Approach to Cognitive Engineering. (vols. 12) Amsterdam: Elsevier Science Publishing Co., Inc.

Rubin, J. (1994). Handbook of Usability Testing. John Wiley & Sons

Santos, L.R., Miller, C.T. & Hauser, M.D. (2003). Representing tools: How two non-human species distinguish between functionally relevant and irrelevant features of a tool. Animal Cognition, 6, (p.269-281).

Spelke, E. S. (1994). Initial knowledge: Six suggestions. Cognition, 50, (p.431-445).

—. (2000). Core knowledge. American Psychologist, 55, (p.1233-1243).

Spelke, E. S. & Tsivkin, S. (2001). Initial knowledge and conceptual change: Space and number. In: M. Bowerman & S.C. Levinson (eds.), Language Acquisition and conceptual Conceptual Development (p. 70-100). Cambridge: Cambridge University Press.

Spelke, E.S. & Kinzler, K. (2007). Core knowledge. Developmental Science, 10, 1, (p.89-96).

Tomitsch, M. (2003). Trends and Evolution of Window Interfaces (2003). Diploma thesis, University of Technology, Vienna.

Trettvik, J. (2006). Affordance, attention and time. Proceedings of the Sixth Danish Human-Computer Interaction Research Symposium, Department of Computer Science (DAIMI), University of Aarhus, Aarhus, (p. 59-60).

Velichkovsky, B.M. (1990). The vertical dimension of mental functioning. Psychological Research, 52, (p.282-289).

—. (2002). Heterachy of cognition: The depths and the highs of a framework for memory research. Memory, 10,5/6, (p.405-419).

Ware, E.A., Uttal, D.H., Wetter, E.K. & DeLoache, J.S. (2006). Young children make scale errors when playing with dolls. Developmental Science, 9, 1, (p.40-45).

Weir, A.A.S. & Kacelnik, A. (2006). A New Caledonian crow (Corvus moneduloides) creatively re-designs tools by bending or unbending aluminium strips. Animal Cognition, Vol. 9, No. 4., (pp. 317-334).

Weiss, D.J. & Santos, L.R. (2006). Why primates? The importance of nonhuman primates for the understanding of human infancy. Infancy, 9, 2, (p.133-146).

Wensveen, S.A.G. (2005). A Tangibility Approach to Affective. Interaction, PhD Thesis, Technische Universiteit Delft.

Whiten, A., Flynn, E., Brown, K. & Lee, T. (2006). Imitation of hierarchical action structure by young children. Developmental Science, 9, 6, (p.574-582).

Whiten, A., Horner, V. & de Waal, F.B.M. (2005). Conformity to cultural norms of tool use in chimpanzees. Nature, 437, (p.737-740).

CHAPTER SEVEN

BEYOND GROOMING: DESCARTES' "COGITO" AND A DARWINIAN NEUROPSYCHOANALYTIC VIEW OF UNIQUELY HUMAN MENTATION

HARRY R. BRICKMAN

Abstract: This paper begins by deconstructing Descartes' Cogito as referring to self-reflective thinking, which it argues to be uniquely human. Further examining Descartes' Meditations, it proposes and illustrates that the thought processes in contemporary neuroscientifically-informed psychoanalytic depth psychotherapy are prime examples of reflexive thinking unique to Homo sapiens. It also postulates a continuum of succorance among mammals, from arched back nursing in rats, through grooming in non-human primates to supportive elements of psychotherapy and psychoanalysis in humans. New research is called for on the selective value of unconscious conflictual neural processes, the acknowledged domain of psychoanalytic depth psychology. Since the paper is not primarily intended for clinicians, no detailed case material will be presented.

Introduction: What Descartes meant by *"Cogito ergo sum"*

A strict reading of Descartes supports a seemingly paradoxical view of the self-reflecting properties of the human mind that this essay proposes as unique among all animals. *Cogito ergo sum,* the Latin translation of *je pense, donc je suis* appeared in his *Discourse on Method* (1637):

"I was then in Germany, attracted thither by the wars in that country, which have not yet been brought to a termination; and as I was returning to the army from the coronation of the Emperor, the setting in of winter arrested me in a locality where, as I found no society to interest me, and was besides fortunately undisturbed by any cares or passions, I remained the whole day in seclusion, with full opportunity to occupy my attention with my own thoughts."

Without claiming to be a Latin scholar, it appears likely to me that the French words *je pense* appear to be most closely derived from the Latin *pensito* – I weigh, weigh out (Lewis and Short 1879). Since Descartes himself translated his French words *je pense* into the Latin *cogito,* it seems arguable that he intended a specific connotation of weighing, pondering, considering, rather than of propositional thinking (E.g.: I think *that….*).

I therefore suggest that – *think* that – Descartes' choice of language did not mean thinking in the sense of "thinking that" or "thinking of" or even "thinking about" .It can be understood as a mental activity of an intransitive nature – the kind of reflective thinking engaged in by philosophers and by poets, who often call it soliloquy. An outstanding example is Shakespeare's *"To be or not to be"* soliloquy in *Hamlet.* The renowned sculptor Auguste Rodin also captured self-reflection in his masterpiece, *The Thinker.*

Reflective thinking is perhaps more a mental state than a mental action (Brickman 1998). I will then be so bold – or so foolhardy—as to suggest that the Latin word *meditatio* would have better approximated Descartes' idea, if indeed he was parsing words at the time. In partial confirmation of my speculation, Descartes entitled his next work *"Meditations"* (1641). My contention, then, is that Descartes' C*ogito* referred to deeper-level cogitation or meditation, and therefore self-reflection. The mind, *qua* mentation, is a loosely defined term referring to the parallel and distributed processing of information within the brain which can include attention, concentration, planning, memory storage and retrieval., rehearsing (within the inner representational world), licking of psychic wounds, and other executive functions as well as dreaming.

Mentation is mainly non-conscious, but in part can also be willingly conscious, and this is also true of self-reflection. Self reflection is an elaboration of theory of mind, the capacity to identify the intentions of conspecifics. Still-controversial reports from many primatologists (E.g. Whiten 1991, DeWaal 1996) indicate that chimpanzees employ theory of mind, or mentalization, in scanning the intentions of conspecifics.

More recently, experiments by Hare, Call, and Tomasello (2001) have refined our understanding of theory of mind behaviours in chimpanzees under competitive conditions which reveal the nuanced nature of that realm of mentalization. Hare et al. have been able to demonstrate the socially adaptive value of theory of mind in subordinates who can, under laboratory circumstances, evade dominant animals in securing food. There are no findings that I am aware of that confirm a capacity to *mentalize one's own thinking* in any non-human animals.

According to Hauser (2000), spider monkeys think, cats think, corvids and scrub jays think, elephants think, dolphins think. According to most dog owners, dogs also think. Humans are known to think, but what other animals beside humans cogitate, self-reflect, meditate, soliloquize, --- *think about thinking*? Regret their thoughts? Experience self-pride or self-contempt? And how can these questions be researched?

The species- significant criterion of self-reflective thinking

Self-reflective thinking fundamentally requires a splitting of self-function--or agency--into two components, the observing self and the observed self (I will be discussing the concretization of mental processes into mental entities presently). Self-reflection is an evolved aspect of theory of mind; often, but not always, implemented in the undisturbed solitude that Descartes described. It is a conversation within one's inner representations of self and of the outer world when the distractions of that outer world do not require the energy for complex mentation and actions devoted to furthering personal and genomic survival. Self-reflection also enables self-narrative formation, again a uniquely human capacity as far as we know. In turn, the role of self-reflection – also known as reflexivity – has been significantly redefined by contemporary psychoanalytic researchers and clinicians who have been influenced by evolutionary biology and psychology (Fonagy et al., 2001) The Fonagy group has built much of its work on the evolutionary arguments of John Bowlby (1969) who located the early emotional development of humans within a continuum of succorance in mammals.

Arguments for a continuum of succorance in mammalian – including human – life

Grooming, licking, and arched-back nursing in laboratory rats increase hippocampal synaptogenesis and promote cognitive development in pups, according to Liu at al. (2000) at McGill. Naturalistic and in-captivity observations of non-human primates provide evidence of what I would call *alloregulation,* by which I mean affect regulation in one individual through specific behaviours of another conspecific. Anthropoid apes and monkeys have evolved such down-regulating practices in the form of grooming behaviours. In the absence of effective auto- or self-regulation of propensities for socially disruptive or destructive behaviours, grooming has evolved to promote individual comfort and subsequent self-control, thereby facilitating group solidarity, so vital to individual survival of social animals in ancestral and contemporary environments. By down-regulating potentially disruptive affects in individuals, grooming also lubricates social exchanges in dominance hierarchies. Concordant behaviours are thereby furthered, signalling reconciliation, ranking acknowledgement, and peacemaking, along with succorance. (DeWaal 1996, esp. pp.40-88 & 176-182). DeWaal acknowledges, however, that the "double-holding" behaviours of rhesus mothers who pick up and briefly hold infants of higher ranking mothers has only been observed at the Wisconsin Primate Center (DeWaal 1996 pp.100-101). This is an example of a researcher acknowledging the limitations of generalizing behaviours observed in captive animals, as Boesch (2007) has underscored more recently.

Similar practices occur in packs of canids in the form of licking and assumption of vulnerable, often sexually receptive, body positions. (Solomon & French 1997) Other, less intense and prolonged, examples are in the necking of horses (who are herd animals in the wild), and analogous alloregulatory behaviour in other non-primates (see Riedman 1982). Similar behaviours are regularly observed in impalas, giraffes and elephants. Mutual face and neck rubbing has been frequently observed in prides of lionesses in the wild (Buechner 1973).

In humans, alloregulation is expressed in handshaking, hugging, holding, and stroking in families and close friendships. It is common in American society to feel "touched" by a particularly poignant event. Within the past decade, the developmental advantages of touch in early human infancy have been researched in an increasing number of academic medical centres (Field 1996). The more intimate grooming represented by kissing occurs not only in humans, but in chimpanzees and bonobos as well (de Waal 1996). On further consideration, evidence of skin-contact succorant behaviours akin to grooming exists across most observed mammalian species. DeWaal (1996 pp.40-45) discusses this behaviour in dogs and whales as well as primates. A patient of mine, temporarily bedridden with severe neck pain, was touched softly on the cheek by his pet cat (after she had eaten).

On comparative psychology and the *sapiens sapiens* of *Homo*

The arguments for a continuum of succorant behaviours would tend to bracket the long-standing nature vs. nurture debate within anthropological circles, revived recently by Boesch (2007), by not claiming validation for either point of view. As proposed earlier in this paper, the uniquely human capacity for self-reflection is postulated, in the absence of empirical studies, to be an outgrowth of theory of mind (TOM). Perhaps Boesch's distinctions between developmentalist and deterministic approaches apply to the reports of evidence of TOM in some chimpanzees, gorillas and orangutans as well as some cetaceans and domesticated dogs, cats, and parrots. Refinement of these findings may well be in order, although it would not negate the argument that self- reflective thinking is limited to our (putatively) doubly wise subspecies.

Evolved grooming in *Homo*

A significant aspect of contemporary medical care, beginning with a history of *curanderos* and other healing functionaries in early societies, includes varying degrees of succorance, including hypnosis, moxibustion, acupuncture, acupressure , chiropractic, and placebo effects. These procedures can be regarded as examples of evolved grooming. The alloregulation (downregulation) of pain and discomfort achieved by such procedures often occurs through the activation of endogenous opioids in the brain. The increasing establishment of alternative medicine departments in several prominent U.S. academic medical centers, i.e. Harvard, UCLA, and Columbia , reflect a new-found respect for the efficacy of these methods.

Avuncular teaching, counselling, and supportive psychotherapy can be considered as types of evolved succorant grooming in humans. This is especially true of massage, cosmetic services, barbering and hairdressing. Earlier in my career, when faced with the task of building a large metropolitan community mental health program, I hired a psychiatrist from another state who had created a smaller local program providing mental health consultation to barbers, bartenders, and hairdressers. All of these serve in most North American communities as front line–although informal–listeners and advisors to their troubled clients and customers. (Brickman 1964)

The roots of psychotherapy in succorant grooming

In the arena of professional caregiving, supportive psychotherapy provides empathy, compassion, instructive advice, and manual-based psychological exercises for those who signal a desire for help. These services can be regarded as evolved grooming consistent with language acquisition in social exchange. Psychotherapy based on depth psychology, such as psychoanalysis, despite denials by many of its practitioners, has been acknowledged by influential contributors in the field to often provide minimal levels of verbal support (Wallerstein 2000). This underlines the evolved roots in grooming of this supposedly 'interpretation-only' therapeutic interchange. It is even arguable, for example, that maintenance of the "frame" of time and space parameters so strongly advocated in conventional analytic circles as essential for establishing a secure base for the patient is also rooted in the alloregulative functions derived from a continuum of grooming behaviours.

Most psychoanalysts and psychodynamic therapists, however, agree that deliberate provision of support in the therapeutic encounter is unwarranted for several reasons, not the least of which is that it can actually impede self-reflection and eventual auto-regulation. On the other hand, many psychoanalytic investigators (E.g. Bion 1963, Winnicott 1965) identify sensations of being "held" as integral to a patient's therapeutic experience. It must be added that such "holding "is strictly figurative in psychoanalytic therapy. It refers to a nurturant type of succorance akin, in the patient's subjective experience, to a marsupial relationship. Most contemporary psychoanalysts probably consider actual holding as likely grounds for ethical complaint, by virtue of the "slippery slope" cautions of analytic ethicists against erotic enactions in therapeutic relationships. In the case of patients who live drastically alienated and isolated lives as a result of early and repeated traumatic experience with caregivers, a carefully titrated amount of literal, but non-erotic, touch by experienced clinicians may be in order with the intent of facilitating trust.

Self-reflective thinking, succorance, and psychoanalytic process

Aside from meditation (and perhaps contemplative prayer), the most intensive patterned self-reflection is instantiated by the psychoanalytic process, which itself can be conceived as a joint meditation. In this interaction, the nominal patient self-reflects verbally in the presence of a presumably trusted self-reflecting other. That joint undertaking, exploring the patient's inner representational world, is intended to reduce or possibly eliminate anxiety, shame, and depression, imprints of early psychic trauma, as well as self-defeating behavioural patterns and social failures resulting from psychopathology. (Psychoanalysis is meant in this essay to include other socially sanctioned psychotherapies that apply psychoanalytic understandings of human subjective experience, development, and behaviour.)

As an essentially verbal interchange, psychoanalytic therapeutic practice constitutes succorant behaviour which goes beyond grooming. It pivots on the uniquely human capacity for self reflection and the use of language. Although its interactions are primarily verbal, an increasing number of psychoanalysts practice, and advocate, acute clinical awareness of prosodic nuances, bodily states ("body language") and neurocirculatory changes, such as blushing, sweating and increased respiratory rate, in their patients (Stern et al. 1998). These non-verbal phenomena are regarded as clues to either conscious or unconscious affect states. In sum, self-reflective thinking is an obligatory portal for seeking psychodynamic help, or succorance, and succorance in turn has a long evolutionary history in pre-verbal mammals mediated through bodily contact and grooming behaviours.

Darwinian neuro-psychoanalysis: accommodating a new synthesis

Freud's depiction of a universal *epistemophilic instinct* (Freud 1909) can be said inductively to energize much of the unique self-reflective capacity of *Homo sapiens sapiens.* In other words, a uniquely human need to assign meaning to subjective experience helps to generate the self-reflected *cogito* that made Descartes famous. The very *non-Cartesian* findings of cognitive neuroscience, by highlighting the emotional foundations of all socially interactive behaviour, identify neuronal plasticity as one of the major neurobiological attributes enabling change under the impact of the analytic process.

Specifically, the encodings of implicit and procedural memory have been found to be modifiable through psychotherapy (Tronick 2001). The approximation of evolutionary biology and cognitive neuroscience amounts to a new synthesis for psychoanalytic theory. This new synthesis has been enhanced, for instance, by the experimental studies in molecular biology by the Nobelist psychiatrist Eric Kandel (1998, 1999), who views psychoanalysis as potentially enriching neurobiology in attempts to understand the vicissitudes of human mentation and behaviour.

It is important to add at this point that brain science is still at a very early stage in its development, and has many years to go to reach its proper maturation. The suppositions underlying many of this essay's examples of reconciled neurobiological and psychoanalytic thinking, despite their identified bases in empirical studies, reflect an expectation that further studies will update and eventually supplant current neurobiological as well as conventional psychoanalytic thinking.

In that vein, most natural and social scientists are unaware of recent advances in psychoanalytic theory and clinical practice that embody significant footnotes and emendations to Sigmund Freud's original instinct theory (Freud 1933). While educated as a neuropathologist, Freud found the localizationist hypotheses of the neurologists of his time thoroughly improbable, and, while maintaining that psychoanalytic theory should be built upon a biological bedrock (Freud 1932), he reluctantly departed from the umbrella of natural science in favour of a pure, largely disembodied psychology (Solms and Saling 1986). ("The" dynamic unconscious, for example, could not be localized in a specific anatomical area of the brain. In actual fact, recent neurobiological understandings of the widespread neural connections involved in all psychological functioning no longer imply that specific non-sensory and non-motor processes can ever be locatable in such a geographic manner).

This disembodied portrayal led to classical Freudian theories of development and psychopathogenesis that portrayed an individual beset with unconscious conflicts between instinctual urges seeking satisfaction and opposing intrapsychic elements attempting to forestall negative and self-defeating behaviours by means of symptomatic compromise formations. While the evolutionary significance of instinctual forces (the"id") and the fundamentally social nature of the human species were implied in the concept of the "super-ego", psychoanalytic theory and clinical process were cast in a mechanistic model of intrapsychic conflict known as metapsychology. Therapy was devised to bring these conflicts to awareness by undoing repression through free association and dream interpretation on the analyst's couch. Because of their lack of conventional empirical verifiability, these theoretical speculations were prime targets for disparagement and scorn from a wide variety of scholars in the natural and social sciences (E.g. Grunbaum 1986) As recently as 2004, a renowned neuroscientific researcher saw fit to refer to "the notoriously lax intellectual standards of Freudian psychology" (Ramachandran 2004 p.8). To a psychoanalyst increasingly informed by Darwinian neuroscience, a prevalent continuing aspect of psychoanalytic conceptualizing can in fact be justly labelled as notoriously lax.

A regrettable effect of Freud's conceptual migration from his neurophysiological roots is understandable in view of the limitations of neurological knowledge of his day. A major consequence has been the intellectually questionable – and increasingly unsupportable – practice among analysts to concretize brain functions into psychic entities. Starting with Freud himself, psychoanalytic discourse has been cast in the grammar of anthropomorphic metaphor, abounding in terms such as: *the* Id, *the* Superego, t*he E*go, *the* self – all struggling on an intrapsychic battleground within "*the* unconscious". These concretizations can be comparable to a theological system incorporating immaterial vectoring entities in unremitting conflict with one another—unseen inner *dybbuks* flaunting the banners of instinctual reward and punitive consequence, polarized forces representing mature versus destructive behaviours. This polarized world view paradoxically reflects a type of regression to pre-Enlightenment medieval ontologies contrary to Freud's embrace of scientific thinking. An increasing tide of neuroscientific research comprehends these intrapsychic phenomena as functions rather than entities, as ever-changing, connecting, and parallel processing neuronal networks within the brain's assemblage of billions of neurons and trillions of synapses (LeDoux 2004). The paramount and ultimate vector of biological existence is not "instinctual" gratification, but propagative survival, according to Darwin (1869) and Dawkins (1976).

While the use of metaphor is unavoidable in any attempt to convey understanding of events in the world, descriptions of brain-based behaviour are better employed by using verbal and adverbial metaphor, rather than resorting to nouns and adjectives. An example is the use of the word "selfing" referring to recursive neuronal circuitry (Edelman 1992) rather than "the self" (Brickman 2008). This grammatical point is less trivial than it may seem, as neurodarwinian impacts on psychoanalytic thought will be illustrated below.

For close to 100 years, psychoanalytic theory has devolved into a variety of competing thought collectives, comparable to competing theories in anthropology, social psychology, and linguistics. While not promising total ideological integration, recent studies in attachment theory and neuroscience have fostered an increasingly discernible drift toward intersubjective, rather than positivistic and mechanistic, thinking. As mentioned above, a major historical figure in psychoanalysis, John Bowlby (1969), has compellingly introduced Darwinian and ecological perspectives on child development and sociality into the field. Accordingly, an anti-Cartesian view of the mind as embodied and rooted in the long history of natural selection of our social species is coming into greater focus. Many psychoanalysts, however, continue to agree with Freud (who obviously was not consistent) that our science is only verifiable through the intensive case study method and should not be judged by empirical perspectives that require non-treated controls and deliberately varied, and frequently unethical, alternative treatment techniques. This was the same Sigmund Freud who later in his career defined the *Weltanschauung* of psychoanalysis as identical to the world view of science in general (Freud 1933).

Nevertheless, recent advances in cognitive and affective neuroscience, accompanied by research in molecular neurobiology and electronic imaging, have contributed to increasing biologization of psychoanalytic theories of development and psychopathogenesis. If psychoanalysis is increasingly, if ponderously, approximating itself with biology, the principles of Darwinian natural selection of behavioural phenotype must inevitably apply. Likewise, since sociality is the midwife of cultural influences on individual behaviour (Fiske 1992, Cosmides & Tooby 2005), a more thorough familiarity with the social sciences has begun to be implanted into psychoanalytic understandings. It is in the spirit of such an accommodationist stance, with postulated connections to both social and neural science, that the view of psychoanalytic theories of pathogenesis and cure can be reconsidered in terms of a continuum of alloregulatory and succorant behaviours. Empirical studies of psychoanalytic theory and practice, however difficult to implement, would be a desirable outcome of such efforts.

Contributions of attachment theory and relational learning theory

In the basically asymmetrical analytically informed therapeutic relationship, the therapist does not promote himself as the authoritative arbiter of "the truth" of the patient's inner representational world. A state of open receptivity encourages warded-off psycho-emotional conflicts to emerge from within the interactive process itself. This praxis reflects a more specific concern with the relational aspects of human behaviour than was the case in analytic technique from the time of Freud until near the end of the 20th Century, when "making the unconscious conscious" was a guiding principle.

Reports of the conceptual foundations and technical implications of this ideological shift are to be found, for example, in the research of Fonagy and his associates at University College London in attachment theory and reflective thinking (2002). Also, Stern and the Boston Process of Change Group (1998), as mentioned above, have thoroughly studied the key role in analytic therapy of procedural learning and non-verbal aspects of interaction within a relational ambience in the consulting room. These theories propose that attachment styles developed in early childhood inform adult relationships including those occurring in the consultation room, and that a non-authoritative co-subjective therapeutic process not relying exclusively on linguistic interaction can bring about modifications in behavioural patterns through new implicit learning.

As mentioned above, increased verbal insight into these processes, while helpful at the cognitive level, takes second place to the acquisition of new procedural knowledge. In contemporary American lingo, it is a matter of not only "talking the talk", but more importantly, "walking the walk". These new findings rely on recent neuroscientific confirmations of continued synaptic plasticity in the adult brain (Braun & Bogerts 2001, (Ansermet & Magistretti 2007). Interestingly, these recent perspectives on procedural learning do not necessarily invalidate the effectiveness of more conventional analytic treatment based on earlier theoretical models. The argument of relational/intersubjective analysts is that, even in those clinicians guided by earlier formulations with their rich vocabulary of anthropomorphic metaphorizing, the "bottom line" effectiveness of psychodynamic therapies seems better explained by the neurodarwinian concepts of procedural learning, including the detoxification of self-defeating attachment styles.

The impact of evolutionary biological thinking on psychoanalytic perspectives

Carrying forward the earlier discussion of the grammar of analytic concepts of unconscious mentation, recent advances in neuro-psychoanalysis suggest a selective advantage of good-enough innate anticipatory neural networks that prepare for the exigencies of human social life. Faulty or self-defeating preparative neural networks and consequent behaviours can seriously disadvantage an individual's social, physical, and genomic survival in many ways. In like manner, a fragmented or totally absent life narrative, even when unconsciously held, deprives an individual of a sense of a robust life trajectory, thereby generating a depressive and anxiety-laden *Weltanschauung*.

A new conceptual triad of psychology, sociality, and evolutionary neurobiology has been increasingly influential in psychoanalytic theory and technique. This conceptual triad is illustrated, for example, by the conclusions of Fonagy and his co-investigators (2001) that the capacity to attune oneself to the intentional states of conspecifics, also known as theory of mind or mentalization, is enhanced significantly by psychoanalytic therapy, and may in fact be the gold standard of therapeutic efficacy. This is further discussed below.

While anticipatory neural networks would seem to be worthy candidates for natural selection, theory of mind has undoubtedly been naturally selected as an advantageous mental process furthering personal and genomic survival. Its universal presence in *Homo sapiens* seems to have been enhanced by the acquisition of language. In turn, the mixed blessing to our species of language acquisition allows for more effective communication as well as more effective deception of one's intentions. Darwin's "The Expression of Emotions in Animals and Man" (1872) was a splendid description of the array of vocal, facial, gestural, postural, and motoric phenomena in all animals that reflect the perception as well as the response to perceived intentions of conspecifics and others. Many of these manoeuvres among non-human animals are deceptive; their persistence indicates their contributions to the organisms' survival through natural selection.

Self deceptive inner conversations, often supported by verbalized beliefs, can be highlighted as a significant factor in a wide range of human behaviours beyond those considered neurotic. For example, the confident skin glow and common belief in the blessings of pregnancy in the expectant young mother-to-be protectively ignore the desperate zero-sum arms race in her uterus between herself and her parasitic fetus The skin glow itself may be an effect of increased blood pressure in the pregnant woman – a neurocirculatory campaign in her unknowing struggle. In extreme cases this internal struggle can lead to fetal death through starvation on one side, or, through eclampsia, to maternal cardiovascular damage, or death through uncontrollable hypertension in the mother.

Linguistic considerations in clinical interaction: toward reconciling three theories

In humans, the self-deceptive sector of language has evolved to prevent the inadvertent communication of self-doubt – a form of undercutting one's effectiveness in social exchange. In ordinary social intercourse, subjects and their relational others rely on language to convey intention and response. In such cases, the inevitable, generally more subtle, non-verbal cues inconsistent with what is actually said and heard, are more or less ignored. This can serve the survival ends of the subject – as long as the intentional counter-currents remain concealed by the spoken word.

The work of the language philosopher J.L. Austin in *How to Do Things with Words* (1975) focuses on his supposition that language can be most usefully understood as action rather than as communication alone. Hence, he regards the behavioural role of spoken language as "*speech acts*", thereby distinguishing the *performative* from the *constative* aspects of verbal utterances. Generally speaking, a constative speech act is a relatively simple one which can be confirmed or negated as a fact, exemplified by a statement such as “It’s raining outside”. One to whom this remark is addressed would tend to either agree or disagree.

A *performative* speech act, always more complex, conveys meaning rather than simple observation. It can reflect various facets of intentionality, including a conscious or unconscious intent to influence the hearer’s attitude or behaviour, or what the speaker perceives as the hearer’s intentional stance or behaviour.

To the extent that speech act theory can serve as an evolved evolutionary foundation suitable for reconciliation with psychoanalytic thought, I suggest that the predominantly verbal track of psychoanalytically-informed psychotherapy, requiring the therapist to decode the meaning-saturated *metalanguage* of the patient’s utterances, potentially illuminates the understanding of performative speech acts. Psychoanalytic therapists are crucially concerned with intentionality in their patients and in themselves. Non-verbal cues have been empirically shown to be major portals of access to the structures and functions of meaning underlying even the most seemingly trivial utterances – even by therapists themselves – provided self-reflective attunement is in play. Austin’s speech act theory, then, is reconcilable with psychoanalytic theory, via a common connection with the evolutionary biologist Robert Triver’s (2002 pp. 271-293) theory of the adaptational functions of deception and self-deception.

In psychodynamically-informed psychotherapeutic exchanges, the therapist must be trained and experienced in picking up non-verbal cues, thereby enabling perception of a more authentic inner narrative than the tale the patient often defensively yet unwittingly tells others and himself. The tool kit of an effective dynamic therapist is enhanced by a capacity for applying linguistic and mentalization theory. In fact the prominent contemporary psychoanalyst, Peter Fonagy, cited above, and his co-investigators, view psychoanalysis and its dynamic offshoots as most successful to the extent that it enhances the patient’s theory of mind by means of “mentalized reflexivity” (Fonagy et al 2002 pp.435-468) This enhancement of theory of mind allows the patient to come to terms more effectively with his own emotional states as well as those of others.. This theory of psychopathogenesis and cure, substantially based on attachment developmental theory, is an impressive example of Darwinian neuro-psychoanalytic thought.

The narrative nature of the human inner world

The work of Llinas and Pare (1998) has demonstrated that no animal with a cerebral cortex has direct and unfiltered sensory access to its physical and social surrounds. Over millions of years, brains have developed a survival-oriented representational function which has co-evolved with the increasing encephalization and corticalization of the brain. At the pre-reptilian and reptilian levels, interaction with the physical and biological surrounds requires no inner representation. The largely reflexive limbic system (amygdala and hippocampus) activities, such as feeding, fighting, freezing, fleeing, or mating, require no cerebral filtering. The increasing complexity of life in social mammals has required the evolution of more complex cortical and subcortical neuronal assemblies for humans to subordinate the foundational limbic system behavioural tendencies with a huge welter of activational and inhibitory neural circuits, synapses and inner world representations. These neural phenomena seem to organize human subjectivity in terms of variations on a self-postulated narrative theme. Much of the psychoanalytic literature portrays these narrative constructs as ongoing unspoken yet influential 'conversations' between a putative self representation and representations of formative others from early in the individual's life.

The survival value of these inner conversations would require their adaptational suitability to interpret the world in closer correspondence to what is real – or, at least, to what is confirmed by valued others as real. In virtual competition with these adaptational inner representations, a persistent unconscious reservoir of conflictual neural processes, conventionally known as "the unconscious", serves as an index of emotionally-drenched memories and their complex neuropsychological innervative connections throughout the brain. The inner representations of the world influenced by these assemblies compete for applicability as models for "here and now" experience. Much of human behaviour, especially in the neurotic band of the spectrum, constitutes what Freud and his followers have called compromise formations – symptoms and actions reflecting, neuropsychologically speaking, a balance between excitatory and inhibitory neural circuits. Again, it is important to add that the neural processes underlying these behaviours are yet to be clearly mapped through neuroimagery studies.

In a manner similar to the body's immune system, neurodynamic systems of conflictual unconscious processing have evolved to segregate noxious memories from conscious declarative and autobiographical memory and to attempt self-healing behaviours. The motivational sources of behaviours generated largely by conflict-derived unconscious functioning are most readily accessed through interpersonal dialogue with trusted others. Psychoanalysts and psychodynamic therapists hope to exemplify such trusted others. Often, successful psychotherapy results can pivot on session-to-session joint resolution of therapists' occasional failures to be trustworthy.

This shimmering, intricately and multiply interconnecting neural reservoir (metaphors fail here), this cerebral immune system, propagates a host of behavioural devices such as denial (largely employed to prevent confessions and actions that would be adverse to the survival of the individual and his genome), self deception, splitting and isolation, projection, intellectualization, dreaming, creative artistic activity, and compensatory symptom formation. These "mental mechanisms" are not always considered psychopathological; at times, for example, denial and self-deception can be advantageous to fitness. The behavioural expressions of these devices are the building blocks of compromise formation discussed above, with the possible exception of many behaviours and inner mental states induced by addictive practices.

The adaptational role of conflictual unconscious processing

A neurodarwinian perspective, therefore, suggests the evolutionary survivability of conflict-derived unconscious functioning as an adaptive functional tool, a neuro-psychological immune system conceivably evolved to manage individual suffering in a social world conceived at times as overwhelming. Dreams, fantasies, "unthought" ideas are among the self-healing activities of that conflict-laden domain of unconscious mentation. Borrowing from naval parlance, the self-healing aims of such an adaptive tool suggests the 'sick bay' function of a ship at sea, where medical personnel promote the goals of the voyage by attending to the injuries of officers and crew so they may remain on partial or full duty until fully healed. A limping, bandaged or depressed member of the ship's company remaining in the succorant care of sick bay personnel is not a critical deterrent to the vessel's continued voyage. Absent an evolved neurobiological capacity for processing the effects of trauma, insecure attachment, abandonment experiences and other hurtful events during early development, destructive and self-destructive behaviours would abound, to the detriment of individual and group survival. Further research is clearly necessary to identify a more detailed evolutionary rationale for this prime domain of psychoanalytic interest.

Conclusion

Beginning with a re-interpretation of Descartes' famous *Cogito,* I argue that the form of contemplative thinking the philosopher had in mind was self-reflective rather than transitive in nature.. Furthermore, self-reflective thinking, as an evolved aspect of theory of mind, appears to be uniquely human, and characterizes psychoanalytically-informed psychotherapy. But a non-Cartesian, thoroughly embodied, psychoanalytic view of psychological and emotional dysfunction is not limited to thought processes when it is informed by evolutionary neurobiology and psychology. It allows us to conceive of psychodynamic therapy as enhancing theory of mind, or mentalization in social contexts. It also allows us to understand the widespread practice of self-deception, and the roots in a continuum of succorant animal behaviours of all forms of psychotherapy. A major implication of such a more contemporary psychoanalytic view is that more research is needed to reveal the selective value of unconscious neuro-psychological processing of intrapsychic conflict. Additional research should also attempt to illuminate the more immediate adaptational function of psychotherapeutic processes based on a reconciliation of Darwinian, neurobiological, and depth psychological perspectives which the author identifies as *Darwinian neuro-psychoanalysis.*

In contrast to mentation in non-human animals, two uniquely human roles of embodied conflict-derived unconscious functioning are evolutionarily understandable: a) as a naturally selected type of self-deception to keep oneself unaware of what is felt to be socially objectionable or dangerous in one's inner representational world, thereby impeding human conspecifics from mentalizing one's hidden intentions; and b) as a self-healing module of the human organism evolved to correct maladaptive predictions, potentially enhanced by the healing effects of the self-reflective, and basically succorant alloregulative *meditations á deux* entailed in psychoanalytically informed psychotherapy.

While the possibilities and problems of addressing these formulations through empirical research are akin to the problems in using observations of present-day hunter gatherer life to scientifically confirm or falsify theories of the nature of human life during ancestral times, a truly scientific theory of conflict-derived unconscious functioning, while being explored at present, is yet to be successfully accomplished (see Luyten et al 1997).

References

Ansermet, F. & Magistretti, P. (2007) Biology of freedom. New York: Other Press

Austin, J.L. (1975) How to do things with words. Cambridge MA: Harvard Univ. Press

Bion, W, R. (1963). Elements of Psycho-Analysis. London: Heinemann.

Bowlby, J. (1969). Attachment and Loss. New York: Basic Books

Braun, A.R. & Bogerts, B. (2001) Experience-guided neuronal plasticity: Significance for pathogenesis and therapy of psychiatric diseases. Nervenarzt 72 (1): 3-10

Brickman, H. R. (1964). Some basic assumptions in community mental health. Am. J. Pub. Hlth. 54:6 890-899

—. (1998). The psychoanalytic cure and its discontents: A Zen perspective on "common unhappiness" and the polarized self. Psychoanalysis and Contemp. Thought 21:1, pp.3-32.

—. (2008). Living within the cellular envelope: subjectivity and self from an evolutionary neuropsychoanalytic perspective. J. Am. Ac. Psychoanal, 36 (2): 317-341.

Buechner, H. K (1973). The sociable "Leo Serengeti". The Quarterly Review of Biology, 1973 – JSTOR.

Cosmides, L. & Tooby, J. (2005). Neurocognitive adaptations designed for social exchange. In D. M. Buss (Ed.), Evolutionary Psychology Handbook. NY: Wiley.

Darwin, C.R. (1859). The Origin of Species. London: John Murray

—. (1872). The Expression of the Emotions in Man and Animals. London: John Murray.

Dawkins, R. (1976). The Selfish Gene. Oxford: Oxford Univ. Press

De Waal, F. (1996). Good natured: the origins of right and wrong in humans and other animals. Cambridge, MA. Harvard Univ. Press

Edelman, G. M. [1992] Bright Air, Brilliant Fire: on the matter of the mind. New York: Basic Books.

Field, T. M. (Ed.) (1996) Touch in early development. New York: Erlbaum Associates.

Fiske, A.P. (1991). Structures of Social Life: The Four Elementary Forms of Human Relations. New York: Free Press.

Fonagy, P., Gergely, G., Jurist, E., & Target, M. (2002). Affect Regulation, Mentalization, and t he Development of the Self. New York: Other Press.

Freud, S. (1905). Three essays on the theory of sexuality. Standard Edition 7: 123-243.

—. (1909). Notes on a case of obsessionality. Standard Edition 10 p.245,

—. (1933). Freud, S. New Introductory Lectures on Psychoanalysis" Lecture XXXV, "The Question of a Weltanschauung," 1933. Standard Edition, Vol. 22, p.160.

Grunbaum, A. (1986) Precis of the Foundations of Psychoanalysis: A Philosophical Critique. Behavioral and Brain Sciences 9: 217-228.

Hare, B., Call, J., & Tomasello, M. (2001) Do chimpanzees know what conspecifics know? Animal Behaviour 61: 1 pp.139-151

Kandel, E. R. [1998] A new intellectual framework for psychiatry Am. J. Psychiatry 155:4 April 1998 pp.457-469

—. [1999] Biology and the future of psychoanalysis: A new intellectual framework for psychiatry revisited. Am. J. Psychiatry 156: 505-524

Le Doux, J, (2002). Synaptic Self: How our brain became who we are.

Lewis, C.T. And Short, C. (1879) A Latin Dictionary. Oxford, Clarendon Press.

Llinas, R., & Pare, D. 1998. The Brain as a Closed System Modulated by the Senses. In Llinas, R. & Churchland, P.S., The mind-Brain Continuum. Cambridge, MA : MIT Press

Luyten, P., Blatt, S. J., & Corvelyn, J. (1997) Minding the gap between positivism and hermeneutics in psychoanalytic research. J. Am. Psychoan. Assn, 54/2 571-610

Ramachandran, V. S. [2004] A Brief Tour of Human Consciousness. New York: Pi Press.

Reidman, M. L. (1982) The evolution of alloparental care in mammals and birds. Quart. Rev. Biol. 57:4. 405-435.

Solms, M. And Saling, M. [1986]. On psychoanalysis and neural science: Freud's attitude to the localizationist tradition. Int. J. Psycho-Anal. 67: 397-428.

Stern, D. and the Boston The Process of Change Study Group:[1998] Non-interpretive mechanisms in psychoanalytic therapy: the 'something more' than interpretation. Internat. J. Psycho-Anal. 79: 903-921.

Trivers, R. (2002) Natural Selection and Social Theory. Oxford: Oxford Univ. Press.

Wallerstein, R.S. (2000) Forty-two lives in treatment: a study of psychoanalysis and psychotherapy: the report of the Psychotherapy Research Project of the Menninger Foundation, 1954-1982, New York: Other Press, 2000.

Winnicott, D.W. (1965). The Maturational Process and the Facilitating Environment. London: Hogarth Press.

CHAPTER EIGHT

WHY THE HUMAN BRAIN IS NOT AN ENLARGED CHIMPANZEE BRAIN

JOHAN DE SMEDT, HELEN DE CRUZ AND JOHAN BRAECKMAN

Abstract: Following Darwin, many comparative psychologists assume that the human mind is a kind of ape mind, differing only in degree from the extant apes – we call this the mental continuity assumption. However, the continuity principle in evolutionary theory does not posit continuity between extant closely related species, but between extant species and their extinct ancestors. Thus, it is possible that some human cognitive capacities have no parallels in extant apes, but that they emerged in extinct hominid species after the human-chimpanzee divergence. Our examination of the case of social cognition from an archaeological and comparative psychological perspective suggests that the human brain is not simply an enlarged chimpanzee brain. Instead, natural selection seems to have favoured a different social cognition in both species.

Introduction

Among comparative psychologists, the belief prevails that the human mind is a kind of ape mind, differing only in degree from the extant apes. An influential formulation of this *mental continuity assumption* comes from Darwin, who probably wrote this under the influence of Thomas Huxley (1871: 105):

> "Nevertheless the difference in mind between man and the higher animals, great as it is, is certainly one of degree and not of kind. We have seen that the senses and intuitions, the various emotions and faculties, such as love, memory, attention, curiosity, imitation, reason, &c., of which man boasts, may be found in an incipient, or even sometimes in a well-developed condition, in the lower animals."

As a consequence, most investigations of nonhuman primate cognition are conducted with an anthropocentric agenda: do chimpanzees seek causal explanations (Povinelli & Dunphy-Lelii, 2001), do they have cultures similar to ours (Whiten et al., 1999), and do they possess a theory of mind (Call & Tomasello, 1999)?

The theory of natural selection compels us to see the history of life like a branching tree rather than a scale. Accepting that there is a diversity of minds, and not a single scale of progressive cognitive development raises several questions. If the human mind is but one evolutionary outcome among many, why then this outcome, and not some other? What cognitive capacities have been selected for during human evolution? Tackling such questions may help us gain insight into how the human mind works, and also promises a better understanding of other animal minds, such as that of the chimpanzee.

The mental continuity assumption is currently a progressive research programme, because it generates a wealth of empirical findings. Yet it is plagued by anomalies. For example, humans behave altruistically, even if it poses a cost to them, yet Silk et al. (2005) found that chimpanzees do not help others, even at no cost to themselves. In the history of science we find two ways of dealing with anomalies. One is to keep on designing experiments, in this case until chimpanzees perform in accordance with the mental continuity assumption, like Warnecken and Tomasello (2006) who set up an experiment in which *infant* chimpanzees spontaneously helped *human* trainers. A second option, presented in this paper, is to treat these anomalies as evidence of fundamental differences between human and ape cognition and to abandon the mental continuity assumption. We examine neuro-scientific, ecological, and developmental psychological evidence that speaks against the widely held belief that the human brain is an enlarged chimpanzee brain. We trace diverging ecological and social contexts in hominid and chimpanzee evolution after their split about 6 million years ago, and show that these have yielded unique social cognitive adaptations in each species.

Neuro-scientific Evidence Against the Mental Continuity Assumption

One must keep in mind that the continuity principle does not apply to extant related species, but to species and their extinct ancestors (see also Pinker, 1994). Since we evolved in differing ecological contexts, it is a contingent fact that some human cognitive capacities have no parallel in extant apes, but that they emerged in extinct hominid species, after the hominid-chimpanzee divergence. Even on a very short timescale, selection can craft cognitive specializations as a result of unique, species-specific pressures. Take the extreme case of differing social cognition in dogs and wolves. Dogs are domestic animals, living in the human cultural niche. They developed adaptations for social interaction with humans, such as understanding pointing gestures and making eye contact (Miklósi et al., 2003). Wolves lack these capacities, although genetic evidence shows that dogs have diverged only 15,000 years ago and genetically differ by only about 0.15 % from contemporary wolves (Savolainen et al., 2002). Thus, the estimated divergence time of 6 million years between humans and chimpanzees (Stone et al., 2002) is in principle long enough to create cognitive specializations in each. Indeed, studies of gene expression in human and chimpanzee brains show that certain brain areas and thus, cognitive capacities have been under divergent evolutionary pressures in both species (Oldham et al., 2006; Cacéres et al., 2007).

Comparative neuro-anatomical studies (e.g., Barton et al., 1995) show that primate brains do not only differ in size, but also in internal organization and structure. Interestingly, this organization reflects a species' ecology and social structure, rather than its cladistic relatedness. For example, woolly monkeys (*Lagothrix poeppigii*), a species of New World monkeys, have an energy-rich diet consisting mainly of fruits and insects. As a result, the internal organization of their brain looks very similar to that of chimpanzees and differs considerably from that of other closer related New World monkeys (De Winter & Oxnard, 2001). Rilling and Insel (1999) compared brains of 44 primate species using magnetic resonance imaging. Their research indicates that the human brain is not simply an enlarged ape brain: some areas have grown allometrically in humans, such as the prefrontal and temporal cortices, which are involved in language and theory of mind, whereas others, such as the cerebellum, which deals with locomotion, are reduced compared to those of orang-utans and gibbons. Interestingly, the corpus callosum, which connects areas of similar function between the hemispheres, is reduced in humans compared to other apes. This reduced connectivity allows for greater autonomy and divergent evolution of different brain regions which may have enabled left-lateralization of cognitive functions such as language and tool-use in humans (Hopkins & Rilling, 2000).

As a matter of empirical fact then, the human brain does not appear to be an enlarged chimpanzee brain. What selective pressures are responsible for this divergent cognitive evolution? Empirical studies of brain size and organization in birds (Burish et al., 2004) and ungulates (Shultz & Dunbar, 2006) indicate that both ecology and social structure are important factors in cognitive evolution. In the next section, we apply this ecological approach to hominid and chimpanzee cognitive evolution, by sketching the different palaeo-ecologies of these lineages since their divergence. Later we will look at the comparative psychological literature for evidence of cognitive specializations that might reflect these differing selective pressures, focusing on social cognition.

Ecological and Social Selective Pressures During Hominid and Chimpanzee Brain Evolution

Like all animals, the extant great apes exhibit dietary adaptations that reflect past selective pressures. During the early Miocene (23-15 million years ago), apes reached their greatest level of diversity, filling a wide range of sympatric niches. Fossil hominoid dentition and gut content (Franzen & Wilde, 2003) indicate that their diet consisted exclusively of vegetable foods. By the late Miocene (11.6-5.3 million years ago) however, they came under pressure from both climate change and competition by the Old World monkeys. The few ape species that survived this competitive sweep did so by adopting unusual ecological niches (Milton, 1999).

Gorillas and orang-utans developed a larger body size and turned to lower quality plant foods, such as mature leaves and bark, thus avoiding intense competition with monkeys. Chimpanzees became dietary specialists of ripe fruits, nuts, young leaves, insects and small vertebrates, which contain higher levels of calories and protein. Because high-quality food is scarce, competition for food is intense. As a result, female chimpanzees with dependent offspring are forced to live and forage solitarily, because their slowness puts them at a competitive disadvantage (Hrdy, 2005). An exception to this general rule is *Pan troglodytes verus* (a West-African chimpanzee subspecies), which exploits locally abundant species of nuts with stone hammers and anvils (Yamakoshi, 2001). Yet even these females compete for food and form linear dominance hierarchies (Wittig & Boesch, 2003). Male chimpanzees cooperate to defend communal ranges, groom, and engage in coalition formation against dominant individuals. However, there is also intense competition between males. Even among maternal brothers, who form the backbone of chimpanzee social groups, meat-sharing is rare (Langergraber et al., 2007). Bonobos, who live in the more food-rich Central African rain forests, rely on fruits, leaves and terrestrial herbaceous vegetation (flowers, young leaves and pith), which is a locally stable and widely available source of protein (Rafert & Vineberg, 1997). This probably enables bonobo females to form cooperative alliances. Cooperation among male bonobos is restricted to periods of relative fruit abundance (White, 1998). Whereas selection has enhanced sociality in chimpanzee males, it seems to have favoured increased sociality in bonobo females. Humans are unique among primates in their obligatory reliance on tools to extract food. Analyses of diets of contemporary hunter-gatherer groups (Kaplan et al., 2000) show that humans prefer food that is hard to obtain – through hunting or extraction – but that is high in energy and nutritive value. Less than 10 % of all food types they exploit are relatively calorie-low vegetable foods that are easy to obtain (fruits, leafy vegetables). In contrast, the bulk of chimpanzee diets consists of easy to obtain food. From this brief survey, it becomes apparent that the apes each occupy a unique ecological niche. These dietary niches also influence social organization. In order to understand how human cognition differs from that of other apes, it is important to reconstruct the environment in which most of the cognitive evolution of these species took place.

Modelling Past Selective Pressures on Cognitive Evolution

Marine sediment sequences provide evidence for stepwise increases in East-African aridity during the last 8 million years (deMenocal, 2004). This dryness was caused both by a global cooling trend, and by the formation of the Rift Valley, which blocked East-Africa from the rains from the Atlantic Ocean (Pickford, 1990). The aridity forced early hominids to extend their daily foraging range, favouring bipedal locomotion, which is more energy-efficient than all other known forms of primate terrestrial locomotion (Leonard & Robertson, 1997).

Bipedalism emerged about 5.2 million years ago (Haile-Selassie, 2001), which is almost immediately after the split between hominids and chimpanzees. It had unexpected consequences for early hominid social interaction, because it made female cooperation practically inevitable. The bipedal ape cannot look at her own offspring while giving birth, and thus cannot assist her young, such as removing mucus from its face or loosening a suffocating umbilical cord. Nonhuman apes habitually help their own offspring in this way, but bipedalism requires assistance (Rosenberg & Trevathan, 2002). In all human cultures, women in labour receive help. Whereas female chimpanzees do not form alliances, female *Ardipithecus* as early as 5.2 million years ago required help during birth. Female alliances were further promoted as early *Homo* developed childhood, a new lifecycle that required alloparenting by other females (Bogin & Smith, 1996). Once hominids developed a brain size that was markedly larger than that of chimpanzees, which occurred about 2 million years ago, they were forced to wean their infants earlier to meet the nutritional demands of the rapidly growing brain. Humans have the shortest lactation period of all apes, weaning infants at about 30 months, whereas chimpanzees are breastfed for about 62.8 months on average, and orang-utan infants are lactated about 7 to 9 years (Kennedy, 2005). Shorter lactation times result in shorter interbirth intervals: traditional hunter-gatherers have interbirth intervals of about 3 to 4 years. However, human children need intensive care until they are 6 or 7. This places heavy demands on their mothers, who often have to take care of several dependent children at the same time. In most pre-industrial societies, women solve this problem by developing extensive networks of alloparenting. In some cases, such as the Aka pygmies, childcare networks comprise over 10 individuals including several nonkin members (Ivey, 2000). Therefore, it seems plausible that early hominids from about 2 million years on showed greater cooperation between females than chimpanzees, perhaps on a par with what we observe in bonobos today.

A second step towards increasing aridity in East-Africa took place around 2.8-2.5 million years ago (deMenocal, 2004), leading to the disappearance of the remaining tropical rainforest (Bonnefille, 1995). Because fruit and young leaves became scarce and seasonally unavailable, hominids were pushed to exploit alternative ecological niches. One clade, early *Homo,* began to rely on meat to compensate for the seasonal unavailability of rich plant food sources. Examinations of cutmarks on fossil bones and traces of wear on stone tools indicate that early hominids obtained most or all of the meat in their diet through scavenging: they cut meat with sharp flakes, and crushed bones to obtain the marrow by using hammer stones (Capaldo, 1997). At least three sabre-tooth genera inhabited Pliocene Eastern Africa. They were relatively small felids with powerful front paws and teeth, which enabled them to kill prey far too large to satisfy their own needs (Lewis, 1997). These partly defleshed carcasses provided the hominids with a stable and reliable food source. Because hyenas, their only competitors, are dangerous gregarious animals, it seems plausible that hominid males cooperated to defend the carcasses, as analyses of tooth marks and cutmarks on bones demonstrate that hominids got to the carcasses before the hyenas (Domínguez-Rodrigo & Barba, 2006). This suggests that male hominid coalitions were at least on a par with those of male chimpanzees.

The evolution of the *Pan* lineages is less well-documented than that of the hominids, because chimpanzees mainly evolved in wooded West- and Central-Africa, where fossil preservation is worse than in dry East-Africa.

Only teeth belonging to one fossil species of the genus *Pan* dating to circa 545,000 years ago have been recovered (McBrearty & Jablonski, 2005). However, it is interesting to note that according to mtDNA studies (Stone et al., 2002), the split between *Pan troglodytes* and *Pan paniscus* (bonobo) occurred about 1.8 million years ago, which coincides with an extreme global cooling and drying event that also gave rise to *Homo ergaster* in East-Africa. Given that East-Africa experienced more climatological instability than West- and Central-Africa, it seems plausible to assume that chimpanzees and bonobos occupy their current niches for at least 1.8 million years.

Differing Social Cognition in Humans and Chimpanzees

These differing ecological conditions resulted in diverging social cognitive adaptations in humans and chimpanzees. A striking difference is that chimpanzee cognition is more tuned to competition whereas that of humans is tuned to cooperation. Povinelli et al. (2000) found that chimpanzees seem oblivious to the attentional state of a person who helps them find food: they are as likely to beg food from a person who looks at them as from someone with a bucket over her head. However, they do better when they are placed in a competitive situation. They can effortlessly find a reward if the experimenter pretends to reach in vain for it, a gesture that looks quite similar to pointing (Hare & Tomasello, 2004) – apparently, once they see the experimenter as a competitor, they can solve the task. Experiments with conspecifics show similar findings. Chimpanzees have difficulties working together to pull a rope to obtain food that is out of reach (Melis et al., 2006). Bonobos outperform chimpanzees in cooperative tasks (Hare et al., 2007), presumably because they have a broader diet and thus are more tolerant.

A host of experiments (see e.g., Fehr & Rockenbach, 2004, for an overview) show that humans across cultures behave exceptionally altruistically; they help others even at some cost to themselves and they are prepared to help people they will never meet again. In our view, human altruism can be explained by the unique ecological conditions in which hominids evolved. As we have seen, they had access to an abundant food-source, scavenged meat and marrow, but obtaining and defending this food required extensive male cooperation. Moreover, hominid females became more social, because bipedalism necessitated assistance at birth, and alloparenting developed when childhood emerged as a new lifecycle.

Human social learning is characterized by a set of distinctive features, such as imitation, teaching and joint attention. Hunter-gatherer children depend on adult members of their community for food. Typically, girls become proficient gatherers during late adolescence, whereas men only become proficient hunters in their thirties (Kaplan et al., 2000). In contrast, chimpanzees acquire all tool-using traditions of their community during their first five years of life. Whereas humans learn mostly from peers and other members in the community, infant chimpanzees learn almost exclusively from their mothers. It is therefore not surprising that humans and chimpanzees differ in the way they acquire cultural behaviours. Laboratory and field studies have revealed a unique kind of social learning in chimpanzees, which Matsuzawa (2007) termed "education by master-apprenticeship". It is characterized as follows: the infant spends most of its time alone with its mother. It is highly motivated to acquire her behaviour. The mother, however, does not teach or provide any positive or negative feedback. In stark contrast to humans whose learning is based on triadic interactions – mother, infant and object – chimpanzees learn from dyadic interactions only: they observe their mothers' behaviour carefully, and this helps them to reconstruct the action for themselves. A detailed comparative study (Tomonaga et al., 2004) shows that chimpanzee and human development are similar during the first months of life: the infants of both species engage in dyadic interactions with their mothers, such as smiling and mutual gazing. However, at about nine months, human infants develop the ability to share their attention for a specific object *with* another person (Tomasello & Carpenter, 2007). This results in a referential triangle of infant, adult and the object upon which they share attention. Crucial for this is that the infant knows that it shares its attention for the object with someone else – it is thus aware of the mental state of the other. This "nine-month revolution" does not occur in chimpanzees. Tomonaga et al. (2004) failed to engage in triadic interactions with infant chimpanzees, despite trying for months. Cultural learning in chimpanzees thus differs fundamentally from that of humans. Tomasello and Rackozy (2003) have proposed that our ability to engage in triadic interactions makes cumulative cultural learning possible. We hypothesize that the technical skills required for making stone tools may have exerted intense selective pressures on human cognition to enable us to share attention over objects.

Even the oldest stone tools of about 2.7 million years old, the Oldowan technology, require extensive cultural learning. Only oblique angled strikes result in razor-sharp flakes with a typical conchoidal fracture pattern. Early *Homo* turned the core around while flaking – indicated by the multiple striking platforms – and chose the most efficient angles to strike the core (Roche et al., 1999). Both a chimpanzee (Kitahara-Frisch,

1993) and a bonobo (Schick et al., 1999) have been extensively tutored to learn how to make Oldowan type stone tools, but both failed to strike the cores at the correct angles, indicating that Oldowan technology is beyond their social learning skills. Next to this, good flakes are only obtained from cores with a dense, fine and even (isotropic) structure. Hominids transported raw materials several kilometres to the sites where they found and butchered the animal carcasses (Plummer et al., 1999). Even when they obtained cobbles from local sources, they were selective in the choice of their raw material: the earliest stone tools from Gona, Ethiopia, were made from locally scarce vitreous volcanic clasts, not from the locally abundant basalt, which is of lower quality and does not yield good flakes (Stout et al., 2005). This anticipatory behaviour in raw material selection has not been observed in extant nonhuman primates.

To facilitate sharing attention and cooperation, humans have evolved unique anatomical and neuropsychological adaptations. For example, the human eye shows a large white sclera on both sides of the iris. This makes it easy to follow our gaze direction. In contrast, all other primates have a pigmented sclera, which makes it difficult to follow their gaze (Kobayashi & Kohshima, 2001). All primates are good at following gaze direction. Yet, in their highly competitive world it is not always good to have others know what one is up to, especially not during competition for food – so it is advantageous to have a dark sclera which camouflages gaze direction. Human white sclera might actually have evolved to facilitate reading eye direction, thus augmenting the possibility to decipher intentions. Through triadic interactions, we still make objects part of our social world. Humans across cultures reason about objects in terms of their intended function (Bloom, 1996), even in cultures with a limited technology (German & Barrett, 2005). This intentional stance for artefacts is apparent in children as young as two years of age (Casler & Kelemen, 2007). It is a direct result of our ability to incorporate objects into triadic interactions: when considering an artefact, we can think about the intentions of the maker, such as the planned function of the object. This would not have been possible if humans, like chimpanzees, engaged in dyadic interactions only. It is therefore not surprising that there is currently no evidence for the intentional stance for artefacts in nonhuman primates.

Conclusion

Although the mental continuity assumption has generated an interesting body of empirical literature, our palaeo-ecological survey suggests that it is fundamentally flawed. If we concentrate too much on human cognition as *the* starting point of investigations of comparative cognition, we risk overlooking many cognitive specializations in other species. There is a real danger that we may never truly understand ape cognition: through the destruction of their natural habitats, most species are threatened with extinction; next to this, many western countries start restricting chimpanzee and bonobo research. Now is the time for comparative psychologists to abandon human cognition as the starting point of their investigations of nonhuman primate cognition, but rather to study the contexts in which these species have evolved. The case of social cognition shows that this approach could turn out to be at least as fruitful as the anthropocentric stance.

Acknowledgements

We wish to express our gratitude to members of the audience at the Human Mind-Human Kind Conference at Aarhus (15-18 August, 2007) for their helpful suggestions and comments. This research was partly funded by grant OZRBOF916 from the Free University of Brussels.

References

Barton, R.A., Purvis, A. & Harvey, P.H. (1995). Evolutionary radiation of visual and olfactory brain systems in primates, bats and insectivores. *Philosophical Transactions of the Royal Society of London B*, *348*, 381–392.

Bloom, P. (1996). Intention, history, and artifact concepts. *Cognition*, *60*, 1–29.

Bogin, B. & Smith, B. H. (1996). Evolution of the human life cycle. *American Journal of Human Biology*, *8*, 703–716.

Bonnefille, R. (1995). A reassessment of the Plio-Pleistocene pollen record of East Africa. In E.S. Vrba, G.H. Denton, T.C. Partridge & Ll.H. Burckle (Eds.), *Paleoclimate and evolution, with emphasis on human origins,* (pp. 299–310). New Haven: Yale University Press.

Burish, M.J., Kueh, H.Y. & Wang, S.S. (2004). Brain architecture and social complexity in modern and ancient birds. *Brain, Behaviour and Evolution*, *63*, 107–124.

Cáceres, M., Suwyn, C., Maddox, M., Thomas, J.W. & Preuss, T.M.. (2007). Increased cortical expression of two synaptogenic thrombospondins in human brain evolution. *Cerebral Cortex*, *17*, 2312–2321.

Call, J. & Tomasello, M. (1999). A nonverbal false belief task: The performance of children and great apes. *Child Development*, *70*, 381–395.

Capaldo, S.D. (1997). Experimental determination of carcass processing by Plio-Pleistocene hominids and carnivores at FLK 22 (*Zinjanthropus*), Olduvai Gorge, Tanzania. *Journal of Human Evolution*, *33*, 555–597.

Casler, K. & Kelemen, D. (2007). Reasoning about artifacts at 24 months. The developing teleo-functional stance. *Cognition*, *103*, 120–130.

Darwin, C. (1871). *The Descent of Man, and Selection in Relation to Sex*. London: John Murray.

deMenocal, P.B. (2004). African climate change and faunal evolution during the Pliocene-Pleistocene. *Earth and Planetary Science Letters*, *220*, 3–24.

De Winter, W. & Oxnard, C.E. (2001). Evolutionary radiations and convergences in the structural organization of mammalian brains. *Nature*, *409*, 710–714.

Domínguez-Rodrigo, M. & Barba, R. (2006). New estimates of tooth mark and percussion mark frequencies at the FLK Zinj site: The carnivore-hominid-carnivore hypothesis falsified. *Journal of Human Evolution*, *50*, 170–194.

Fehr, E. & Rockenbach, B. (2004). Human altruism: Economic, neural, and evolutionary perspectives. *Current Opinion in Neurobiology*, *14*, 784–790.

Franzen, J.L. & Wilde, V. (2003). First gut content of a fossil primate. *Journal of Human Evolution*, *44*, 373–378.

German, T.P. & Barrett, H.C. (2005). Functional fixedness in a technologically sparse culture. *Psychological Science*, *16*, 1–5.

Haile-Selassie, Y. (2001). Late Miocene hominids from the Middle Awash, Ethiopia. *Nature*, *412*, 178–181.

Hare, B. & Tomasello, M. (2004). Chimpanzees are more skilful in competitive than in cooperative cognitive tasks. *Animal Behaviour*, *68*, 571–581.

Hare, B., Woods, V., Melis, A.P., Hastings, S. & Wrangham, R. (2007). Tolerance allows bonobos to outperform chimpanzees on a cooperative task. *Current Biology*, *17*, 619–623.

Hopkins, W.D. & Rilling, J.K. (2000). A comparative MRI study of the relationship between neuroanatomical asymmetry and interhemispheric connectivity in primates: Implication for the evolution of functional asymmetries. *Behavioral Neuroscience*, *114*, 739–748.

Hrdy, S.B. (2005). On why it takes a village. Cooperative breeders, infant needs, and the future. In R.L. Burgess & K. MacDonald (Eds.), *Evolutionary Perspectives on Human Development,* 167–188. Thousand Oaks, Cal.: Sage.

Ivey, P.K. (2000). Cooperative reproduction in Ituri forest hunter-gatherers: Who cares for Efe infants? *Current Anthropology*, *41*, 856–866.

Kaplan, H., Hill, K., Lancaster, J. & Hurtado, A.M. (2000). A theory of human life history evolution: Diet, intelligence, and longevity. *Evolutionary Anthropology*, *9*, 156–185.

Kennedy, G.E. (2005). From the ape's dilemma to the weanling's dilemma: Early weaning and its evolutionary context. *Journal of Human Evolution*, *48*, 123–145.

Kitahara-Frisch, J. (1993). The origin of secondary tools. In A. Berthelet & J. Chavaillon (Eds.), *The Use of Tools by Human and Non-human Primates*. (pp. 239–246). Oxford: Clarendon Press.

Kobayashi, H. & Khoshima, S. (2001). Unique morphology of the human eye and its adaptive meaning: Comparative studies on external morphology of the primate eye. *Journal of Human Evolution*, *40*, 419–435.

Langergraber, K.E., Mitani, J.C. & Vigilant, L. (2007). The limited impact of kinship on cooperation in wild chimpanzees. *Proceedings of the National Academy of Sciences of the USA*, *104*, 7786–7790.

Leonard, W.R. & Robertson, M.L. (1997). Rethinking the energetics of bipedality. *Current Anthropology*, *38*, 304–309.

Lewis, M.E. (1997). Carnivoran paleoguilds of Africa: Implications for hominid food procurement strategies. *Journal of Human Evolution*, *32*, 257–288.

Matsuzawa, T. (2007). Comparative cognitive development. *Developmental Science*, *10*, 97–103.

McBrearty, S. & Jablonski, N.G. (2005). First fossil chimpanzee. *Nature*, *437*, 105–108.

Melis, A.P., Hare, B. & Tomasello, M. (2006). Engineering cooperation in chimpanzees: Tolerance constraints on cooperation. *Animal Behaviour*, *72*, 275–286.

Miklósi, Á., Kubinyi, E., Topál, J., Gácsi, M., Virányi, Z. & Csányi, V. (2003). A simple reason for a big difference: Wolves do not look back at humans, but dogs do. *Current Biology*, *13*, 763–766.

Milton, K. (1999). A hypothesis to explain the role of meat-eating in human evolution. *Evolutionary Anthropology*, *8*, 11–21.

Oldham, M.C., Horvath, S. & Geschwind, D.H. (2006). Conservation and evolution of gene coexpression networks in human and chimpanzee brains. *Proceedings of the National Academy of Sciences of the USA*, *103*, 17973–17978.

Pickford, M. (1990). Uplift of the roof of Africa and its bearing on the evolution of mankind. *Human Evolution*, 5, 1–20.

Pinker, S. (1994). *The Language Instinct. The New Science of Language and Mind*. London: Allen Lane.

Plummer, T., Bishop, L.C., Ditchfield, P. & Hicks, J. (1999). Research on late Pliocene Oldowan sites at Kanjera South, Kenya. *Journal of Human Evolution*, *36*, 151–170.

Povinelli, D.J., Bering, J.M. & Giambrone, S. (2000). Toward a science of other minds: Escaping the argument by analogy. *Cognitive Science*, *24*, 509–541.

Povinelli, D.J. & Dunphy-Lelii, S. (2001). Do chimpanzees seek explanations? Preliminary comparative investigations. *Canadian Journal of Experimental Psychology*, *55*, 185–193.

Rafert, J. & Vineberg, E.O. (1997). Bonobo nutrition - Relation of captive diet to wild diet. In J. Mills, G. Reinartz, H. De Bois, L. Van Elsacker & B. Van Puijenbroeck (Eds.), *Bonobo Husbandry Manual: The Care and Management of Bonobos in Captive Environments,* (pp. 3.1–3.18). Milwaukee: Zoological Society of Milwaukee County.

Rilling, J.K. & Insel, T.R. (1999). The primate neocortex in comparative perspective using magnetic resonance imaging. *Journal of Human Evolution*, *37*, 191–223.

Roche, H., Delagnes, A. Brugal, J.-P., Feibel, C., Kibunjia, M., Mourre, V. & Texier, P.-J. (1999). Early hominid stone tool production and technical skill 2.34 myr ago in West Turkana, Kenya. *Nature*, *399*, 57–60.

Rosenberg, K. & Trevathan, W. (2002). Birth, obstetrics and human evolution. *British Journal of Obstetrics and Gynaecology*, *109*, 1199–1206.

Savolainen, P., Zhang, Y., Luo, J., Lundeberg, J. & Leitner, T. (2002). Genetic evidence for an East Asian origin of domestic dogs. *Science*, *298*, 1610–1613.

Schick, K., Toth, N., Garufi, G., Savage-Rumbaugh, E.S., Rumbaugh, D. & Sevcik, R. (1999). Continuing investigations into the stone tool-making and tool-using capabilities of a bonobo (*Pan paniscus*). *Journal of Archaeological Science*, *26*, 821–832.

Shultz, S. & Dunbar, R. (2006). Both social and ecological factors predict ungulate brain size. *Proceedings of the Royal Society B*, *273*, 207–215.

Silk, J.B., Brosnan, S.F., Vonk, J., Henrich, J., Povinelli, D.J., Richardson, A.S., Lambeth, S.P., Mascaro, J. & Schapiro, S.J. (2005). Chimpanzees are indifferent to the welfare of unrelated group members. *Nature*, *437*, 1357–1359.

Stone, A.C., Griffiths, R.C., Zegura, S.L. & Hammer, M.F. (2002). High levels of Y-chromosome nucleotide diversity in the genus *Pan*. *Proceedings of the National Academy of Sciences of the USA*, *104*, 3907–3912.

Stout, D., Quade, J., Semaw, S. Rogers, M.J. & Levin, N.E. (2005). Raw material selectivity of the earliest stone toolmakers at Gona, Afar, Ethiopia. *Journal of Human Evolution*, *48*, 365–380.

Tomasello, M. & Carpenter, M. (2007). Shared intentionality. *Developmental Science*, *10*, 121–125.

Tomasello, M. & Rakoczy, H. (2003). What makes human cognition unique? From individual to shared to collective intentionality. *Mind & Language*, *18*, 121–147.

Tomonaga, M., Tanaka, M., Matsuzawa, T., Myowa-Yamakoshi, M., Kosugi, D., Mizuno, Y., Okamoto, S., Yamaguchi, M.K. & Bard, K.A. (2004). Development of social cognition in infant chimpanzees (*Pan troglodytes*): Face recognition, smiling, gaze, and the lack of triadic interactions. *Japanese Psychological Research*, *46*, 227–235.

Warneken, F. & Tomasello, T. (2006). Altruistic helping in human infants and young chimpanzees. *Science*, *311*, 1301–1303.

White, F.J. (1998). Seasonality and socio-ecology: The importance of variation in fruit abundance to bonobo sociality. *International Journal of Primatology*, *19*, 1013–1027.

Whiten A., Goodall, J., McGrew, W.C., Reynolds, N.T., Sugiyama Y., Tutin, C.E., Wrangham, R.W. & Boesch, C. (1999). Cultures in chimpanzees. *Nature*, *399*, 682–685.

Wittig, R.M. & Boesch, C. (2003). Food competition and linear dominance hierarchies among female chimpanzees of the Taï national park. *International Journal of Primatology*, *24*, 847–867.

Yamakoshi, G. (2001). Ecology of tool use in wild chimpanzees: Towards a reconstruction of early hominid evolution. In T. Matsuzawa (Ed.), *Primate origins of human cognition and behaviour* (pp. 537–556). Tokyo & Berlin: Springer.

SECTION II

HUMAN SOCIALITY, MORALITY AND RELIGIOSITY

CHAPTER NINE

DEVELOPMENT OF SOCIAL RELATIONSHIPS WITH SIGNIFICANT OTHERS FROM 3- TO 8-YEAR-OLD JAPANESE CHILDREN: NATURE AND NURTURE

KEIKO TAKAHASHI, MARIKO INOUE, KAYOKO YAMAKAWA, REIKO SHIBATA

Abstract: Based on the key assumption about human evolution that human beings have survived by being members of social networks, this paper examines the nature and contents of close relationships among young Japanese children. According to conventional wisdom, young children are supposed to develop mainly close mother-oriented relationships. However, as we analysed our longitudinal data from children between 3 to 8 years, we found that, contrary to our expectations, Japanese children as young as three years old constructed their own personal frameworks of social relationships that consisted of multiple significant others, including non-family members. A corollary of this assumption is that the human mind from early on reaches purposefully out towards other minds, and so the child finds its assigned place in a network and moves on to refine and revise its terms of "membership" according to how its needs develop.

Introduction

Today we know that from birth infants are drawn to and seek interaction with other human beings to establish emotional bonds with caregivers. But why are they so sociable from the beginning of life? For a long time, many developmental psychologists who studied close relationships in infancy and early childhood struggled to explain this why; this question of the origins of social relationships. Although most developmental psychologists agree that nature and nurture are equally necessary for human development, when confronted with research findings that amply indicated neonates' initial capacities to form and maintain relationships without any previous experience, they endeavoured to find an alternative explanation of this why without depending on conventional environmental learning theories. Thanks to insights from evolutionary literature, two groups of theorists have successfully shown that the origins of close relationships have an evolutionary explanation, i.e. infants are born with pre-programmed tendencies to respond to human beings.

Attachment Theorists

A paragon of the first group is the Attachment theory proposed by Bowlby (1969/1982). He constructed his unique and influential Attachment theory, in which he hypothesized an innate predisposition in human infants to "ask" the primary caregiver (the mother) to protect them for their survival and security. From an evolutionary perspective, Bowlby claimed that to lose touch with the mother is tantamount to death for her baby. After observing miserable orphans in the aftermath of World War II, he concluded that maternal deprivation endangered a child's life (Bowlby, 1951). His evolutionary hypothesis that accompanied his observations of children deprived of maternal care, urged him to emphasize and focus on the role of the mother-figure in child development. That is, Bowlby insisted that the infant possessed a pre-programmed tendency of attachment to the mother, which was triggered by maternal care. Accordingly, attachment theorists assert that the origin of the infant's attachment to the mother must be innate because of its adaptive and survival value. In addition to an evolutionary slant, the attachment theory was strongly influenced by psychoanalytic theories, that both claim an indispensable role of the mother-figure and lasting influences of experiences with her in early childhood on future development. Thus, the attachment theory highlights a specific and narrow aspect of close relationships, mostly with the mother as primary caregiver. As a natural consequence, this perspective on close relationships emphasizes and focuses on the role of the mother-figure (Ainsworth, 1989; Main, 1999). Recently, beyond the 'monotropy' (the seeking of one figure, the mother-figure) hypothesis by Bowlby (1969/1982), some attachment theorists have assumed a plural form of attachment and 'implicitly' suggested that there are other attachment figures than the mother (e.g. Cassidy, 1999; Kobak, 1999). Some theorists discussed 'an attachment hierarchy' from the assumption that the mother-figure is primary and other figures secondary in the hierarchy (e.g. Cassidy, 1999, Main, 1999). However, according to their rationale, the attachment theory mainly stresses the role of the mother-figure. In fact, even for adulthood, attachment theorists try to evaluate the present state of attachment of adults by having them recount how their mothers treated them in early childhood (George, Kaplan& Main, 1984-1996; Main & Goldwyn, 1998).

Social Network Theorists

An alternative evolutionary explanation has been posed by so-called social network theorists, including the present authors. These theorists have also witnessed the accumulated empirical findings that indicate that from the cradle human beings are social by nature. In contrast to attachment theorists, who focus on the mother-figure, social network theorists emphasize that humans need 'multiple' significant others, including non-family members for their survival (e.g. Kahn & Antonucci, 1980; Lewis, 1982; Lewis & Takahashi, 2005; Takahashi, 1974; 1990; 2004; 2005; Takahashi, Ohara, Antonucci & Akiyama, 2002). Similar to the attachment theorists, the social network theorists posit an important role for the mother-figure, especially at the beginning of the social relationship. However, at the same time, they hypothesize that human infants are born in a social arena: Many different humans around them will naturally attract and interact with the newcomers. In fact, mothers of 11-months-old infants reported that many people interacted with them; the father as a partner in playful situations, such as bathing and playing; the grandmother as a surrogate caregiver when the mother was absent; and siblings whose behaviour was the most interesting to watch (Takahashi, 1990). A record of a baby's first years described how a 14-months-old boy had multiple figures: the mother as a caregiver, his sibling as a play partner, the grandmother as a generous presenter of special candy and the grandfather as an expert at repairing toys (Yamada, 1982).

Thus, if we accept this multiplicity of social relationships as a given in nature, we are naturally led to another assumption: i.e. there is a functional differentiation among the multiple significant others. Namely, each child assigns different psychological functions to each of these multiple social figures. Here, psychological functions means sub-goals through which we achieve the main purpose of having significant others, that is, to maintain our survival and feeling of security and to enhance our well-being. Moreover, if we accept the fact that individuals consciously choose the most appropriate figures for themselves and assign one or more psychological functions to each of the figures, it is naturally assumed that there are individual differences in the configurations of a person's social relationships that can be described in terms of figure-function pairs. Thus, there must be individual patterns of social relationships. It is assumed that in young children this is not an exception. We further hypothesize that such tendencies will appear from birth.

Based on social network theory, this study has specifically aimed to describe the nature of close relationships in early childhood. In particular, our question for this paper is whether children as young as three years old have already constructed their own frameworks of close relationships, just as older subjects have. In particular, this study aims to confirm: (1) Multiplicity: Children as young as 3 have multiple significant others, including non-family members such as friends and teachers; (2) Articulations: Children construct their own frameworks of social relationships, assigning several psychological functions of close relationships among the multiple significant others. Such a framework has a hierarchical structure, as previous studies have clearly indicated among older subjects (e.g. Antonucci, Akiyama & Takahashi, 2004; Levitt, 2005). That is, a focal figure will satisfy almost all of the psychological functions of close relationships, especially the most critical functions for the children's survival and well-being. In addition to the focal figure, there would be a limited but sufficient number of significant others to satisfy a variety of psychological functions; and finally, (3) Individual patterns: There would be individual patterns of frameworks, because each child would choose the figures that were the most appropriate and available for them to fulfil each of the functions.

Method

Participants

When the children were three and a half years old, a total of 83 families were asked to participate in our ongoing longitudinal study on the development of social relationships. The reason this study began at age 3 ½ is that our preliminary research indicated that these children were the youngest subjects capable of reporting their close relationships in face-to-face interviews.

The sample included in this study consisted of 83 children aged 3 1/2 (42 girls and 41 boys, with a mean age of 3:65, SD = 2.97 months); 83 aged 4 (39 girls and 44 boys, M = 4:33, SD = 2.08 months); 81 aged 5 (40 girls and 41 boys. M = 5:12, SD = 2.02 months); 73 aged 6 (38 girls and 35 boys, M = 6:21, SD = 1.92 months); 64 aged 7 (33 girls and 31 boys, M = 7:22, SD = 3.55 months), and 61 aged 8 (28 girls and 33 boys, M = 8:31, SD = 3.87 months). Almost all were from middle-class families in

Tokyo. Initially this study consisted of two-parent families; 86% were nuclear families with working fathers, full-time mothers (78%) and families with one child (64%), two children (32%) or three children (4%). Seventy-three percent of the girls and 68% of the boys were the first-born child.

Assessing Close Relationships: The PART

To describe the personal framework of close relationships, which consists of multiple significant others, a series of self-report assessment instruments for young children up to elderly people, the Picture Affective Relationships Test (PART) has been constructed (Takahashi, 1978-2000). PART focuses on what we have termed 'affective relationships', the core and relatively stable close relationships which are assumed to be essential for the survival and well-being of human beings (Takahashi, 1990; 2004; Takahashi & Sakamoto, 2000). PART can effectively describe and also summarize how figures and functions are related to each other in individual frameworks of close relationships. That is, PART is designed to assess the strength of the need felt toward each of the figures by asking children to differentiate between which of the functions they assign to each figure.

Each set of PART for young children and for elementary school children consists of two sets of 15 cards, one for girls and one for boys. Each card illustrates a daily life situation in which affective behaviours toward another person may be induced (see Figure 1). Five psychological functions are measured. Each of the cards depicts an imaginary situation presenting one of the following five functions:

1. *Seeking proximity* (e.g. when you play at home, who would you like to play with?)
2. *Receiving emotional support* (e.g. who would make you feel secure by being with you?)
3. *Receiving reassurance for behaviour and/or being* (e.g. if you were sick, who would you want to be with?)
4. *Receiving encouragement and help* (e.g. if you found a flower that you do not know the name of, who would you want to ask to find out?)
5. *Sharing information and experience* (e.g. if something pleasant happened to you, who would you like to share it with?)

During the test session, the child was instructed to suppose that the major figure in the picture was her/himself and to choose the best partner in each situation.

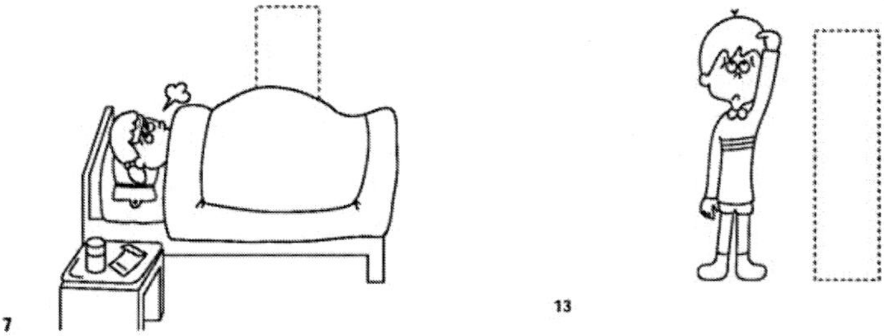

Figure 1. Example of PART for young children
Card 7: If you were sick, who would you want to be with (Girl's card)
Card 13: If something sad happened to you, who would you like to be with? (Boy's card)
Note: Part cards are available at the website: http:// www.keiko-takahashi.com

With the 15 cards (three cards representing each of the five functions), PART is designed to yield two kinds of information about social frameworks: the total frequency of each figure, which indicates who the most important figure is; and the frequency of each function for each figure, which indicates how and to what extent each figure is important. Moreover, we identified each individual pattern of close relationships in terms of the figure with the highest total frequency, the one who was most often named throughout the 15 cards. This was based on our previous studies, which indicated that by knowing who the top figure was, we could neatly summarize the nature of each framework of social relationships among older subjects (e.g. Inoue & Takahashi, 2000; Takahashi, 2004).

Operationally, a child who answered "myself", "anybody" or "nobody" for more than half of the cards, was designated a Lone wolf-type. Among the remaining subjects, the figure named most frequently was designated as the dominant figure, and the child was labelled accordingly. For example, if a child selected the mother most often, the child was identified as a Mother-type. When as many as seven cards had been shown and a child either did not name any particular figure, or chose two or more figures, s/he was identified as a Multi-focus type, because it was difficult to identify the primary figure in such cases. However, if children nominated "mommy and daddy" instead of the mother or the father for more than half of the cards, or the total number of nominations of parents was more than 8 out of 15 cards, we specifically identified them as Parents-type, because to these children the parents seemed to be a unit.

Results

The Multiplicity of Significant Others

As predicted, children as young as 3 named several different figures as the best partner throughout the 15 hypothetical situations: parents, sibling, grandparents, friends, teacher and others. As Figures 2a and 2b indicate, children from ages 3 to 8 most often nominated family members, but they included non-family members from age 3. The nominations of the figures to PART were subjected to a 4 (Figures: family, friend, others, and myself) x 6 (Age group: at ages 3, 4, 5, 6, 7, and 8) x 2 (Gender) ANOVA with figures and age group as within-subject variables and gender as a between-subjects variable. As ANOVAs with repeated measures automatically excluded those participants who had not undergone all of the assessments for each age, these ANOVAs were actually calculated among the intact sample. The main effects of Figure, F (3, 51) = 204.95, $p < .001$, and of Age group, F (5, 51) = 6.42, $p < .01$, were significant. Because the ANOVAs did not suggest the differences between genders, further analyses for each figure were conducted with girls and boys together. A main effect of the nomination of a friend was significant, F (5, 51) = 2.50, $p < .05$, but not as significant to that of age, F (1, 51) = 6.69, $p < .19$. Post-hoc analyses by LSD（least significant difference）among age groups indicated that at ages 7 and 8, that is, after entering elementary school, children nominated a significantly greater number of friends than at ages 3 and 6. As for nominations of other figures such as a cousin and an aunt, the main effect of age was significant, F (5, 51) = 6.69, $p < .05$, and a post-hoc analysis by LSD indicated at age 3 that children more often nominated other figures than in later years. This is probably because 3-year-old children are not mature enough to articulate clearly the various psychological functions served by significant others. There were no significant changes to the nominations of family members, F (5, 51) = .77, $p < .57$ or of references to oneself, F (5,51) = .50, $p < .77$. However, it is noteworthy that, while the total was less than 10%, at least some children in every age period reported “myself”, “anybody” or “nobody” for each of the cards as we found among participants in later age groups from adolescents to elderly people (Takahashi, 2004). Thus, children from ages 3 to 8 consistently and most often nominated the mother as an important figure, but at the same time had expanded their relationships to non-family members, especially friends.

Among girls

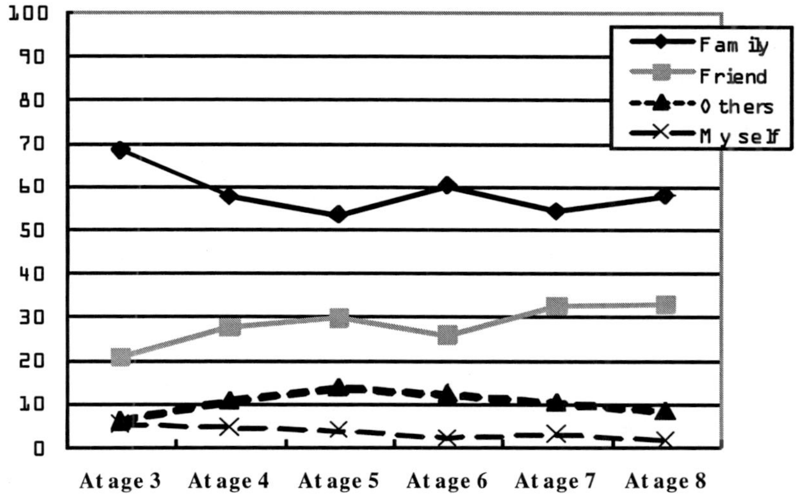

Among boys

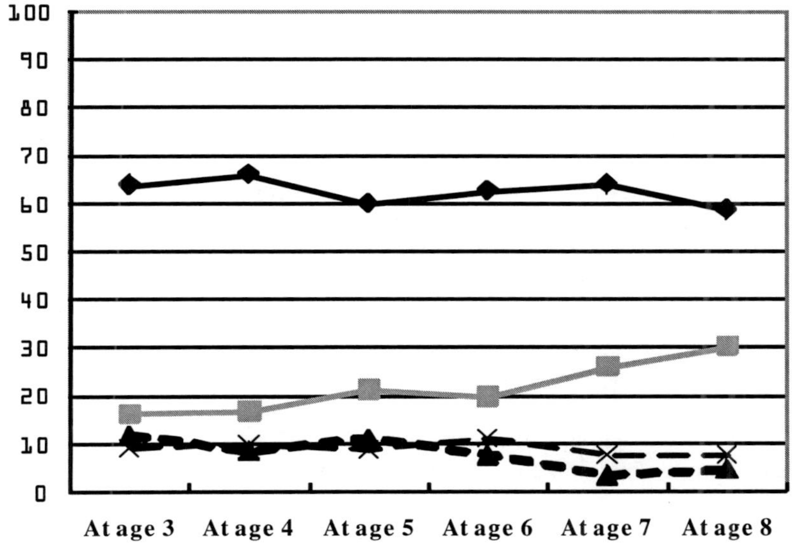

Figures 2a & 2b. Percentage of Nominations: Family vs. Friend vs. Others vs. Myself

Then, how many different figures did each child nominate throughout the 15 hypothetical situations? Figure 3 shows the average numbers of different kinds of social figures nominated by each child. The data was subjected to a 2 x 6 (Gender x Age group) ANOVA, with gender as a between-subject variable and an age group as within-subject variable. The

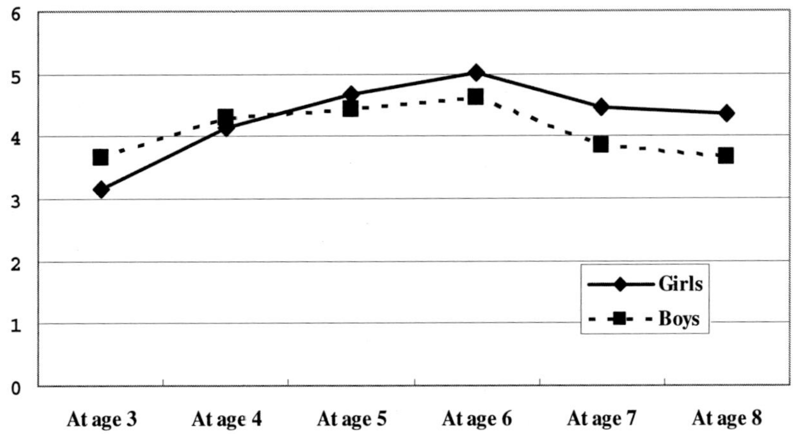

Figure 3. Different Types of Figures Nominated at Ages 3-8

main effect of age group was significant, F (5, 51) = 10.33, p < .01, but not significant to gender, F (5, 51) = 1.76, p < .19. Post-hoc analyses by LSD among age groups indicated that the number of nominations of significant figures at age 3 was significantly lower than at later ages (*ps* < .01), and was significantly higher at age 6 than at ages 3, 7, or 8 (*ps* < .01). That is, at age 3, the children of both genders generally named three or more different people on average (for girls M = 3.17, SD = 1.20, *Range* = 1-5; for boys M = 3.66, SD = 1.32, *Range* = 1-6) as important, and as shown in Figure 3, the average numbers increased as the children turned 6, but decreased after they entered elementary school at age 7. No statistical differences were found between ages 7 and 8, i.e., between first graders and second graders in elementary school. Thus, we have come to realize that Japanese children as young as 3 1/2 already have multiple significant others apart from the parents, and continue to expand their social relationships to new figures while growing up. However, the present data indicate that while children include a limited number of different figures as close partners, they definitely expand their social figures. There must be some limit to the number of people with whom children interact closely.

This limitation was also reported among adolescents and adults (e.g. Antonucci & Akiyama, 1987; Antonucci, Akiyama & Takahashi, 2004).

Normative Trends of Articulations of Psychological Functions among Multiple Figures

The percentage of the nomination of figures fulfilling each of the five functions at each age, among girls and boys separately, as well as articulations of psychological functions for the children are shown in Figures 4a and b. The data shows the normative trends of how children assign psychological functions to each of their multiple significant figures.

First, the data at each age was subjected to 2 (Gender) x 4 (Figure: mother, father, sibling, friend) x 5 (Function) ANOVAs with gender as a between-subject variable. The main effects of gender were significant only at age 3: F (1, 92) = 8.44, $p < .01$, and interactions of figure x function were significant throughout ages 3 to 8, respectively: F (12, 92) = 15.29, $p < .01$; F (12, 92) = 17.15, $p < .01$; F (12, 92) = 26.76, $p < .01$; F (12, 92) = 34.29, $p < .01$; F (12, 92) = 32.28, $p < .01$; F (12, 92) = 33.32, $p < .01$. As the ANOVA suggested gender differences at age 3, further analyses were conducted of girls and boys separately at each age.

Among girls, the mother was most frequently and significantly nominated for the most critical functions to the child's well-being, i.e. for Function 3 (*Receiving reassurance for behaviour and/or being*) the mother was most frequently nominated at each age from 3 to 8 [F (3,44) = 16.61; F (3,44) = 11.01 ; F (3,44) = 15.88; F (3,44) = 23.04; F (3,44) = 27.32; F (3,44) = 22.53 ($ps < .01$)], and for Function 2 (*Receiving emotional support*), at ages 5 to 8, respectively [F (3,44) = 4.19, $p < .05$; F (3,44) = 4.81, $p < .05$; F (3,44) = 18.30, $p < .01$; F (3,44) = 18.85, $p < .01$.] As children mature and develop their articulation of the psychological functions that the various figures fulfil, the mother's functions become distinct to many of the children. As Figure 4a indicates, when the children were young, quite a few of them nominated the mother even for Function 4 (*Receiving encouragement and help*) and the two parents for Function 5 (*Sharing information and experience*). However, generally speaking, beginning at age 6 (the last year of kindergarten), the more active experiences with friends the children had, the less important the mother was for these functions. Friends were consistently and significantly more important social figures as playmates (Function 1, *Seeking proximity*) from ages 3 to 8, respectively [F (3,44) = 5.09; F (3,44) = 16.45; F (3,44) = 42.06; F (3,44) = 31.88; F (3,44) = 29.12; F (3,44) = 22.75 ($ps < .01$)]. As Figure 4a shows, many of the children preferred friends to parents.

Starting at age 5, among girls, friends were more often assigned the role of instrumental support (Function 4) than other figures, including parents [F (3,44) = 8.94; F (3,44) = 11.26; F (3,44) = 22.30; F (3,44) = 12.58 (ps < .01), from ages 5 to 8, respectively]. The father was nominated by some children for the most critical functions (Functions 2 & 3) at all ages, and at older ages the father or other figures seem to be a partner for having fun with (Function 5). In Function 5, each child chose its own preferred partner. In other words, the normative trend in the selections of figures was not apparent in Function 5. Moreover, the analysis of boys' data shown in Figure 4b indicated similar tendencies to those of girls, suggesting a general tendency towards functional articulations among social figures.

It is noteworthy that even among young children, for whom attachment to the mother is commonly looked upon as the norm, some named the father or even a friend as best suited to fulfil critical functions. The data clearly indicated that no one, not even the mother, was an omnipresent, omnipotent figure to these young children even in so-called mother-oriented developmental periods. Children as young as 3 1/2 years old certainly discriminated between psychological functions and assigned each function to whoever seemed to be the appropriate figure to them. These individual differences persisted even in situations where it has long been considered 'normal' for one particular figure to fulfil the function. Thus, while the influence of these normative trends continues to be felt, we can certainly assume the existence of individual patterns in selections of appropriate social figures.

Individual Patterns of Close Relationships

In terms of the figure who was most often nominated throughout the 15 situations shown on the cards, we identified individual patterns of close relationships. In other words, although the children nominated multiple significant others, they had a focal figure, a person who was most often nominated for their survival and security, especially in the cards that depicted situations of critical functions (Functions 2 & 3).

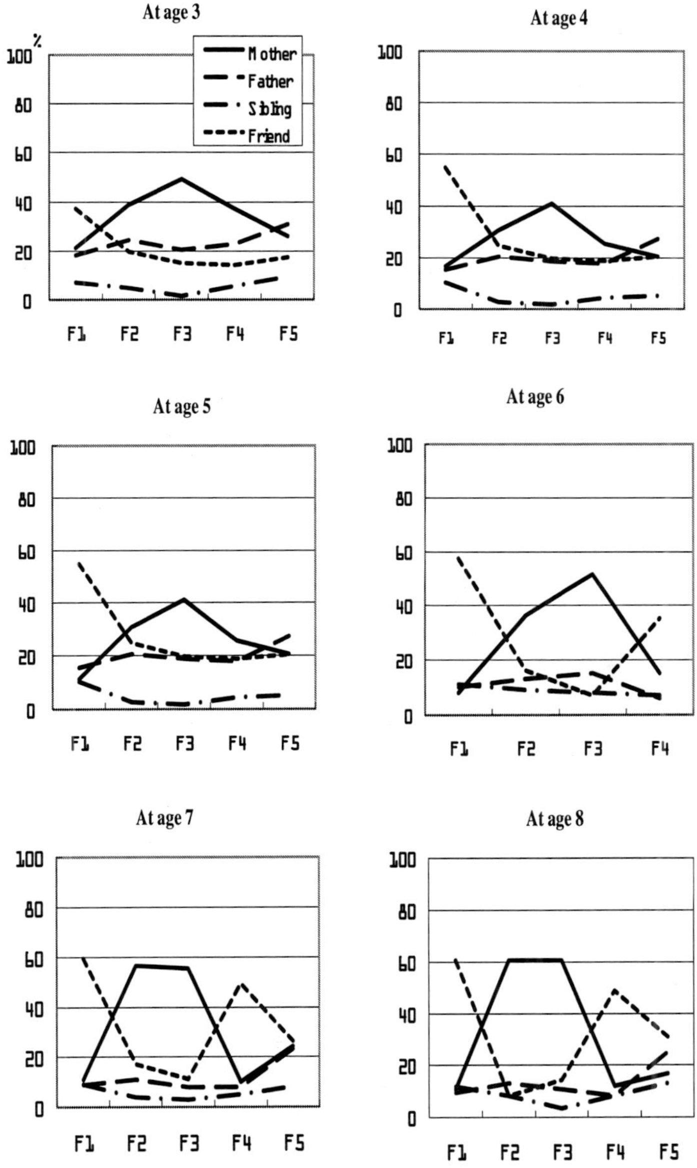

Figure 4a. Percentage of Nomination of Each Figure for Each Function among Girls

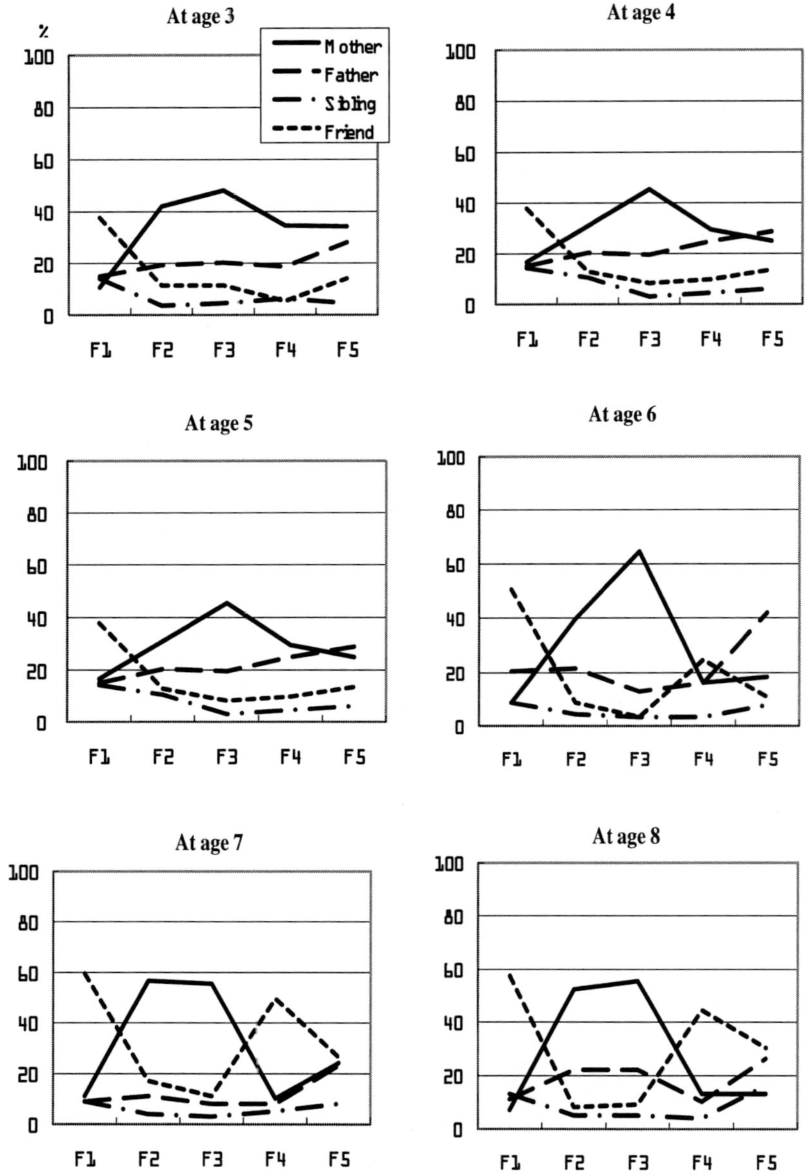

Figure 4b. Percentage of Nomination of Each Figure for Each Function among Boys

As shown in Figure 5, we found different types of relationships among the participants. Even at age 3, 90% of the children were identified as Single-focus type in that they largely named just one of their significant others as the focal figure. In addition to family-dominant types (Mother-, Father-, Parents- and Sibling-types), there were Friend-type children (13%). There were no significant gender differences in the appearance of types except at age 4: We found a larger number of Parents-type ($\chi2(6, N = 83, p < .05)$ among girls.

The percentage of appearance of the Mother-type was small but stable (10-15%) throughout, except at age 3 (24%). The father was significantly more often nominated as a focal figure at age 3 (12%) than at later ages (2-8%). We identified 14 % (at age 3) to 29% (at age 4) children as Parents-types. These children nominated the parents as an inseparable unit. The father seemed to be a focal figure together with the mother, but not so independently. At age 8, the percentage of Parents-types significantly decreased. Friend-types, that is, Non-family-types, had already appeared at age 3 and increased up to age 5. The number of Family-types at ages 6 and 7 was still large like at age 3. However, children at ages 7 and 8, by now in elementary school, were significantly more frequently identified as Friend-

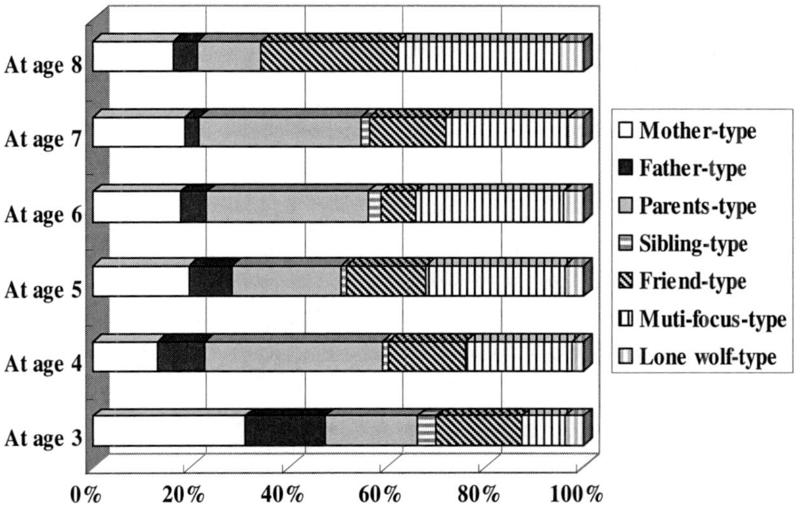

Figure 5. Proportion of Appearance of Affective Relationship Types

type than at age 6. As they grew and matured, there were Multi-focus types, who reflected their expansions of social relationships to a wider range of people. Finally, from ages 3 to 8, two to three percent of the children showed so little concern with social figures that they were designated as Lone wolf-type.

Functional articulations among figures were most apparent when we compared nominations of focal figures with those of other figures, because the focal figures were specifically assigned the most critical function or functions. In order to refine our understanding of this, the average number of nominations of the mother and of the closest friend in Function 3 (the critical function that the focal figure fulfils: *Receiving reassurance for behaviour and/or being*) among both the Mother-types and Friend-types were calculated as shown in Figure 6. They were subjected to a 2 x 2 (Type x Figure) ANOVA at each age, with type as between-subject variables and figure as within-subject variables, to affirm differences of the nominations in Function 3. Results indicated that all interactions from ages 3 to 8 were highly significant. That is, interactions of Type x Figure: $F\,(1,37) = 100.38$; $F\,(1,22) = 29.37$; $F\,(1,27) = 14.62$; $F\,(1,16) = 12.63$; $F\,(1,20) = 28.13$; $F\,(1,25) = 75.34$, $ps < .01$, at ages 3 to 8, respectively. Thus, Mother-type children more often assigned Function 3 to the mother and not to a friend, whereas Friend-type counterparts nominated the friend, not the mother, for this function.

Looking at Figure 5, the appearances of types at each age seem to be coherent. Although we found the above-mentioned significant changes that accompanied their development, the proportion of types was quite similar. As shown in Figure 7, an analysis of the person-centred trajectories of types from ages 3 to 8 indicated that 60% of the girls and 63% of the boys were stable as to type (in at least four out of the six interviews). From ages 3 to 8, for 41% of the girls and 57% of the boys, the parents were consistently the dominant figures. However, 5% of the girls and 3% of the boys consistently identified as Non-family type, that is, Friend-type. Moreover, 41% of the girls and 37% of the boys were unstable as to types; they showed fluctuations in their selections of the dominant figure. No child was consistently identified as a Lone wolf-type throughout all 6 interviews, but 14% of the boys were identified as Lone wolf-type at least once.

Thus, there were individual differences in the trajectories of development of social relationships in early childhood. Contrary to popular expectation, only half of the young children were identified as consistent Family-types. There were consistent Non-family types, and the

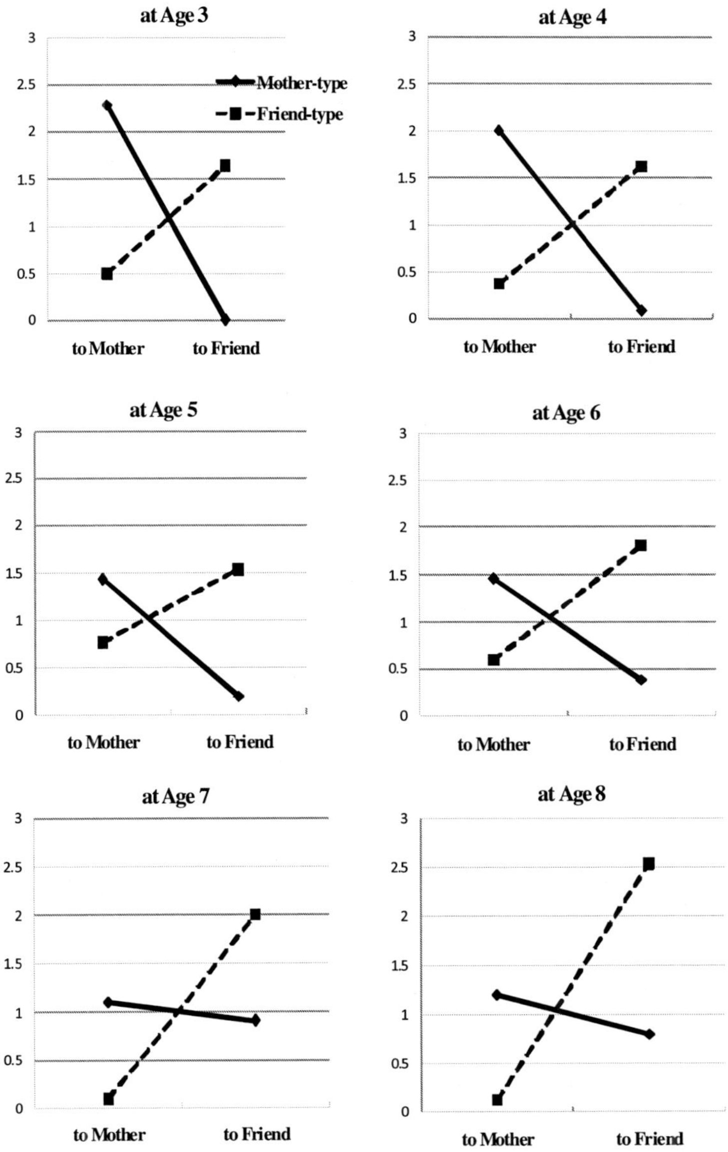

Figure 6. Nominations of mother and friend in Function 3:Mother-type vs. Friend-type

remaining 40% of the children appeared to be exploring to find the best source of psychological support suitable to each social situation.

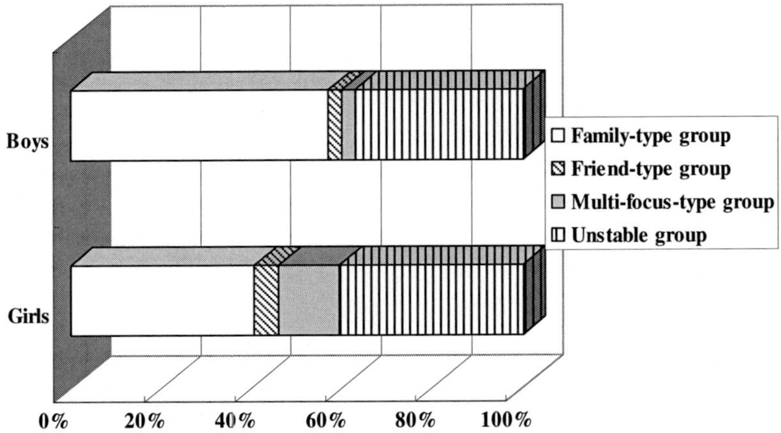

Figure 7. Person-centred Trajectories of Affective Relationship Types

Discussion

The Nature of Children's Social Relationships

The present findings indicate that children as young as 3 1/2 years old, who lived and were taken care of by family members clearly constructed their own personal frameworks of social relationships consisting of multiple figures, including non-family members (3.3-4.7 different people on the average). They discriminated between who they felt was the most appropriate figure for them in each situation. Namely, in contrast to conventional expectation and the attachment theory that young children develop mainly close mother-oriented relationships, and therefore these children were not invariably tied to their family members. Rather, the children chose the mother only when they thought that she was the most appropriate figure to fulfil a particular function. It is noteworthy that Mother-type children assigned the mother the most critical function that characterized the nature of their focal figure. In contrast, Friend-type children reported their close friend to fulfil this function.

Nature and Nurture in Personal Constructions of Social Relationships

The present results undoubtedly suggest that from the beginning of life, humans are social by nature. As social network theorists have anticipated, humans seek interaction with multiple social figures concurrently from the beginning rather than seeking one omnipotent caregiver as attachment theorists have asserted. As there are a number of humans in the social arena, infants must also possess a pre-programmed ability to choose their favourite people and to construct their own personal framework. As children grow older, they continuously expand their options of possible figures and examine who the best partner is in a given situation. For example, once they were in kindergarten, many of the children chose a teacher as a competent supporter in a cognitive task such as a paper-folding exercise.

Our findings suggest that from early on the human mind reaches purposefully out towards other minds. That is, there is a universal tendency in children towards establishing close relationships that support both their survival and well-being at various dimensions. They choose multiple social figures, articulate psychological functions among them, and in so doing construct personal frameworks. In this regard, it appears that environmental differences may not only be a determining factor, as researchers have often expected.

At the same time, however, this study suggests that one can speak of social situations of relationships; that is, we found that within this universal tendency to form social relationship frameworks, there were not only age differences but also both individual and gender differences in these frameworks. In further analysis of our data set, we hope to find environmental factors that influence these differences. We should be able to explain why some young children claimed, ‘I prefer a friend to my parents’ even in a critical situation, and even which factors push a small number of children to be separated from others to the degree that they are identified as Lone wolves.

Cultural Universality vs. Diversity

Finally, can we apply the present findings to young participants in other cultures? Our answer is Yes. We believe that human beings from birth universally live and interact with multiple social figures concurrently and construct their own social networks for their survival and well-being. In fact, studies among young children in the United States indicate the multiplicity of their social networks. However, the selection of multiple significant figures in a given function was affected by customs and values surrounding children in their daily situations (e.g. Levitt, 2005; Takahashi & Hirai, 2007). In this vein, we can assume that the proportion of appearances of types will vary with and be influenced by cultural expectations. However, in any society, there are bound to be a number of children who can clearly be identified as Family-type, Friend-type, or Lone wolf-type, as this study indicates.

References

Ainsworth, M. D. S. (1989). Attachment beyond infancy. *American Psychologist, 44*, (p.709-716).

Antonucci, T. C., & Akiyama, H. (1987). An examination of gender differences in a social support among older men and women. *Sex Roles, 17*, (p.737-749).

Antonucci, T. C., Akiyama, H., & Takahashi, K. (2004). Attachment and close relationships across the life span. *Attachment & Human Development, 6*, (p. 353-370).

Bowlby, J. (1951). Maternal care and mental health. *WHO Monograph Series No. 2*. Geneva: World Health Organization.

—. (1969/1982). *Attachment and Loss: Vol. 1*. Attachment. New York: Basic Books.

Cassidy, J. (1999). The nature of the child's ties. In: J. Cassidy & P. R. Shaver. (eds.). *Handbook of Attachment: Theory, Research, and Clinical Applications* (p. 3-20). New York: Guilford.

George, C., Kaplan, N. & Main, M. (1984-1996). *Adult Attachment Interview Protocol* (3rd. ed.). Unpublished manuscript, University of California, Berkeley.

Inoue, M., & Takahashi, K. (2000). Patterns of social relations and psychological adjustment among elementary school children: Using the Picture Affective Relationships Test. *Japanese Journal of Educational Psychology, 48*, (p.75-84). (in Japanese).

Kahn, R. L. & Antonucci, T. C. (1980). Convoys over the life course: Attachment, roles, and social support. In: P. B. Baltes & O. B. Brim (eds.), *Life-span Development and Behavior, Vol. 3* (p. 253-268). New York: Academic Press.

Kobak, R. R. (1999). The emotional dynamics of disruptions in attachment relationships: Implications for theory, research, and clinical intervention. In: J. Cassidy & P. R. Shaver (eds.). *Handbook of*

Attachment: Theory, Research, and Clinical Applications (p.. 21-43). New York: Guilford.

Levitt, M. J. (2005). Social relations in childhood and adolescence: The convoy model perspective. *Human Development, 48*, (p.28-47).

Lewis, M. (1982). The social network model. In: T. M. Field, A. Huston, H. C. Quary, L. Troll, & G. E. Finley (Eds.), *Review of Human Development* (p 180-214). New York: Wiley.

Lewis, M. & Takahashi, K. (2005).Beyond the dyad: conceptualization of social networks. *Human Development, 48*, 1-2.

Main, M. (1999). Attachment theory: Eighteen points with suggestions for future studies. In: J. Cassidy & P. R. Shaver eds.). *Handbook of Attachment: Theory, Research, and Clinical applications* (p. 845-887). New York: Guilford.

Main, M. & Goldwyn, R. (1998). *Adult attachment scoring and classification systems*. Unpublished scoring manual. University of California, Berkeley.

Takahashi, K. (1974). Development of dependency among female adolescents and young adults. *Japanese Psychological Research, 16*, (p.179-185).

—. (1978-2000). *The measurement of affective relationships among young children.* Unpublished manuscript.

—. (1990). Affective relationships and their lifelong development. In: P. B. Baltes, D. L. Featherman, & R. M. Lerner (eds.), *Life-span Development and Behavior Vol. 10* (p. 1-27). Hillsdale, NJ: Erlbaum.

—. (2004). Close relationships across the life span: Toward a theory of relationship types. In: F. R. Lang, & K. L. Fingerman (eds.), *Growing Together: Personal Relationships Across the Lifespan* (p. 130-158). New York: Cambridge University Press.

—. (2005). Toward a life span theory of close relationships: The Affective relationships model. *Human Development, 48*, (p.48-66).

Takahashi, K. & Hirai, M. (2007). *Children in their mothers' social networks in the United States and Japan*. Unpublished manuscript.

Takahashi, K., Ohara, N., Antonucci, T. & Akiyama, H. (2002). Commonalities and differences in close relationships among the Americans and the Japanese: A comparison by the Individualism/ Collectivism concept. *International Journal of Behavioral Development, 26*, (p.453-465).

Takahashi, K. & Sakamoto, A. (2000). Assessing social relationships in adolescent and adults: Constructing and validating the Affective Relationships Scale. International *Journal of Behavioral Development, 24*, (p.451-463).

Yamada, Y. (1982). Expressions of request and rejection and the emergence of self in infancy. *Japanese Journal of Educational Psychology, 30*, (p.38-48) (in Japanese).

CHAPTER TEN

THE ADAPTIVE PROBLEM OF ABSENT THIRD-PARTY PUNISHMENT

GORDON P. D. INGRAM, JARED R. PIAZZA, JESSE M. BERING

Abstract: Language is a uniquely human behaviour, which has presented unique adaptive problems. Prominent among these is the transmission of information that may affect an individual's reputation. The possibility of punishment of those with a low reputation by absent third parties has created a selective pressure on human beings that is not shared by any other species. This has led to the evolution of unique cognitive structures that are capable of handling such a novel adaptive challenge. One of these, we argue, is the propositional theory of mind, which enables individuals to model, and potentially manipulate, their own reputation in the minds of other group members, by representing the beliefs that others have about the first party's intentions and actions. Support for our theoretical model is provided by an observational study on tattling in two preschools, and an experimental study of giving under threat of gossip in a dictator game.

Language makes humans unique. Other animals employ complex systems of communication (Hauser, 1997), but their communicative signals are indexical in reference (Deacon, 1997). As with non-verbal communication in humans, the reference of animal signals inheres in the drawing of attention to the presence of a stimulus. The stimulus referred to may be internal, such as dominance/submission displays in dogs (Lorenz, 1966, Figure 3) or emotional signals in humans (Ekman, 1999). Or the stimulus may be external, such as predator alarm calls in vervet monkeys (Cheney & Seyfarth, 1990), the waggle dance in honeybees (Dyer, 2002), or pointing in humans (Kita, 2003). Alone among animals, humans employ an additional, symbolic mode of communication in the form of language: human communicative signals do not always refer to stimuli directly, but may refer to other combinations of symbols (Deacon, 1997).

The use of recursive symbols means that, in addition to drawing attention to a present state of affairs in the world, humans can use language to describe past states of affairs, predict future states, and fabricate counterfactual states. This important design feature of language is called *displacement* (Crystal, 1997, pp. 400–401, following Hockett, 1960).

Language has doubtless helped humans to produce great cultural achievements: information about past innovations, for example in the production of stone tools, can be stored in the form of language and fed into the development of new innovations. Less obvious, perhaps, are the adaptive problems that language has created. Displacement means that individuals have to be concerned not only with the reactions of direct witnesses to their actions, but also with the potential reactions of many other individuals to whom the witnesses may communicate information about their actions. In this paper, we present the hypothesis that another uniquely human competence – our highly developed theory of mind – evolved in response to this adaptive problem.

We argue that language enabled the development of systems of indirect reciprocity, in which absent third parties punished individuals for negative behaviour towards others. We briefly present two recent studies by the authors that offer supporting evidence for this hypothesis. Young children's language when they report other children's behaviour shows design features, such as honesty and negative bias, which ensure that absent third-party punishment is effective. And in adults, selfish behaviour in the dictator game is inhibited by the threat of gossip about such behaviour. We surmise that a sophisticated, propositional theory of mind is responsible for implementing this inhibition.

Gossip and indirect reciprocity

Recently, two important areas of scientific research have opened up in relation to the evolution of uniquely human behaviour. One area of research has focused on the role of language in the expansion of human group sizes; the other has examined the theoretical conditions necessary for the evolution of cooperation in large groups. Our main aim in this section is to show that these two independent areas of research are systematically related. As Nowak and Sigmund (2005, p. 1295) put it: "The co-evolution of human language and cooperation by indirect reciprocity is a fascinating and as yet unexplored topic."

Dunbar (1993; 2004a; 2004b) proposes that a propensity towards a specific kind of language – namely, gossip about other social agents – enabled the evolution of complex human societies. The size of other primate societies, according to Dunbar, is limited by the amount of time that they can devote to the practice of social grooming, which serves to reinforce social bonds and communicate emotional information between allies. In humans, a selective pressure towards larger group sizes led to the evolution of language as an adaptation which could perform similar functions to grooming, in maintaining group cohesion, but which was better suited to being deployed in a distributed manner across a large social group. This was partly because language enabled our ancestors to communicate social information to more than one individual at a time, and partly because it enabled our ancestors to communicate strategic information about the behaviour of absent third parties. In this chapter we focus on the latter activity, which we consider to be broadly synonymous with gossip.

Communication about absent third-parties allowed our ancestors to stay in touch with group members who were physically distant, enabling early human groups to spread out over far wider areas than were occupied by other primates, and enabling individuals to maintain wider networks of contacts. Such communication would also have helped to encourage cooperation within groups by spreading information about free riders and other norm violators (see also Dunbar, 1999). In Dunbar's words:

> [Language] allows us to exchange information about other people, so short-circuiting the laborious process of finding out how they behave. For monkeys and apes, all this has to be done by direct observation. I may never know that you are unreliable until I see you in action with an ally, and that opportunity is likely to occur only rarely. But a mutual acquaintance may be able to report on his or her experiences of you, and so warn me against you – especially if they share a common interest with me. Friends and relations will not want to see their allies being exploited by other individuals, since a cost borne by an ally is ultimately a cost borne by them too. (2004b, p. 79).

A similar argument is made by Boehm (1999; 2000) in relation to the evolution of egalitarianism in humans. Consider the case of a strong individual living in a group of social primates that bullies a weaker group member, for example by snatching food from its grasp. In most species the weaker individual has only two possible avenues of resistance: (i) it may withhold future cooperation from the aggressor; (ii) it may attempt to alert other individuals *in the vicinity* with a vested interest in its fitness, which may be powerful enough to deter or punish the aggressor.

Language dramatically expanded the scope of the second form of resistance. Early humans did not have to rely on a powerful third party directly witnessing an act of aggression: the victim themselves, or any other witness, could spread information about the aggression around an entire social network. Furthermore, third parties who heard about the aggression did not have to punish the aggressor single-handedly, but could use language to recruit others to support them, as well as to justify to others the action that they took. Boehm (1999; 2000) argues that the use of language to punish or deter aggression led to the evolution of human (hunter-gatherer) societies that were highly egalitarian compared to the rigid hierarchies of chimpanzee societies.

Another way of looking at the same process is in terms of *indirect reciprocity* (Alexander, 1987; Nowak & Sigmund, 2005). Systems of direct reciprocity (Trivers, 1971) rely on the return of a positive or negative payback for an action to the individual *A* who performed that action, by the recipient *B* of the action. The potential of direct reciprocity for enhancing cooperation is limited by the scope for future interactions between these two individuals *A* and *B*. In contrast, in systems of indirect reciprocity payback may be returned either by the recipient *B*, by those with whom *B* shares genetic material, or even by unrelated group members who learn about the incident (Alexander, 1987, p.85). Indirect reciprocity can therefore lead to cooperation even in cases where *A* and *B* are unlikely to meet again. While direct reciprocity may be exhibited by some (though perhaps not many) other animals, indirect reciprocity appears to be exhibited only by humans (Alexander, 1987; Richerson, Boyd, & Henrich, 2003, p. 379). It seems plausible that systems of indirect reciprocity are responsible for the unique ability of humans to cooperate in large groups of genetically unrelated individuals (but see Richerson, et al., 2003, for a different view). We contend that the novel possibilities offered by indirect observation through language greatly expanded the scope of indirect reciprocity for maintaining cooperation.

Surprisingly, theorists have often neglected the importance of language in potentiating systems of indirect reciprocity. For example, Alexander (1987, p. 77) writes that: "Indirect reciprocity develops because interactions are repeated, or flow among a society's members, and because information about subsequent interactions can be gleaned from observing the reciprocal interactions of others." Yet plenty of non-human animals are capable of distinguishing between individual group members and engaging in repeated interactions; so this does not explain why other species have not evolved cognitive mechanisms for deciding whether conspecifics would make suitable cooperative partners. The key word in the preceding quotation is "observing," referring to the observation of an interaction between *A* and *B* by a third party, *C*. Alexander fails to note that in most cases, only a small and not necessarily representative proportion of *A* and *B*'s interactions will be directly observed by *C* (Nowak & Sigmund, 2005). In humans, however, *C*'s observation is often indirect, based on the testimony of *A* or *B* (or indeed a direct observer, *D*). Linguistic testimony greatly increases the observability of interactions, and hence the scope of indirect reciprocity.

Indirect reciprocity is generally mediated by *A*'s *reputation* within a social network – where reputation may be defined as the information that is known about *A*'s behaviour and personality at second hand, independently of direct experience. Over the last decade there has been increasing interest in formal models and experimental paradigms in which reputation encourages the evolution of cooperation (e.g., Nowak & Sigmund, 1998; Semmann, Krambeck, & Milinski, 2004; Wedekind & Braithwaite, 2002; see Nowak & Sigmund, 2005, for a review). These models incorporate the important point that individuals in a society of cooperators are generally in competition with each other through their reputations (see especially Barclay & Willer, 2007). For instance, in Nowak and Sigmund's (1998) model an individual agent's "image score" – which is analogous to a simple reputation – controls that agent's access to mutually beneficial interactions. Researchers in this area sometimes note in passing that reputations are constructed and disseminated via language (but see Enquist & Leimar, 1993, and Nowak & Sigmund, 2005, for fuller treatments). Yet for modelling purposes, such research often treats image scores as though they were stamped on individuals' foreheads, ready for potential exchange partners to read instantly and unambiguously. Of course, in the real world, reputations are often manipulated via exaggeration, innuendo or downright deception, and getting to the truth of the actions and personalities behind them can be very difficult. Humans constantly use language to tell people about our own and others' actions; and we frequently use language to make value judgements about our own and others' actions (Foster, 2004).

Another line of research (e.g., Boyd, Gintis, Bowles, & Richerson, 2003; Fehr & Gächter, 2002; Fehr & Fischbacher, 2004; Fowler, 2005) has focused on altruistic punishment as a key mechanism in the evolution of human cooperation, and as an alternative to reputation-based mechanisms. However, the dichotomy between punishment and reputation may be illusory, because in human societies those who punish a transgression are not always (or even often) direct witnesses to that transgression. Most punishment of transgressors – including punishment administered by the legal systems of complex societies – relies on the spread of information by witnesses via language. In other words, it relies on second-hand information that lowers a suspect's reputation. It is important to remember that indirect reciprocity can be negative as well as positive, which means that "altruistic punishment" may be considered to be a form of negative indirect reciprocation, and one which may be expected to bring its own rewards through positive indirect reciprocity (i.e., it is also a form of costly signalling; see Smith & Bliege Bird, 2005).

As with models of indirect reciprocity, the importance of language has been neglected or even denied in models of altruistic punishment. Richerson and colleagues (2003) argue that indirect reciprocity, as mediated by language, could not have been solely responsible for the evolution of large-scale cooperation in humans, because it is not easy to distinguish true gossip from false:

> It is not obvious that language potentiates indirect reciprocity. Whereas superficially language may seem to promote the exchange of high-quality information required for indirect reciprocity to favour cooperation, this addition merely changes the question slightly to one of why individuals would cooperate in information sharing; language merely recreates the same public goods dilemma. Lies about hunting success, for example, are difficult to check and often ambiguous. Among the Gunwinggu (Australian foragers), members of one band often lied to members of other bands about their success to avoid having to share meat (Altman & Peterson, 1988). (Richerson et al., 2003, p. 257, n. 3)

Richerson and his colleagues refer to this line of reasoning as the "cheap talk" argument. However, their argument has a number of weaknesses. Firstly, verbal deception is only cheap if it goes undetected; if detected, it may incur penalties that are just as severe as those for stinginess or other forms of cheating. For example, in a recent experimental study, Tyler, Feldman and Reichert (2006) found that frequency of lying by confederates is inversely related to judgements of likeability that participants make about the confederates. Therefore lying may foreclose opportunities for cooperation by advertising the liar's unreliability as a cooperative partner. Secondly, while some lies may be hard to detect, others are not. If three or more people witnessed an event, and one person's story differs from everyone else's, that person risks being labelled a liar. Since it is not always easy to know exactly how many people witnessed something happen, or who they may tell and get to repeat the story, honesty is usually the safest policy. Thirdly, as the example of the Gunwinggu shows, deception may be adaptive between groups but not within groups. In general, the likelihood of deception increases where there are conflicts of interest, such as between mimics and their predators (Alexander 1987, pp.73 – 75; Dawkins & Krebs, 1978). In an extremely social species such as humans, conflicts of interest are likely to be far stronger between groups than within groups. And while one hunting band may know little about another band's exploits, it would be much more difficult for an individual hunter to conceal success or failure from other members of the same band. Some degree of honest communication is surely necessary to coordinate the activities of a group such as a hunting band in the first place.

Despite its flaws, Richerson et al.'s (2003) argument does help to focus attention on one of the main characteristics that gossip needs to possess in order to support systems of indirect reciprocity. If gossip is to promote cooperation, it must be reliable: either it should be mostly trustworthy, or it should be easy to distinguish between trustworthy and untrustworthy gossip. The second possibility – that humans have reliable cognitive mechanisms for distinguishing between truthful and deceptive gossip – was investigated by Hess and Hagen (2006). In a series of experiments using social psychological methods, they found that participants were more likely to believe gossip if it came from multiple sources, and less likely to believe negative gossip if the source was someone who was in competition with the subject of the gossip. Their results suggest that people may be equipped with strategies for marking gossip as either reliable or untrustworthy, just as they are equipped with mechanisms for spotting cheaters in the first place (Cosmides & Tooby, 1992). Taken together, these two sets of cognitive mechanisms would seem to encourage the growth of indirect reciprocity, by ensuring the extraction and reliable transmission of information about antisocial activities.

The evolution of language and the evolution of cooperation through indirect reciprocity are intimately linked, as theorists have recently begun to acknowledge (Nowak & Sigmund, 2005). Research into altruistic punishment reminds us that indirect reciprocity can be negative as well as positive. In fact, it seems that in certain contexts, the use of language to support indirect reciprocity may be biased towards achieving punishment rather than rewards. The next section presents evidence from one such context, namely social cognitive development in children.

Tattling: the Reporting of Negative Behaviour by Young Children

The development of gossip in young children has been little studied, but it is generally accepted that the behaviour of absent others is not nearly as common a conversational topic for children under the age of 5 as it is for adults (Engel & Li, 2004; Fine, 1977). Yet children do frequently report to adults the negative behaviour of other children who are close at hand – a kind of informal linguistic testimony known as *tattling*. Tattling is the act of informing on a peer's activities, with the intention that this will lead to his/her punishment. This linguistic behaviour can be seen as an attempt to secure absent third-party punishment, even though the audience is not completely absent, because the adult has often not witnessed the behaviour described, or at least was not paying attention to it.

Tattling is a frequent (and frequently annoying, for teachers) occurrence in the classroom – much more common than the reporting of positive behaviour by peers, which has to be actively encouraged by teachers if it is to take place with any regularity (Skinner, Cashwell, & Skinner, 2000). Young children's general bias towards reporting negative behaviour is consistent with Harris and Núnez's (1996; see also Núnez & Harris, 1998) finding that three-to-four-year-olds' deontic reasoning about actions that would breach a permission rule is superior to their indicative reasoning about actions that would breach a description rule. Harris and Nunez (1996, Experiment 4) found that their participants were better at identifying a picture where a child was "doing something naughty" than a picture where a child was "doing something different". As they point out, this task is closely related to Cosmides and Tooby's (1992) cheater detection task. Cosmides and Tooby famously demonstrated that performance on the Wason selection task is greatly enhanced by framing the task in terms of a possible breach of social contract (e.g. checking a drinker's ID before serving them alcohol). When performing this sort of reasoning about social norms, fast and frugal cognitive mechanisms are activated that are not usually deployed when reasoning about abstract symbols. It seems that heightened reasoning about norm violation is reflected in young children's everyday behaviour, as they are more likely to report breaches of the teachers' rules than unusual actions by other children that are not in breach of a rule.

One of the few studies to focus on tattling itself is by Ross and den Bak-Lammers, who carried out a longitudinal observational study of tattling in the home environment between Canadian sibling dyads aged 2 and 4 (den Bak & Ross, 1996) and 4 and 6 (Ross & den Bak-Lammers, 1998). The authors found that: (1) reports of a sibling's negative (that is, rule-violating) behaviour are much more frequent than reports of positive or neutral behaviour combined; and (2) tattling is overwhelmingly truthful. They also found that the proportion of negative reports decreased significantly as children grew older, suggesting important differences between tattling in children and the more inhibited forms of gossip practised by adults.

One of us (GI) recently conducted an observational study of tattling in the very different social setting of the pre-school. GI carried out research in two pre-schools in Belfast, with one group of 16 children and another group of 24 children. Both groups were aged 3–4 years, with mean ages of 4 years 1 month and 4 years 2 months respectively. A mixture of participant observation and behavioural ecological techniques (event sampling, social network analysis, and focal follows) were used to analyze the content and context of children's reports of peers' behaviour. These two types of methodologies complemented each other well: participant observation provided qualitative insights into the meaning and motivation of tattling, while the quantitative techniques allowed for a statistical analysis of the defining characteristics of tattling events.

GI spent a total of 66 hours engaged in event sampling in the two schools, recording 391 events in which children reported the behaviour of peers (Ingram, 2007; Ingram & Bering, 2008). Mean percentage results for the valence, and truth value of tattling in both pre-schools are summarised in Table 1, with sibling data from Ross and den Bak-Lammers (1998) for comparison:

In all six contexts, the clear majority of young children's behavioural reports concern some sort of proscribed behaviour. The most common types of transgressions reported are property disputes and instances of physical harm – although in classroom settings, where there are many more or less arbitrary rules, breaches of social convention were frequently reported too. The finding that the content of peer reports is more likely to be negative in the pre-school setting than in the home setting could be due (at least in part) to sampling bias, because complaints about negative behaviour would be easier to notice than comments about positive or neutral behaviour in the noisy environment of the pre-school.

Table1: Comparison of tattling between siblings and in the two preschools

	Proportion of negative reports		Proportion of true reports[1]	
	Mean	*SD*	*Mean*	*SD*
Preschool A ($n = 13$)	0.92	0.16	0.91	0.28
Preschool B ($n = 21$)	0.94	0.08	0.88	0.23
2-year-old on 4-year-old siblings	0.87	*N/A*	0.96	0.13
4-year-old on 2-year-old siblings	0.80	*N/A*	0.91	0.19
4-year-old on 6-year-old siblings	0.70	*N/A*	0.96	0.08
6-year-old on 4-year-old siblings	0.56	*N/A*	0.94	0.13

The truthfulness of children's behavioural reports is even more striking. It is noteworthy that most of the tiny minority of false reports recorded by GI seemed to be accidentally rather than deliberately false. It was impossible to know for sure whether a child was lying, but false tattling seemed to be more often the result of one of three different types of mistake: a misidentification of the child who was reported on; a bias towards attributing hostile intentions (e.g. "He pushed me" when another child accidentally bumped into the tattler – see Orobrio de Castro, Veerman, et al., 2002); or a misuse of language (e.g. saying "He punched me" but meaning "He pushed me"). There were fewer than 5 clear-cut instances of deception in all 391 recorded verbal reports.

The experimental literature on deception by children is equivocal (Reddy, 2007), with some studies suggesting that children as young as 2 or 3 are capable of intentional deception, whereas others indicate that it is not consistently achieved until the age of 4 or 5 (both sets of studies reviewed in Carlson, Moses, & Hix, 1998). In an attempt to resolve this contradiction, Carlson, Moses and Hix (1998) compared 3-year-old children's abilities to deceive by pointing with deception using a novel pictorial cue. They found that the children had much more difficulty with deceptive pointing, and argue that this is because children were unable to inhibit the accurate use

[1] Truth values are given for negative reports only, since Ross and den Bak-Lammers (1998) do not report the percentages of non-negative reports that were true.

of pointing as an action that has been reinforced through years of interaction with caregivers (and may even be instinctive).

A similar argument might be made for indicative uses of language to describe social events. Young children may not be able to inhibit their impulse to provide an accurate description of events, much as they might like to get a rival into trouble. It is interesting that the children in our study frequently *did* attempt deception in other contexts, for example by making vigorous (but plainly false) denials when they themselves were accused of a transgression (see also Lewis, Stanger, & Sullivan, 1989, who observed high rates of deception when 3-year-old children were asked whether they had peeked in a box that they were not supposed to look in). Perhaps the urge for self-preservation is enough to inhibit the impulse to tell the truth in such cases. Or it could be that preschool children do not tend to fabricate accounts of others' transgressions because their primary motivation in making such reports is not to get others in trouble without good reason, but to draw an authority figure's attention to authentic rule violations. A third possibility is that the executive demands of deception by simple denial of an accusation are much weaker than those involved in fabricating a story from scratch. Controlled experimentation would be required to test between these competing explanations.

Regardless of the proximate causes for the lack of deceptive tattling by children, it is tempting to postulate an ultimate cause. In small-scale societies, young children mainly communicate with members of their kin group, and this presumably reflects conditions in the ancestral evolutionary environment. The early preponderance of communication within the kin group means that there are few conflicts of interest between tattlers and their audience. As Alexander (1987, pp.73–75) notes, a high frequency of deception is expected in any communication where there are significant conflicts of interest between the communicative partners (such as mimics deceiving predators; Dawkins & Krebs, 1978). Congruence of interests is easy to understand in the case of siblings tattling to parents, who are naturally interested in making sure that neither sibling is unfairly exploited by the other, because they have a huge amount of resources invested in both. Since young children are not prepared by evolution for life in the institutional environment of the preschool, it is reasonable to suppose that they treat staff at the preschool as surrogate parents, or at least as senior members of their kin group. Moreover, children use language not primarily to communicate useful information to arbitrary others within the social group, but far more often to communicate their own needs and desires to adults who look after them – and this is reflected in the egocentric nature of tattling, in which around 75% of reports were made about a transgression of which the tattler was the victim. It would make little sense for tattlers to lie in these circumstances: doing so could lead to a "cry wolf" type of scenario, in which the tattler is ignored just when they really need help. For language to be adaptive for a young child, it has to be an honest signal of need. This helps to ensure that reports of peers' behaviour are reliable and are not undermined by the problems with "cheap talk" discussed by Richerson et al. (2003).

In summary, young children frequently report the behaviour of other children, and show a bias towards reporting negative, rule-violating behaviour. Tattling is adaptive at the individual level because it allows the absent-third party punishment of aggressors and transgressors who would not be susceptible to direct punishment by weak individuals such as children. Two important properties of children's tattling – truthfulness and negative bias – are robust across varying social contexts, and would favour the development of systems of indirect reciprocity incorporating punishment of offenders by a group (or by those with authority within a group).

Experimental Evidence of Reputation Management under Threat of Gossip

The children in the observational study were too young to show much behavioural inhibition in response to the likelihood of tattling (although they did occasionally desist from an activity if directly threatened with tattling). In contrast, we hypothesized that an adult individual's selfish impulses are often inhibited by the risk that negative social information about his or her actions will be transmitted to absent third parties. As a preliminary test of this hypothesis, we employed an economic experiment called the *dictator game* (Camerer, 2003, ch. 2; Kahneman, Knetsch, & Thaler, 1986), in which each participant is asked to divide a monetary endowment between themselves and one other participant. Previous findings with the dictator game suggest that even in anonymous settings, equitable allocations are often the mode (Camerer, 2003, Table 2.4). However, selfish allocations tend to become more frequent as the "social distance" (or social isolation) between players increases, in highly anonymized experimental paradigms (Hoffman, McCabe, & Smith, 1996). Conversely, Bohnet and Frey (1999) found that dictators who could visually identify their recipients were less selfish with their allocations than dictators who could not identify their recipients, and selfishness decreased further when recipients were required to disclose their names. Even showing generic eyespots on a computer while dictators are making their decisions (Haley & Fessler, 2005) – or in a naturalistic setting, displaying an image of a pair of eyes rather than an image of flowers above a canteen "honesty box" (Bateson, Nettle, & Roberts, 2006) – can significantly increase allocations. The possibility of identifying the donor also has a strong effect on generosity. Burnham (2003) found that dictators' allocations were at similarly high levels – significantly higher than in an anonymized control – (i) when they were shown a photo of the recipient, and (ii) when they were told that their own photo would be shown to the recipient.

Given the importance of anonymity and identification in this research, we decided to investigate the role of language – that is, gossip – in facilitating the identification of individual actors and the spread of information about their actions. One of us (JP) used a modified dictator game to investigate the effect of two variables on altruistic behaviour (see Piazza & Bering, 2008). The first variable (*reputational concern*) indicated whether or not a third party could identify the participant. This variable was manipulated using a confederate. Before being instructed about the dictator game, participants either did or did not meet and disclose personally identifying information (e.g., their name, place of study, place of residence) to a female confederate in private. The second variable (*threat of gossip*) indicated whether or not participants were told that the person directly affected by their economic decision (the receiver) would be discussing their economic decision with a third party. For some participants, this third party was the confederate with whom they had disclosed personally identifying information. For others, this third party could not identify them. The experiment thus had a 2 x 2 (*reputational concern* x *threat of gossip*) factorial, between-subjects design with *allocation scores* (i.e., number of tokens allocated to the receiver) as the main dependent measure.

The game itself entailed having participants unilaterally distribute an endowment of ten tokens between themselves and an anonymous second party (the Receiver). Each individual token represented a chance to have one's name entered into a lottery drawing to win £100. Participants were instructed that they would never meet the Receiver and that they were free to distribute the endowment however they wished – the only constraint was that the distribution total had to equal 10. Measures were also taken to reduce the motivation for building a reputation with the experimenter, in that participants' economic decisions were written down and enclosed in opaque envelopes to conceal them from the experimenter.

Table 2: Mean allocation scores (and standard deviations) for each condition ($n = 18$) of dictator game

	No threat of gossip		Threat of gossip	
	Mean	*SD*	*Mean*	*SD*
No reputational concern	4.22	1.73	4.67	1.09
Reputational concern	4.17	2.23	5.44	1.15

The results were that participants who received the threat of gossip variable were significantly more generous with their allocations to the Receiver (see Table 2, and Piazza & Bering, 2008). However, simple-effects analysis revealed that this was only the case for participants who had previously disclosed personally identifying information to the recipient of the gossip (the third party). When the recipient of the gossip was unable to identify the participant, participants in this condition were no more generous with their allocations than control participants (i.e., participants without the threat of gossip). This suggests that the threat of gossip can promote generous economic decisions, but only when an individual's reputation is placed at risk by the threat.

These findings are consistent with the hypothesis that selective pressure to present the self strategically (i.e., to promote a rewarding reputation) increased during human evolution with the emergence of language. This was accomplished by inhibiting selfish behaviour in order to prevent negative social information from being disseminated to absent third parties (i.e., potential mates and coalition partners). In the experiment, the threat of gossip appeared to inhibit selfish behaviour and encourage prosocial behaviour, but only when the threat was backed by real concerns about one's reputation. When the recipient of the gossip could identify the participant, the gossip prime led to a significant increase in allocations.

Perspective-taking and the Management of Strategic Social Information

While the two empirical studies described in this paper used very different methodologies, their results are highly complementary. Our observational study shows that young children are biased towards the communication of truthful negative information about peers' activities, and that they use this to compete for adult attention. Tattling frequently results in punishment of those whose antisocial activities are reported. However, at the age of 3–4 years, the risk of absent third-party punishment does not seem to inhibit antisocial behaviour in general. This finding contrasts with Piazza and Bering's (forthcoming) results for the dictator game. In adults, the mere possibility of someone talking to a third party about a protagonist's activities, combined with the ability of the third party to identify the protagonist, is enough to increase prosocial behaviour in the dictator game. (Note that the sort of behaviour that the threat of gossip inhibits in the dictator game has a natural counterpart in the preschool environment, where a refusal to share is a frequent subject of tattling.) The question naturally arises: what are the developmental mechanisms by which the threat of absent third-party punishment encourages the inhibition of antisocial behaviour, as children grow older?

Given the conflict between the risks of impairing one's reputation by acting selfishly and the costs associated with behavioural inhibition, early humans must have been under intense selective pressure to find a way of balancing these two forces. The solution, we argue, was to develop improved theory of mind (ToM) skills. Great apes and some other animals, such as social carnivores, may have a rudimentary ToM in the sense that they can predict behaviour by anticipating the intentions of other social agents (Premack & Woodruff, 1978; Tomasello, Call, & Hare, 2003). ToM in humans is vastly more sophisticated, however. It is possible that the divergence of improved human ToM skills is seen at a very simple level, in terms of an awareness that seeing leads to knowing (O'Neill, 1996), since evidence that chimpanzees possess this awareness is mixed (compare Hare, Call, Agnetta, & Tomasello, 2000, with Povinelli & Eddy, 1996; see also Povinelli & Bering, 2002). As discussed in the previous section, humans tend to behave more altruistically when observers are present (Bateson, Nettle, & Roberts, 2006; Burnham, 2003; Haley & Fessler, 2005). Linguistic testimony increases the number of potential observers of an individual's behaviour, thereby increasing the risk of punishment. There is no reason for a chimpanzee to inhibit its behaviour in response to the observation of a subordinate individual, and therefore little selective pressure for chimpanzees to respond to general cues of observation (rather than, for example, the presence of a dominant individual). Humans, on the other hand, must be sensitive to the presence of *any* possible observers of their behaviour, since it is quite possible for these observers to tell *anyone* else. Awareness of the link between seeing and knowing may have been the first stage in the development of improved perspective-taking skills in humans.

Whether or not humans show increased sensitivity to the link between seeing and knowing, it seems that a fully propositional theory of mind – which may logically require language – is unique to humans (Heyes, 1998). Propositional ToM is also known as the belief/desire model (Wellman, 1990), and includes a capacity to model false beliefs (Flavell, 1999; Wellman, Cross, & Watson, 2001; Wimmer & Perner, 1983). The development of reasoning about beliefs and desires created the potential for us to model our reputations in the eyes of others, because in order to assess whether other people will see us in a good or bad light, we have to represent both what others believe to be the case about our actions, and what they desire our actions to be. In fact, because a representation of the actor's intentions is important for evaluating action, a fully developed model of reputation may require second-order theory of mind (Liddle & Nettle, 2006; Perner & Wimmer, 1985; Perner, 1988) – that is, an ability to think about what others believe our own intentions to be.

Recent research has highlighted the links between theory of mind, inhibition, and executive control (e.g. Carlson & Moses, 2001; Perner & Lang, 1999). All of these cognitive functions are associated with the prefrontal cortex (Gallagher & Frith, 2003; Stone, Baron-Cohen, & Knight, 1998), which is one of the brain areas that has shown the greatest expansion during human evolution (Deacon, 1997; Povinelli & Preuss, 1995; Schoenemann, 2006). It seems likely that propositional ToM is a vital part of the conscious mind, where it acts as a sort of "cognitive override" exerting inhibitory control over an individual's actions (Alexander, 1987, pp. 107–110; Bering & Shackelford, 2004). When deciding whether to behave selfishly or unselfishly, humans use ToM skills to assess the likelihood of discovery, and also to estimate other people's approval or disapproval of their behaviour. They are thereby modelling the likely effects of their actions on their reputations – a process that may occur either consciously or unconsciously.

It has previously been suggested that ToM played a key role in the evolution of human cooperation (e.g., Bjorklund, Cormier, & Rosenberg, 2005; Dunbar, 2003; Povinelli & Preuss, 1995). The novel part of our argument is the drawing of an explicit connection between the evolution of language and the evolution of ToM. Language was not just a cognitive pre-requisite for propositional ToM to appear (as might be inferred from the fact that language is developed to a high degree in children well before they pass the false belief task[2]). Language was an *adaptive problem* for the

---

[2] See Astington, 2006, for a recent account of the complex interrelationship between language and theory of mind in child development.

individual because it frequently led to absent third-party punishment. This adaptive problem was solved by new adaptations, in the form of highly developed ToM and perspective-taking skills.

The position set out in this chapter is superficially at odds with the views of Nowak and Sigmund, who write that:

> indirect reciprocity seems to require a 'theory of mind'. Whereas altruism directed towards kin works because similar genomes reside in different organisms, reciprocal altruism recognizes that similar minds emerge from different brains. It is easy to conceive that an organism experiences as 'good' or 'bad' anything that affects the organism's own reproductive fitness in a positive or negative sense. The step from there to judging, as 'good' or 'bad', actions between third parties, is not obvious. The same terms 'good' and 'bad' that are applied to pleasure and pain are also used for moral judgements: this linguistic quirk reveals an astonishing degree of empathy, and reflects highly developed faculties for cognition and abstraction. (2005, p. 1291)

However, as discussed above, theory of mind is not limited to a propositional modelling of false beliefs, but should be considered as a complex portfolio of skills. Empathizing is one such skill, and probably arises earlier in development than does the representation of false beliefs (Flavell, 1999, citing Baldwin & Moses, 1994; Hoffman, 2000; Nichols, 2004, ch. 2). Some form of empathy may even be present in non-human primates (Preston & de Waal, 2002), dolphins (Kuczaj, Tranel, Trone, & Hill, 2001), and elephants (Douglas-Hamilton, Bhalla, Wittemyer, & Vollrath, 2006). In any case, early forms of indirect reciprocity may have been supported by the tattling of immature individuals to their parents, and it is doubtful that a fully empathic response is needed in order for a mother to defend her young. It is not difficult to imagine that early humans, possessing sophisticated enough brains to be capable of some sort of proto-language, would also be capable of enough empathy (or something like it) to judge actions between third parties – especially where kin or close associates are involved – as 'good' or 'bad.'

Finally, an important prediction of our evolutionary hypothesis is that ToM eventually feeds back into the linguistic behaviour by which strategic social information is spread. This is because of recursive, higher-order ToM skills: the ability to say to oneself, for instance, "she thinks that I think that..." (Perner & Wimmer, 1985). Using these skills strategically, it becomes clear to most people (starting in middle childhood) that open, unrestricted boasting and tattling are hardly likely to win many friends: other people will tend to see through such testimony for the reputational competition that it is. The first signs of these higher-order feedback effects are perhaps seen as early as six years of age, for although the six-year-olds in Ross and den Bak-Lammers's (1998) study tattled about as often as the four-year-olds, they engaged in positive or neutral talk about their siblings far more frequently. This suggests that by the age of six, children have already learned the value of exhibiting a positive attitude to their siblings in front of their parents. Second-order effects on the transmission of strategic information through language are likely to gain in strength through late childhood and into adolescence, as children identify more with their peer group and come to see adult authority figures as out-group members. This accounts for the inverse relationship between tattling frequency and social dominance found by Friman et al. (2004) among teenage boys in a residential care program, where the least popular boys were perceived to do the most tattling. Further research, both experimental and observational, is needed to elaborate the precise developmental mechanisms by which ToM helps to support reputation-oriented behaviour.

Conclusion

In this chapter, we have presented the hypothesis that one of the reasons why propositional theory of mind evolved in humans was to inhibit behaviour in response to the selective pressure of absent third-party punishment. Participants in our dictator game experiment (Piazza & Bering, 2008) used ToM to model (i) the carrying of two different pieces of information (the identity of the giver, and the amount that he or she gave) by two individuals, and (ii) the combined effect of this information in the mind of the recipient as a result of communication between recipient and confederate. The three-to-four-year-old children in our observational study (Ingram & Bering, 2008) either did not possess such highly developed mind-reading abilities, or they were not yet accustomed to using these abilities strategically, to inhibit the selfish behaviour that would once have been adaptive in a world without language, in favour of the cooperative behaviour that would now be adaptive. In support of the latter interpretation, it is notable that when confronted directly with the possibility of absent third-party punishment (when an aggrieved peer said something like "I'm going to tell the teacher on you") children sometimes reversed their selfish behaviour, e.g. by handing over a toy that they had previously refused to share. Despite this, they showed little anticipation of the likelihood of tattling unless directly threatened with it. This suggests that they are missing a crucial competence which would allow them to model the spread of information within their social group – a fully developed propositional theory of mind. However, the linguistic reports of preschoolers possess important properties – truthfulness and negative bias – that would support the development of systems of indirect reciprocity and altruistic punishment. Without wanting to argue that ontogeny recapitulates phylogeny, it is tempting to speculate that early humans might at some stage during our ancestral past have made use of similar kinds of language in their reports of conspecifics' behaviour.

Recent developmental research has emphasized that language and theory of mind are inextricably linked (e.g., Astington, 2006, and chapters in Astington & Baird, 2005). We have presented one out of many possible evolutionary hypotheses which helps to explain why these two unique human competences are linked. The hypothesis that ToM evolved as an adaptive response to the threat of absent third-party punishment, inflicted on individual humans because of the spread of strategic social information via language, is no "just so story." Our hypothesis is falsifiable, and we hope that future research will attempt to disprove it. For example, neuropsychological work needs to be done to flesh out the links between the neural pathways supporting ToM and executive control and those supporting language and social cognition. If neural pathways supporting ToM are shown to be more evolutionarily ancient than those supporting language, this would pose a major problem for our account. A similar problem would arise if animal species lacking language are shown to possess higher-order ToM skills or rudimentary systems of indirect reciprocity: there may be animals, other than humans, for which observing an individual behave antisocially towards a conspecific (which is not closely related to the observer) sometimes causes the observer to withhold cooperation from the antisocial individual. We hope that our hypothesis will inspire future research in these and other areas.

References

Alexander, R. D. (1987). *The biology of moral systems*. New York: Aldine de Gruyter.

Altman, J., & Peterson, N. (1988). Rights to game and rights to cash among contemporary Australian hunter-gatherers. In T. Ingold, D. Riches, & J. Woodburn (Eds.), *Hunters and gatherers*. Vol. 2. *Property, power and ideology* (pp. 75–94). Oxford, England: Berg.

Astington, J. W. (2006). The developmental interdependence of theory of mind and language. In N. J. Enfield & S. C. Levinson (Eds.), *Roots of human sociality* (pp. 179-206). Oxford, England: Berg.

Astington, J. W., & Baird, J. A. (Eds.). (2005). *Why language matters for theory of mind*. Oxford, England: Oxford University Press.

Baldwin, D., & Moses, L. (1994). Early understanding of referential intent and attentional focus: Evidence from language and emotion. In C. Lewis & P. Mitchell (Eds.), *Children's early understanding of mind: Origins and development* (pp. 133–156). Hove, England: Lawrence Erlbaum.

Barclay, P., & Willer, R. (2007). Partner choice creates competitive altruism in humans. *Proceedings of the Royal Society B, 274*, 749-753.

Bateson, M., Nettle, D., & Roberts, G. (2006). Cues of being watched enhance cooperation in a real-world setting. *Biology Letters, 2*, 412-414.

Bering, J. M., & Shackelford, T. K. (2004). The causal role of consciousness: A conceptual addendum to human evolutionary psychology. *Review of General Psychology, 8*, 227-248.

Bjorklund, D. F., Cormier, C. A., & Rosenberg, J. S. (2005). The evolution of theory of mind: Big brains, social complexity, and inhibition. In W. Schneider, R. Schumann-Hengsteler, & B. Sodian (Eds.), *Young children's cognitive development: Interrelationships among executive functioning, working memory, verbal ability, and theory of mind* (pp. 147-174). Mahwah, NJ: Lawrence Erlbaum.

Boehm, C. (1999). *Hierarchy in the forest: The evolution of egalitarian behavior.* Cambridge, MA: Harvard University Press.

—. (2000). Conflict and the evolution of social control. In L. D. Katz (Ed.), *Evolutionary origins of morality: Cross-disciplinary perspectives* (pp. 79-148). Thorverton, England: Imprint Academic.

Bohnet, I., & Frey, B. S. (1999). The sound of silence in prisoner's dilemma and dictator games. *Journal of Economic Behavior and Organization, 38*, 43-57.

Boyd, R., Gintis, H., Bowles, S., & Richerson, P. J. (2003). The evolution of altruistic punishment. *Proceedings of the National Academy of Sciences (USA), 100*, 3531-3535.

Burnham, T. C. (2003). Engineering altruism: A theoretical and experimental investigation of anonymity and gift-giving. *Journal of Economic Behavior and Organization, 50*, 133-144.

Camerer, C. F. (2003). *Behavioral game theory: Experiments in strategic interaction*. Princeton, NJ: Princeton University Press.

Carlson, S. M., & Moses, L. J. (2001). Individual differences in inhibitory control and children's theory of mind. *Child Development, 72*, 1032-1053.

Carlson, S. M., Moses, L. J., & Hix, H. R. (1998). The role of inhibitory processes in young children's difficulties with deception and false belief. *Child Development, 69*, 672-691.

Cheney, D. L., & Seyfarth, R. M. (1990). *How monkeys see the world: Inside the mind of another species.* Chicago: University of Chicago Press.

Cosmides, L., & Tooby, J. (1992). Cognitive adaptations for social exchange. In J. H. Barkow, L. Cosmides, & J. Tooby (Eds.), *The adapted mind: Evolutionary psychology and the generation of culture* (pp. 163-228). Oxford, England: Oxford University Press.

Crystal, D. (1997). *The Cambridge encyclopedia of language* (2nd ed.). Cambridge, England: Cambridge University Press.

Dawkins, R., & Krebs, J. R. (1978). Animal signals: Information or manipulation? In J. R. Krebs & N. B. Davies (Eds.), *Behavioural ecology: An evolutionary approach* (pp. 282-309). Oxford: Blackwell Scientific.

Deacon, T. (1997). The symbolic species: The co-evolution of language and the human brain. London: Penguin.

den Bak, I. M., & Ross, H. S. (1996). 'I'm telling!' The content, context, and consequences of children's tattling on their siblings. *Social Development, 5*, 292-309.

Douglas-Hamilton, I., Bhalla, S., Wittemyer, G., & Vollrath, F. (2006). Behavioural reactions of elephants towards a dying and deceased matriarch. *Applied Animal Behaviour Science, 100*, 87-102.

Dunbar, R. I. M. (1993). Co-evolution of neocortex size, group size and language in humans. *Behavioral and Brain Sciences, 16*, 681-735.

—. (1999). Culture, honesty and the freerider problem. In Dunbar, R. I. M., C. Knight, & C. Power (Eds.), *The evolution of culture: An interdisciplinary view* (pp. 194-213). Edinburgh, Scotland: Edinburgh University Press.

—. (2003). The social brain: Mind, language, and society in evolutionary perspective. *Annual Review of Anthropology, 32*, 163-181.

—. (2004a). Gossip in evolutionary perspective. *Review of General Psychology, 8*, 100-110.

—. (2004b) [1996]. *Grooming, gossip and the evolution of language* (2nd ed.). London: Faber and Faber.

Dyer, F. C. (2002). The biology of the dance language. *Annual Review of Entomology, 47*, 917-949.

Ekman, P. (1999). Basic emotions. In T. Dalgleish & M. J. Power (Eds.), *Handbook of cognition and emotion* (pp. 301-320). Chichester, England: John Wiley & Sons.

Engel, S., & Li, A. (2004). Narratives, gossip, and shared experience: How and what young children know about the lives of others. In J. M. Lucariello, J. A. Hudson, R. Fivush, & P. J. Bauer (Eds.), *The development of the mediated mind* (pp. 151-174). Mahwah, NJ: Lawrence Erlbaum.

Enquist, M., & Leimar, O. (1993). The evolution of cooperation in mobile organisms. *Animal Behaviour, 45*, 747-757.

Fehr, E., & Fischbacher, U. (2004). Third-party punishment and social norms. *Evolution and Human Behavior, 25*, 63-87.

Fehr, E., & Gächter, S. (2002). Altruistic punishment in humans. *Nature, 415*, 137-140.

Fine, G. A. (1977). Social components of children's gossip. *Journal of Communication, 27*, 181-185.

Flavell, J. H. (1999). Cognitive development: Children's knowledge about the mind. *Annual Review of Psychology, 50*, 21-45.

Foster, E. K. (2004). Research on gossip: Taxonomy, methods, and future directions. *Review of General Psychology, 8*, 78-99.

Fowler, J. H. (2005). Altruistic punishment and the origin of cooperation. *Proceedings of the National Academy of Sciences, 102*, 7047-7049.

Friman, P. C., Woods, D. W., Freeman, K. A., Gilman, R., Short, M., McGrath, A. M., et al. (2004). Relationships between tattling, likeability, and social classification: A preliminary investigation of adolescents in residential care. *Behavior Modification, 28*, 331-348.

Gallagher, H. L., & Frith, C. D. (2003). Functional imaging of 'theory of mind'. *Trends in Cognitive Sciences, 7*, 77-83.

Haley, K. J., & Fessler, D. M. T. (2005). Nobody's watching? Subtle cues affect generosity in an anonymous economic game. *Evolution and Human Behavior, 26*, 245-256.

Hare, B., Call, J., Agnetta, B., & Tomasello, M. (2000). Chimpanzees know what conspecifics do and do not see. *Animal Behaviour, 59*, 771-785.

Harris, P. L., & Núñez, M. (1996). Understanding of permission rules by preschool children. *Child Development, 67*, 1572-1591.

Hauser, M. D. (1997). *The evolution of communication*. Cambridge, MA: MIT Press.

Heyes, C. M. (1998). Theory of mind in nonhuman primates. *Behavioral and Brain Sciences, 21*, 101-114.

Hockett, C. F. (1960). The origin of speech. *Scientific American, 203*, 88-96.

Hoffman, E., McCabe, K., & Smith, V. L. (1996). Social distance and other-regarding behavior in dictator games. *American Economic Review, 86*, 653-660.

Hoffman, M. L. (2000). Empathy and moral development: Implications for caring and justice. Cambridge, England: Cambridge University Press.

Ingram, G. P. D. (2007). *Tattling among pre-school children: The development of strategic social cognition.* Paper delivered at the 37th annual meeting of the Jean Piaget Society, Amsterdam, June 2007.

Ingram, G. P. D., & Bering, J. M. (2008). *Children's tattling: A developmental precursor of gossip?* Manuscript under review.

Kahneman, D., Knetsch, J. L., & Thaler, R. H. (1986). Fairness and the assumptions of economics. *The Journal of Business, 59*, 285-300.

Kita, S. (Ed.). (2003). *Pointing: Where language, culture, and cognition meet.* Mahwah, NJ: Lawrence Erlbaum.

Kuczaj, S., Tranel, K., Trone, M., & Hill, H. (2001). Are animals capable of deception or empathy? implications for animal consciousness and animal welfare. *Animal Welfare, 10*, 161-173.

Liddle, B., & Nettle, D. (2006). Higher-order theory of mind and social competence in school-age children. *Journal of Cultural and Evolutionary Psychology, 4*, 231-244.

Lorenz, K. (1966). *On aggression* (M. Latzke trans.). London: Methuen.

Nichols, S. (2004). *Sentimental rules: On the natural foundations of moral judgment.* Oxford, England: Oxford University Press.

Nowak, M. A., & Sigmund, K. (1998). Evolution of indirect reciprocity by image scoring. *Nature, 393*, 573-577.

Nowak, M. A., & Sigmund, K. (2005). Evolution of indirect reciprocity. *Nature, 437*, 1291-1298.

Núñez, M., & Harris, P. L. (1998). Psychological and deontic concepts: Separate domains or intimate connection? *Mind and Language, 13*, 153-170.

O'Neill, D. K. (1996). Two-year-old children's sensitivity to a parent's knowledge state when making requests. *Child Development, 67*, 659-677.

Perner, J. (1988). Higher-order beliefs and intentions in children's understanding of social interaction. In J. W. Astington, P. L. Harris & D. R. Olson (Eds.), *Developing theories of mind* (pp. 271-294). Cambridge, England: Cambridge University Press.

Perner, J., & Lang, B. (1999). Development of theory of mind and executive control. *Trends in Cognitive Sciences, 3*, 337-344.

Perner, J., & Wimmer, H. (1985). "John *thinks* that Mary *thinks* that...": Attribution of second-order beliefs by five- to ten-year-old children. *Journal of Experimental Child Psychology, 39*, 437-471.

Piazza, J. R., & Bering, J. M. (2008). Concerns about reputation via gossip promote generous allocations in an economic game *Evolution and Human Behavior, 29*, 172-178.

Povinelli, D. J., & Bering, J. M. (2002). The mentality of apes revisited. *Current Directions in Psychological Science, 11*, 115-119.

Povinelli, D. J., & Eddy, T. J. (1996). What young chimpanzees know about seeing. *Monographs of the Society for Research in Child Development, 61*(3).

Povinelli, D. J., & Preuss, T. M. (1995). Theory of mind: Evolutionary history of a cognitive specialization. *Trends in Neurosciences, 18*, 418-424.

Premack, D., & Woodruff, G. (1978). Does the chimpanzee have a theory of mind? *Behavioral and Brain Sciences, 1*, 515-526.

Preston, S. D., & de Waal, Frans B. M. (2002). Empathy: Its ultimate and proximate bases. *Behavioral and Brain Sciences, 25*, 1-72.

Reddy, V. (2007). Getting back to the rough ground: Deception and 'social living'. *Philosophical Transactions of the Royal Society of London, Series B: Biological Sciences, 362*, 621-637.

Richerson, P. J., Boyd, R., & Henrich, J. (2003). The cultural evolution of human cooperation. In P. Hammerstein (Ed.), *The genetic and cultural evolution of cooperation* (pp. 357-388). Cambridge, MA: MIT Press.

Ross, H. S., & den Bak-Lammers, I. M. (1998). Consistency and change in children's tattling on their siblings: Children's perspectives on the moral rules and procedures of family life. *Social Development, 7*, 275-300.

Schoenemann, P. T. (2006). Evolution of the size and functional areas of the human brain. *Annual Review of Anthropology, 35*, 379-406.

Semmann, D., Krambeck, H., & Milinski, M. (2004). Strategic investment in reputation. *Behavioral Ecology and Sociobiology, 56*, 248-252.

Skinner, C. H., Cashwell, T. H., & Skinner, A. L. (2000). Increasing tootling: The effects of a peer-monitored group contingency program on students' reports of peers' prosocial behaviors. *Psychology in the Schools, 37*, 263-270.

Smith, E. A., & Bliege Bird, R. (2005). Costly signaling and cooperative behaviour. In H. Gintis, S. Bowles, R. Boyd, & E. Fehr (Eds.), *Moral sentiments and material interests: The foundations of cooperation in economic life* (pp. 115-148). Cambridge, MA: MIT Press.

Tomasello, M., Call, J., & Hare, B. (2003). Chimpanzees understand psychological states – the question is which ones and to what extent. *Trends in Cognitive Sciences, 7*, 153-156.

Trivers, R. L. (1971). The evolution of reciprocal altruism. *Quarterly Review of Biology, 46*, 35-57.

Tyler, J. M., Feldman, R. S., & Reichert, A. (2006). The price of deceptive behavior: Disliking and lying to people who lie to us. *Journal of Experimental Social Psychology, 42*, 69-77.

Wedekind, K., & Braithwaite, V. A. (2002). The long-term benefits of human generosity in indirect reciprocity. *Current Biology, 12*, 1012-1015.

Wellman, H. M. (1990). *The child's theory of mind.* Cambridge, MA: MIT Press.

Wellman, H. M., Cross, D., & Watson, J. (2001). Meta-analysis of theory-of-mind development: The truth about false belief. *Child Development, 72*, 655-684.

Wimmer, H., & Perner, J. (1983). Beliefs about beliefs: Representation and constraining function of wrong beliefs in young children's understanding of deception. *Cognition, 13*, 103-128.

CHAPTER ELEVEN

THE ROLE OF COGNITIVE AND AFFECTIVE EMPATHY IN ALTRUISTIC MOTIVATION: AN EVOLUTIONARY APPROACH

JELLE DE SCHRIJVER

This research was funded by the Research Foundation Flanders

Abstract: The moral sense is a quintessential part of human nature. As such, empathy is one of the most important instruments in the toolbox of the human mind. It is the trigger of altruistic motivation. However, from an evolutionary point of view, altruistic behaviour is costly. Kin altruism is worth the effort when it adds to the individual's indirect fitness. In contrast, the benefit of altruistic behaviour towards strangers depends on the reciprocation of the other person. Helping a non-reciprocator lowers fitness. Therefore control mechanisms should be in place to fine-tune the amount of empathy displayed towards strangers, whereas this is less necessary for kin. I propose that two types of altruistic motivation occur. The first is primarily based on (maternal) attachment and affective empathy and is observed in many social mammals. The second depends on cognitive empathy and is only observed in primates and humans.

Altruism and Empathy

Darwin observed that 'of all the differences between Man and the lower animals, the moral sense or conscience is by far the most important' (Darwin, 1871). The moral sense is commonly regarded as a quintessential part of human nature. Although morality does not simply boil down to altruistic behaviour, few doubt that pro-social behaviour lies at its core. Pro-social behaviour is a form of helping behaviour with no immediate advantage to the individual. Yet, from an evolutionary perspective, the coexistence of altruism and the process of evolution - that seemed to promote selfish behaviour - seem hard to reconcile.

Ever since the problem was acknowledged in the middle of the 20th century, several theories have been proposed to resolve this Darwinian paradox, ranging from kin selection to reciprocal altruism (Hamilton, 1964; Nowak, 2006; Trivers, 1971). Kin selection is the process whereby natural selection can favour cooperation if the donor and the recipient of an altruistic act are genetic relatives. The closer the relative, the more it pays to help him or her. Haldane's remark 'I will jump into the river to save two brothers or eight cousins' illustrates this idea (Nowak, 2006). Trivers suggested that cooperation could be favoured between non-relatives, in reciprocal interactions. The central idea is that individuals can take turns in helping each other: "you scratch my back and I scratch yours". Individuals can choose to help those who have helped them in the past. Trivers termed this 'reciprocal altruism' (Trivers, 1971).

From an evolutionary point of view, altruistic behaviour is costly. Helping without receiving any benefits may reduce the chances of survival or offspring. Time and food invested in others do not benefit one's own survival or offspring. Once kinship has been determined, there is no need for further pondering whether or not to help, as helping kin is beneficial or profitable as it helps to spread the shared genes. In contrast, altruistic behaviour toward strangers may rely more on contextual evaluation, because one's fitness is only increased if, in general, helping behaviour is reciprocated. Therefore mechanisms should be in place to point out whether the recipient of help is a reliable and trustworthy party.

Kin selection and reciprocal altruism focus on the ultimate causation of altruistic behaviour. The proximate question is concerned with the psychological mechanism(s) that motivate(s) us to help. Hume already argued that our 'propensity to sympathize with others (…) forms our chief source of moral distinctions', implying that our motivation to help stems from inner sentiments (Hume, 1888/1990). Generally, empathy is referred to as the proximate explanation of helping behaviour. In fact, Batson formulated the empathy-altruism hypothesis: Feeling empathy for a person in need evokes altruistic motivation to help that person (Batson, 1991). Ample experimental evidence, wherein the degree of empathy felt by a subject was measured and manipulated, demonstrates a causal link between the experience of empathy and subsequent helping behaviour (Batson, 1991; Eisenberg et al., 2002; Eisenberg, Miller, Shell, Mcnalley, & Shea, 1991).

Yet there are probably as many definitions of empathy as there are researchers studying the phenomenon. Still, one of the most embraced is Hoffman's definition of empathy as 'an affective response more appropriate to someone else's situation than to one's own' (Hoffman, 2000). The perception of a person in pain for instance, can elicit a feeling of pain in the observer. It is striking that at least some of the brain regions involved in experiencing pain are also activated when someone else's pain is observed (Avenanti, Bueti, Galati, & Aglioti, 2005; Jackson, Meltzoff, & Decety, 2005; Singer et al., 2004). This is probably true for empathy in general: The perception of emotions or actions in another individual activates corresponding representations and brain networks that are associated with that state of mind or behaviour in an observer (Carr, Iacoboni, Dubeau, Mazziotta, & Lenzi, 2003; Gallese, Keysers, & Rizzolatti, 2004; Preston & de Waal, 2002). The neural system involved in this type of resonance system is called 'the mirror system' (Gallese et al., 2004).

Despite the activation of common neural networks in experiencing and perceiving pain, some critical differences between the mental state of the observer and the observed must occur in order to be motivated to help the person in pain. If not, perceiving a person in distress may only elicit an egoistic response, whereby both the victim and the observer are in distress, but the observer is not inclined to help (called 'empathic distress'). But the observer's affective response to facing another's distress may result in a motivation to help this person (called 'altruistic motivation' or 'empathic concern') (Batson, 1991; Decety & Lamm, 2006; Eisenberg et al., 1994).

Roughly two types of models can be discerned on the topic of altruistic motivation. Whereas the first type primarily emphasizes bottom-up processes and automatic reactions associated with the perception of the (painful) stimuli, the second type stresses the importance of cognitive and regulative processes.

The Perception-Action-Model (PAM) on empathy by Preston and de Waal is an illustration of a model where the bottom-up processes are accentuated. The PAM states that 'perception of the object's state automatically activates the subject's representations of the state (…) and that activation of these representations automatically primes or generates the associated autonomic and somatic responses'. This means that perception automatically primes the individual to react with distress or concern upon the person in need. It is suggested that the underlying mechanism rests primarily on affective mirroring of the emotional state of the person perceived to be in distress. Though the inhibiting role of the prefrontal brain areas is acknowledged, it is suggested that 'covert responses may still occur, even outside awareness' (Preston & de Waal, 2002). Similarly, Nichols looks for the minimal mental architecture necessary to induce altruistic motivation. He constructs an account of altruistic motivation that draws on as little 'mind-reading' as necessary to accommodate the core cases of altruism. 'Mind-reading' or 'Theory of Mind' (ToM) is the capacity to attribute mental states (feelings, thoughts, motives) to other agents. In his view the ability to experience affective empathy combined with a minimal contribution of ToM is sufficient for altruistic motivation to occur (Nichols, 2004).

When subjects are asked to imagine themselves in the shoes of a person in distress, they are much more likely to help this person than when they were not asked to do so (Batson, 1991). Actively taking the perspective of someone else enhances empathy and the motivation to help. In the top-down models it is primarily this type of higher cognitive perspective-taking processes that are emphasized to explain the nature of altruistic motivation. Apart from an important role attributed to ToM, the ability to take someone else's perspective, the role of executive control and regulation of the primary affective responses are stressed. Decety, for instance, emphasizes that empathy requires some form of active inhibitory mechanism to regulate and tone down the self-perspective to allow the evaluation of the other-perspective (Decety & Chaminade, 2003; Decety & Jackson, 2004). He suggests this is necessary because the prepotent self-perspective, driven by the automatic link between perception and action, is the default mode. Thus cognitive empathy is suggested to enable helping behaviour.

However, rather than with two mutually exclusive models, we are confronted with differences in emphasis. The role of affective empathy, emotional mirroring and influence of ToM are acknowledged by both. Yet, the bottom-up model emphasizes the automaticity and intuitivity of the reactions, whereas the top-down models focus on cognitive control, a higher role for theory of mind and emotional regulation. In fact, both the bottom-up and the top-down model are involved in empathic processes (Goubert et al., 2005).

Evolutionary psychology is the science of reverse-engineering psycho-biological structures. Confronted with some kind of mental machinery, one wonders why the design is as it is. A central question is how both models are related to each other. To what degree are the modelled processes necessary to induce altruistic motivation? And how may the interaction of both processes guarantee the apt helping reactions? A model will be presented that unites both the bottom-up and the top-down empathic processes and an evolutionary rationale will be offered as to why these two processes occur.

Two Pathways to Altruistic Motivation, A Unifying Model

Evolutionary psychologists suggest that most of our current behaviour is the output of adaptive psychological mechanisms that proved helpful for the survival or reproduction of our ancestors (Buss, 2004). Feeling empathy with someone else's distress may be helpful. Depending on the situation both a reaction of personal distress and altruistic motivation can be adaptive reactions when confronted with a person in pain.

On the one hand, vicarious distress when witnessing people in pain may cause a danger-avoidance response (Prinz, 2007). For instance, when you observe a person in distress after touching an electric wire, you may associate the wire with someone else's pain and subsequently learn about the imminent dangers. This is a learning process called 'vicarious conditioning'. Personal distress, a simple form of empathy, thus may prove beneficial to the observer. Furthermore, a heightened alertness resulting from someone else's distress may help you to detect troubles and be on your guard for further danger.

On the other hand, if the individual in distress is a close kin or a possible reciprocator, it may benefit you to feel altruistic motivation and to help. Yet, there is a different adaptive choice to be made when cooperating with kin than when helping strangers. Helping kin is evolutionarily reasonable in itself (if the relative cost is not too high). Confronted with strangers, being on your guard is adaptive. What is the impression this stranger makes? Will he be a reliable reciprocator?

The act of helping may seem similar whether you help kin or strangers. Still, both types of helping are the result of specific adaptive pressures: Either kin selection or reciprocal altruism. I suggest that these acts rely on partly overlapping, but different, neural pathways. More specifically, I argue that the bottom-up process, with an emphasis on affective empathy, is an old brain system that humans share with many other mammals and that plays a role in helping kin, or helping those one is attached to. The top-down process plays a role when strangers are encountered. This last model allows for more modulation, and assessment of the person one is interacting with. This means the reaction can be changed easier and is more dependent on contextual factors. The top-down model with its emphasis on perspective taking and ToM allows one to assess the reliability of the person in distress. After a favourable assessment, the affective empathic response is activated through higher cognitive processes.

Figure 1: The response to a person in distress consists of several steps. (1) Observation of a person in pain may either lead to indifference or elicit activation of brain areas involved in pain experience (mirroring), leading to affective empathy accompanied by personal distress. (2) If the observer is attached to the suffering person, the personal distress is semi-automatically converted to altruistic motivation. (3) If the observed person is a stranger, the observer may react with distress, unless the activation of brain areas related to ToM, perspective taking and emotional regulation allow the observer to react with concern.

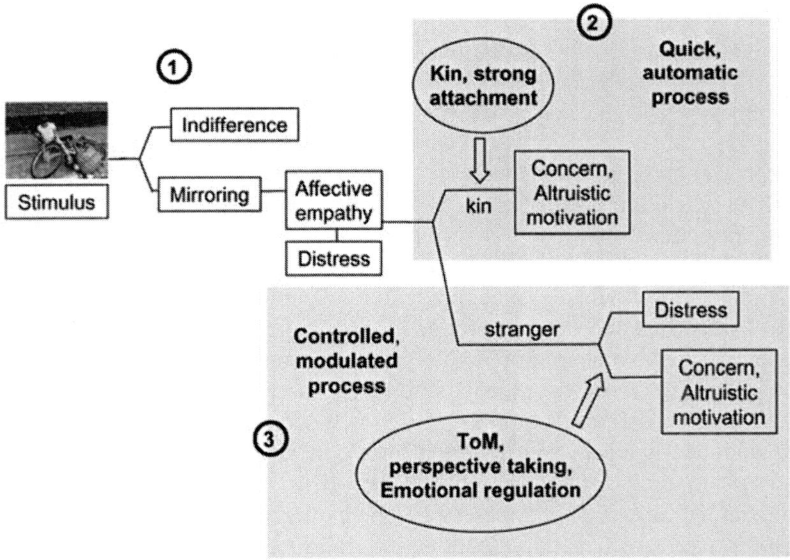

In fact, the model (figure 1) contains three steps, in order for altruistic motivation to occur after witnessing a person in pain. (1) In the first step, a subject may respond either with indifference or with affective empathy to someone else's distress. This last reaction is due to the activation of similar brain areas in the observer, as well as, in the observed (mirroring). Affective empathy or emotional mirroring leads to personal distress, which means the emotions between both are shared (Decety & Grezes, 2006). This step is regarded as a fairly automatic reaction based primarily on intuitive processes. (2) If the observer is emotionally attached to the suffering person, the personal distress is converted into altruistic motivation. This process depends merely on bottom-up affective reactions in combination with the feeling of attachment. (3) If the observed person is unfamiliar, the extent of perspective taking, emotional regulation and the assessment of the sufferer as a trustworthy person, will determine whether this person in distress receives help or not. If the observer does not take the other's perspective, he or she will probably react with a self-centred flight-reaction. When a high amount of perspective taking in the observer leads to the judgment of the distressed as a reliable reciprocator, this may result in a reaction of empathic concern and helping behaviour.

Neurological Underpinnings and Psychological Illustrations

If the model is correct, we expect that altruistic behaviour can occur without perspective taking (if the helper is attached to the helpee). This implies that if a person were attached to the person in distress, we would expect another form of empathy to occur: less activation of ToM and higher cognition related brain areas, but a higher activation of affective empathy networks.

(1) How can mice mother their young if ToM is necessary for helping? Despite the absence of cognitive perspective taking abilities, care for offspring occurs in many (lower) social mammals. Studies demonstrated some rudimentary forms of pro-social behaviour such as emotional mirroring in rodents when cage mates where in pain (Church, 1959). It is argued this type of helping behaviour is mediated by affective attachment. Mothers and infants of many social mammals show vigorous social bonding (Panksepp, 2005). Oxytocin is one of the hormones mediating these bonding processes. A reduction of the level of oxytocin in mother rats significantly reduces their nursing behaviour towards juveniles (Pedersen & Boccia, 2002). Evidence is mounting that the neural circuitry and neurochemistry implicated in studies of lower mammals also facilitate human and non-human primate interpersonal bonding (Pedersen, 2004). This means that other mammals and humans share some basic forms of brain circuitry involved in parenting behaviour and helping of kin.

Nichols (2004) argues that perspective taking is not necessary to perform altruistic acts. The study of autistic children has provided evidence for his thesis. Autistic children - who are believed to have a deficit in perspective taking - still respond to the distress of others and even engage in comforting behaviours (Bacon, Fein, Morris, Waterhouse, & Allen, 1998). Psychopaths – who have a deficiency in their affective functioning that causes a lack of remorse and empathy – are less likely to exhibit altruistic behaviour (J. Blair, Marsh, Finger, Blair, & Luo, 2006). Thus, Nichols shows it is not perspective taking, but the affective condition that underlies altruistic motivation. The fact that very young children (of 23 months), who have not yet developed the capacity of perspective taking, care for their parents and even help familiar others, underlines this thesis (Zahn-Waxler, Radke-Yarrow, Wagner, & Chapman, 1992).

(2) Bartels and Zeki undertook an fMRI study to look for the neural correlates of maternal love. In the study, the mothers viewed pictures of their own and of acquainted children. Mothers responded to pictures of their own children with a higher activation of brain regions and with a high distribution of attachment-mediating neurohormones, such as the brain's reward system at the ventral striatum. This activation pattern was similar, but distinct from results in an earlier study focusing on the neural correlates of romantic love, pointing at a common neural activation for attachment. It is interesting that both romantic and maternal love seemed to deactivate a common set of regions associated with negative emotions, social judgment and mentalizing (middle prefrontal cortex, superior temporal sulcus medial prefrontal cortex and paracingulate cortex). These are the same brain areas that are active when ToM is activated or perspective taking is performed. In fact, Bartels and Zeki propose human attachment probably employs a push-pull mechanism that overcomes social distance by deactivating networks used for critical social assessment and negative emotions. At the same time, it binds individuals through the involvement of reward circuitry (Bartels & Zeki, 2004). This study illustrates that brain areas involved in Theory of Mind and perspective taking (forms of cognitive empathy) are deactivated when attachment occurs. This illustrates the fact that attachment is associated more with affective and less with cognitive forms of empathy, suggesting attachment and affective empathy are sufficient for altruistic motivation to occur. Furthermore, the study demonstrates a reduced activation of the amygdala when viewing familiar faces. The amygdala is involved in the experience of fear and unrest (Bauman, Lavenex, Mason, Capitanio, & Amaral, 2004; R. J. R. Blair, 2007). Individuals are less on their guard when viewing familiar faces allowing a reaction of concern rather than personal distress. However, this observation does not entail that brain regions involved in cognitive empathy are completely deactivated. It seems probable that the activation of the bottom-up system may trigger the top-down system and vice versa. This may explain why Leibenluft found an activation of the superior temporal sulcus, a brain region involved in cognitive empathy, in addition to an activation of emotion related brain areas when mothers

viewed pictures of their children (Leibenluft, Gobbini, Harrison, & Haxby, 2004).

(3) Further, an interesting hypothesis can be drawn from the model. On the one hand, through affective empathy people may help other people because they are attached to them. On the other hand, cognitive empathy may contribute more to the decision to help strangers. So, we might expect differences in help behaviour when individuals display different forms of empathy, because not everyone has the same type of empathy. The Interpersonal Reactivity Index of Davis is often used to measure the empathy of subjects (Davis, 1983). This scale consists of different subscales: Perspective taking, empathic concern, personal distress and fantasising. Those scoring higher on perspective taking and cognitive empathy, can be expected to show stronger help reactions towards strangers, whereas those scoring higher on affective empathy – measured by personal distress – would more often help kin and people they are attached to. Experimental research may shed further light on this matter.

Dual Processing and Altruistic Motivation

LeDoux discerned two systems for processing emotional stimuli (LeDoux, 1996). The experience of fear, for instance, rests on both a quick, fast and automatic pathway leading to an immediate avoidance reaction and a slower, cognitive reaction. Panksepp observes that 'we must remember that core emotional systems have low-level, subcortical perceptual inputs as well as high-level cortico-cognitive ones' (Panksepp, 2005). A perception-action model implies that the same brain-areas are involved in perception as in the action or subsequent reaction or motivation (Preston & de Waal, 2002). I suggest that this dual process approach also applies to altruistic motivation. On the one hand, there is a quick and automatic altruistic response that depends on attachment and affective mirroring. On the other hand, higher cognitive processing may lead to a motivation to help. This type of process requires more perspective taking and ToM. Probably both types of processes are at work whenever altruistic motivation occurs. Yet, one or the other can be considered as the trigger for the motivation.

Arguments for this dual process model rely on (1) the selective deactivation of one of the two systems in clinical populations, (2) a different pattern of brain activation when these types of altruistic motivation are at work, (3) the observation that one system is much older and shared with lower mammals. (1) Persons with autism who lack a ToM still engage in helping behaviour. Characteristic for these people is that they react appropriately towards people they are familiar with, yet fail when they meet strangers. This demonstrates that despite a lack of ToM, a basic mechanism is still at play that regulates altruistic behaviour. Yet, as is characteristic for people with autism spectrum disorder, this is mainly directed toward kin and close ones (Nichols, 2004). Besides, if this basic system is absent, such as with people diagnosed with psychopathy, a general concern for others is absent (R. J. R. Blair, 2005). The observation of selective deactivation emphasizes the idea that independent pathways may be present. (2) Bartels and Zeki demonstrated that different types of attachment or love are associated with different brain region activation. Romantic love suppresses brain regions involved in ToM. A different brain activation pattern associated with forms of attachment suggests that in subsequent helping behaviour different systems are involved (Bartels & Zeki, 2004). (3) The presence of rudimentary forms of helping behaviour in lower mammals suggests (cfr. Church, 1959) that a primitive system was present that was more elaborated on in human helping behaviour.

However, these two models presented above do not have to be mutually exclusive, but can be complementary to each other. Preston and de Waal compare empathy to breathing. Breathing usually occurs automatically, but by focusing on it, we can control or influence it (Preston & de Waal, 2002). Similarly, we can control the feelings of empathy if we focus on them.

The ultimate reason why these two types of systems contribute to empathy and altruistic motivation depends on different types of adaptations. Kin selection requires a quick, automatic and immediate response to a kin member (or someone one is attached to) in need. Reciprocal altruism, however, requires some kind of assessment of the person in need. Higher cognitive processes have to be deployed. Darwin made clear that the difference between humans and other animals is gradual rather than categorical. Yet, apart from the common characteristics many animals share, human nature is constituted of a specific collection of mental adaptations. This can be illustrated by means of the motivations involved in helping behaviour. I suggest we share a basic emotional system (affective empathy) with many other social mammals, but that we may differ with regard to the cognitive control and regulation of it.

It should be pointed out that thus far, the dual processing model is a hypothesis that may need further corroboration by experimental research, yet the collected evidence seems to point at this direction. Furthermore, it should be stressed that the model presented may not cover all the types of helping behaviour that occur. One can, for instance, help another because one has explicitly and rationally chosen to do so. In this case, none of the basic or emotional processes will play a role. The elements in the model presented thus far rely mainly on subconscious processing. Moreover, the model does not imply that in real life situations an automatic empathic response is solely directed at close kin and is never directed towards strangers. It may well be that the system evolved to cope with kin problems is 'hijacked' by strangers. Babies, young faces of puppies elicit sympathy and concern without a necessary close kin connection. Nevertheless, explaining the diversity of helping behaviour and its different foundations remains a major challenge. Perhaps it is because altruism is one of those central elements of our human nature that we will cherish forever.

References

Avenanti, A., Bueti, D., Galati, G., & Aglioti, S. M. (2005). Transcranial magnetic stimulation highlights the sensorimotor side of empathy for pain. *Nature Neuroscience, 8*(7), (p.955-960).

Bacon, A. L., Fein, D., Morris, R., Waterhouse, L., & Allen, D. (1998). The responses of autistic children to the distress of others. *Journal of Autism and Developmental Disorders, 28*(2), (p.129-142).

Bartels, A., & Zeki, S. (2004). The neural correlates of maternal and romantic love. *Neuroimage, 21*(3), (p.1155-1166).

Batson, C. D. (1991). *The Altruism Question: Toward a Social-psychological Answer*. Hillsdale, NJ: Erlbaum.

Bauman, M. D., Lavenex, P., Mason, W. A., Capitanio, J. P., & Amaral, D. G. (2004). The development of mother-infant interactions after neonatal amygdala lesions in rhesus monkeys. *Journal of Neuroscience, 24*(3), (p.711-721).

Blair, J., Marsh, A. A., Finger, E., Blair, K. S., & Luo, J. (2006). Neuro-cognitive systems involved in morality *Philosophical Explorations, 9*(1).

Blair, R. J. R. (2005). Applying a cognitive neuroscience perspective to the disorder of psychopathy. *Development and Psychopathology, 17*(3), (p.865-891).

Blair, R.J.R. (2007). The amygdala and ventromedial prefrontal cortex in morality and psychopathy. *Trends in Cognitive Sciences, 11*(9), (p.387-392).

Buss, D. M. (2004). *Evolutionary Psychology: The New Science of the Mind* (2nd ed.). Boston, MA: Pearson

Carr, L., Iacoboni, M., Dubeau, M. C., Mazziotta, J. C., & Lenzi, G. L. (2003). Neural mechanisms of empathy in humans: A relay from neural systems for imitation to limbic areas. *Proceedings of the National Academy of Sciences of the United States of America, 100*(9), (p.5497-5502).

Church, R. M. (1959). Emotional reactions of rats to the pain of others. *Comparative Physiology and Psychology 52*, (p.132-134).

Darwin, C. R. (1871). *The Descent of Man, and Selection in Relation to Sex*. London: John Murray.

Davis, M. H. (1983). Measuring Individual-Differences in Empathy - Evidence for a Multidimensional Approach. *Journal of Personality and Social Psychology, 44*(1), (p.113-126).

Decety, J., & Chaminade, T. (2003). Neural correlates of feeling sympathy. *Neuropsychologia, 41*(2), (p.127-138).

Decety, J., & Grezes, J. (2006). The power of simulation: Imagining one's own and other's behavior. *Brain Research, 1079*, (p.4-14).

Decety, J., & Jackson, P. L. (2004). The functional architecture of human empathy. *Behavioral and Cognitive Neuroscience Reviews, 3*(2), (p.71-100).

Decety, J., & Lamm, C. (2006). Human empathy through the lens of social neuroscience. *The scientific world journal, 6*, (p.1146-1163).

Eisenberg, N., Fabes, R. A., Murphy, B., Karbon, M., Maszk, P., Smith, M., et al. (1994). The Relations of Emotionality and Regulation to Dispositional and Situational Empathy-Related Responding. *Journal of Personality and Social Psychology, 66*(4), (p.776-797).

Eisenberg, N., Guthrie, I. K., Cumberland, A., Murphy, B. C., Shepard, S. A., Zhou, Q., et al. (2002). Prosocial development in early adulthood: A longitudinal study. *Journal of Personality and Social Psychology, 82*(6), (p.993-1006).

Eisenberg, N., Miller, P. A., Shell, R., Mcnalley, S., & Shea, C. (1991). Prosocial Development in Adolescence - a Longitudinal-Study. *Developmental Psychology, 27*(5), (p.849-857).

Gallese, V., Keysers, C., & Rizzolatti, G. (2004). A unifying view of the basis of social cognition. *Trends in Cognitive Sciences, 8*(9), (p.396-403).

Goubert, L., Craig, K. D., Vervoort, T., Morley, S., Sullivan, M. J. L., Williams, A. C. D., et al. (2005). Facing others in pain: the effects of empathy. *Pain, 118*(3), (p.285-288).

Hamilton, W. D. (1964). The genetical evolution of social behaviour. *Journal of Theoretical Biology 7*, (p.1-52).

Hoffman, M. L. (2000). *Empathy and Moral Development: Implications for Caring and Justice*. Cambridge: University Press.

Hume, D. (1888/1990). *A Treatise of Human Nature*. Oxford: Clarendon Press.

Jackson, P. L., Meltzoff, A. N., & Decety, J. (2005). How do we perceive the pain of others? A window into the neural processes involved in empathy. *Neuroimage, 24*(3), (p.771-779).

LeDoux, J. E. (1996). *The Emotional Brain* New York: Simon & Schuster.

Leibenluft, E., Gobbini, M. I., Harrison, T., & Haxby, J. V. (2004). Mothers' neural activation in response to pictures of their children and other children. *Biological Psychiatry, 56*(4), (p.225-232).

Nichols,S. (2004). *Sentimental Rules: On the Natural Foundations of Moral Judgment*. New York: Oxford University Press.

Nowak, M. A. (2006). Five rules for the evolution of cooperation. *Science, 314*(5805), (p.1560-1563).

Panksepp, J. (2005). Affective consciousness: Core emotional feelings in animals and humans. *Consciousness and Cognition, 14*(1), (p.30-80).

Pedersen, C. A. (2004). Biological aspects of social bonding and the roots of human violence. In: *Youth Violence: Scientific Approaches to Prevention* (Vol. 1036, p. 106-127).

Pedersen, C. A., & Boccia, M. (2002). Oxytocin links mothering received, mothering bestowed and adult stress responses. *Stress 5, 5*, (p.259–267).

Preston, S. D., & de Waal, F. B. M. (2002). Empathy: Its ultimate and proximate bases. *Behavioral and Brain Sciences, 25*(1), 1-+.

Prinz, J. (2007). Is morality innate? In: W. Sinnott-Armstrong (ed.), *The Evolution of Morality* (Vol. 1, p. 367-406). Cambridge: MIT Press.

Singer, T., Seymour, B., O'Doherty, J., Kaube, H., Dolan, R. J., & Frith, C. D. (2004). Empathy for pain involves the affective but not sensory components of pain. *Science, 303*(5661), (p.1157-1162).

Trivers, R. L. (1971). The evolution of reciprocal altruism. *Quarterly Review of Biology 46*, (p.35-57).

Zahn-Waxler, C., Radke-Yarrow, M., Wagner, E., & Chapman, M. (1992). Development of concern for others. *Developmental Psychology 28*, (p.126–136).

CHAPTER TWELVE

'PRIMATES AND PHILOSOPHERS': RECIPROCITY AND DISTANT PEOPLE IN NEED

KRISTIAN HØYER TOFT

Abstract: In a recent book primatologist Frans De Waal discusses the evolution of morality. A prime target of De Waal's criticism is so-called Veneer Theory (VT) according to which morality is "a cultural overlay, a thin veneer hiding an otherwise selfish and brutish nature" (De Waal, 2006).

It is discussed whether evolutionary theory ranging from reciprocal altruism to gene-culture coevolution can accommodate altruism towards distant strangers in need. De Waal exemplifies such an extension of our sentiments to regard the entire human race by The Geneva Convention of 1949 and states: "But we all know how fragile an effort this is" (ibid. p. 53).

It is argued that universal morality must invoke features of VT since evolutionary theory of altruism does not sufficiently account for a concern for distant strangers in need.

Introduction

It is becoming increasingly clear that evolutionary theory is now capable of explaining more complex cases of human altruism. Widening the scope of altruism from close-knit relations of kin and group to also include strangers within larger communities challenges traditional conceptions of morality. According to traditional conceptions of morality, the features of altruism are not fully compatible with evolution, since human motivation for acting altruistically in its purest form is regarded as an 'other-regarding' act. Now evolutionary theory has also set out to explain what are regarded as purely self-sacrificing altruistic acts – acts in which nothing is expected in return in terms of money, status, friendship, or even reproductive fitness. According to such strands of evolutionary theory as, for instance, multi-level selection theory (Sober & Wilson,

1998) and gene-culture coevolution (Boyd & Richerson, 2005), individual acts of altruism can be explained within the naturalist framework of evolution. However, it may still be debatable whether altruism in these recent theories is not simply ‘apparent altruism’, since it is still ultimately the reproductive fitness of the group or species that determines whether ‘altruism’ is selected over time. This distinction between ‘apparent’ and ‘genuine’ altruism provides a prism for understanding a possible difference between evolutionary theory and conceptions of altruism found in moral philosophy. Robert Trivers’ seminal article on reciprocal altruism (1971, p. 51) points to the fact that humans have psychological capacities for detecting genuine altruists and distinguishing these from cheaters who merely give the appearance of being altruists. However, the notion of genuine altruism might also point out that a genuine altruist is capable of being motivated by purely moral reasons and that this is the feature that distinguishes the genuine altruist. The influential tradition within deontological moral philosophy emphasises that duty should be done based on purely imperative and normative reasons. According to Immanuel Kant, a moral reason is distinctively different from a hypothetical reason that rationalizes how to optimize a good, and he states that a moral reason is based on pure reason, i.e. the capacity to judge impartially. It is clear that the aspect of genuineness of an altruistic act or motive is perceived from very different perspectives in deontological theory and evolutionary theory.

To investigate the possible tension between the two views on altruism – the evolutionary and the Kantian, the case of distant strangers in need is discussed. It seems that this particular case is pertinent since it not only represents one of the most pressing global political and humanitarian problems; it also exemplifies a case that might not easily fit into the evolutionary emphasis on reciprocal altruism and group loyalty. In that sense the felt obligation to relieve global poverty presents a challenge to the evolutionary theory of morality. The structure of the discussion is first to address the case of distant strangers in need. It is not straightforward to define such people or to define to what extent they pose a challenge to evolutionary theory. Does evolutionary theory account for such a wide scope of justice? And if not, should it do so? One consequence of a restricted scope of justice is that our felt obligations towards distant strangers in need would lack explanation, and moreover, it might have to appeal to an alternative non-evolutionary theory of morality such as Kant’s moral philosophy.

However, to judge whether the traditional conception of morality needs to be invoked to save the morality of concerns for distant strangers in need, mainstream evolutionary theory is first assessed. The discussion will start out with the theory of reciprocal altruism, the theory of indirect reciprocal altruism and the theory of costly signalling, as well as, altruistic punishment. Then the theories of group-selection and gene-culture coevolution are examined. Finally, Frans De Waal's account of morality and his so-called Veneer Theory as an example of the traditional conception of morality is discussed.

Distant Strangers in Need

It is beyond dispute that human poverty and the suffering related to it makes a serious case for any view on morality today. The number of people that live on an amount equivalent to that of $1 a day, the number of children dying from diseases caused by malnutrition and the number of people who are victims of totalitarian regimes are in the millions (Pogge, 2001). The sad fact of global poverty gives rise to a global divide between rich and poor. Responsibility for this divide and the proper level of foreign aid is often debated. Across the populations in rich countries, one finds disagreements over the level of aid. However, most people in these countries, irrespective of their political outlook, agree that some basic level is required, i.e. the obligation to relieve poverty and to aid distant strangers in need is actually broadly acknowledged. One major reason for taking on global responsibility is the increasing tendency towards globalization. Global poverty, climate changes and global media all create a political situation in which individuals are confronted with issues of distant strangers. But are the features of the case of distant strangers in need relevant to moral theory, evolutionary or not? The morally salient and relevant feature is that there are facts of need and that we, the affluent ones, could help to alleviate. Moreover, the needs are not comparable to the needs found on average in rich countries, in fact, they are vital needs – needs of basic subsistence. Some moral theorists would argue that needs as such do not make a moral reason. We are obliged to help only because they are humans found in dire circumstances. The particular status of humans is the ultimate reason for help. Hence, similar obligations are not owed to non-human animals. A final relevant feature of distant strangers in need would be that they pose some demand on us. One aspect of morality is that it is demanding – it is imperative to help, regardless of whether it is costly or not. Besides the moral aspects of distant strangers in need, they also exhibit evolutionary relevant aspects. Importantly, distant strangers are not part of kin, group or any other reciprocal relation. Hence, it should be obvious that distant strangers in need pose a challenging case to evolutionary theory of altruism. The challenge persists exactly because people generally and actually hold intuitions about a duty to help distant strangers in need. Other cases might be suggested. The point is such cases constitute 'hard cases' that exhibit features generally acknowledged as genuine moral demands, which are revealed in situations where the helper is in a significantly stronger position compared to those in need. It seems that a profound asymmetry between donor and recipient might be a core criterion of a genuine moral situation.

Evolutionary Theory and the Scope of Altruism

If asymmetry is to be considered an important feature of a genuine moral situation, how does evolutionary theory of altruism respond to this? The theory of reciprocal altruism introduced by Robert Trivers in 1971 is path-breaking since it introduces a theory of altruism that encompasses instances of altruism towards non-kin and strangers. Before Trivers the dominant view of altruism relied on kin-selection (Hamilton, 1964). However, certain conditions will constrain acts of reciprocal altruism. The constraints predict that altruists should have a reasonable probability of having their favour reciprocated at a later point in time (Trivers, p. 37). This implies that the stranger must be part of the group or at least a group within some physical proximity. Moreover, the reciprocator must have the resources to return the favour (ibid. p. 36, 45). Hence, the relationship between helper and recipient must be reciprocal or symmetrical in order for altruism to be an evolutionarily stable strategy (ESS). Trivers also concedes that individuals are inclined to invite strangers into reciprocal relationship with the aim of future beneficial mutual exchange (ibid. p. 52). Though the scope of reciprocal altruism may provide cues as to why people are inclined to be altruistic towards strangers, it does not cover a scope wide enough also to include distant strangers. Several theories have elaborated on reciprocal altruism to widen the scope of explaining altruism. Indirect reciprocal altruism (Nowak & Sigmund, 1998) stipulates situations characterized by only one encounter between the altruist and the recipient. Instead of receiving a return of favour, the altruist receives an image score in the group that is equivalent to high status since the group considers the altruist to be a reliable partner in cooperation. A similar concept is found in the theory of costly signalling that was introduced by Thorstein Veblen already in 1899 in his 'Theory of the Leisure Class'. Veblen assumed that the affluent class in society would flag their high status by a conspicuous consumption of very expensive and luxurious goods. Altruistic acts may be very expensive and have no apparent returns and in that sense resemble conspicuous consumption. Recent proponents of costly signalling (Gintis, Smith & Bowles, 2001; Zahavi & Zahavi, 1997) point out that it remedies the theory of indirect reciprocity in the sense that costly signalling merely invokes the self-interest of the other group members that admire the altruist. Altruists are admirable people that everybody wants to interact with. Costly signalling explains cases of altruism where the altruistic act is publicly visible. Hence, it is possible to take this type of theory and apply it to the case of foreign aid, since it predicts that nations seek to create an image that other nations admire (Gilady, 2003). However, the problem with indirect reciprocal altruism and costly signalling is that they presuppose the publicity of altruism. This is in conflict with intuitions saying that genuine altruism should not depend on whether or not it is perceived by the public. In fact, genuine altruism might be performed in pure anonymity. And, in fact, much foreign aid is contributed by private individuals who donate anonymously. Moreover, both theories are certainly in contrast to an understanding of morality that stresses unselfish motives as opposed to vanity. This complaint might not be at the level of scientific theory since it invokes a normative understanding of morality that is insulated from scientific facts of human motives, though it raises the general issue of whether a scientific theory should take notice of subjective views.

However, the shortcomings in the above theories of altruism could be summed up as inability to account for "cooperation [..] among genetically unrelated people, in non-repeated interactions when gains from reputation are small or absent" (Fehr & Gächter 2002, p. 137). To remedy this, 'altruistic punishment' is suggested since it predicts that people who cheat and defect in cooperation will be punished regardless of the relative cost, and absence of benefit, to the one giving out the punishment. Altruistic punishment therefore removes focus from positive motives for altruism and focuses instead, on breaking the rule of altruism. Altruistic punishment has attracted many adherents (Boyd & Richerson 2005, p. 204) since it is a promising theory that may explain how larger groups in which the members are strangers to each other cooperate. The constant threat of future punishment provides the cue for an internalization of a moral conscience and therefore works pre-emptively. The problem with altruistic punishment, in regard to the case of distant strangers, is that it does not account for acts of altruism that are not part of the compliance system of a stable society. Helping distant strangers in need is not strictly required for most societies to function stably. Perhaps the citizens of such states, that do donate to distant strangers or support political parties that argue for a raised level of foreign aid, feel that their less generous fellow citizens are defecting and free-riding on their contribution. In so doing the generous citizens may try to punish them in public by engaging in political activism. This is what actually happens in most rich countries. So, altruistic punishment might be invoked to explain a tendency towards a pressure to contribute to distant strangers in need, though it does not fully explain why, in the first instance, the generous became generous enough to care about those far away. Altruistic punishment seems to work on the basis of already established moral rules, and in particular rules that are essential in securing social stability in society. Adding to this - any arbitrary rule might be supported by the mechanism of altruistic punishment, such as the rule of wearing a tie (Boyd & Richerson, 2005, p. 201).

Expanding the Scope of Altruism: Group Selection

Since the above evolutionary theories of altruism all exhibit shortcomings in explaining why people feel they are obligated to help distant strangers in need, other types of theory might be required. Except for the theory of altruistic punishment, the theories did not look at group-level selection. However, if altruistic acts by individuals are detrimental to those individuals, but at the same time benefit the group to which they belong, altruism may find an explanation. Natural selection not only works at the individual level but also at group-level between groups (Wilson & Wilson, 2007). The theory of multi-level selection predicts equilibrium between the many levels of life – from the cell, the organism and the group. According to group-selection theory, the group can be perceived as an organism. The main advantage of moving focus to also include the group in regard to the case of distant strangers is that such strangers may be distant members of the same group. Certainly the theory of group-selection is not designed to explain why we feel the tendency to provide for foreigners, since foreigners are by definition not members of our group. Hence, it seems difficult to encompass distant strangers within group-selection theory. However, the scope of the group might be expanded in order to allow some distant strangers to be part of the group. The widest expansion would be equivalent to the universal scope of human kind – the scope advocated by human rights proponents. The notion of a 'global village' has been invoked to say that in the age of globalization we are all members of the same group. However, such extensions are contradictory to the scope of any theory of group-selection. Less ambitious extensions of scope are also optional, for instance, it could include only countries to which we are committed by international trade and common international law. From the perspective of multi-level selection, it could be argued that the international society resembles a pyramidal structure with global institutions like the UN at the top.[1] It is of course, not unlikely that the present type of rather weak UN global governance may show resilience over time, and in that sense global duties have some evolutionary explanation.

Moreover, the problem with group-selection theories of the international order is that they tend to prioritize the state over the individual and the region over the state and so forth. Even if the interests of individuals are part of a global multi-level selection process, such process does not necessarily provide any guarantee that individuals are secured by any sort of basic minimum assurance. So, even if multi-level selection provides a framework for understanding why altruism could have a global scope, it does not take the level of individuals sufficiently into consideration. This becomes clear in comparison to our present culture based on human rights, according to which the individual, regardless of national identity, is acknowledged certain basic rights. In conclusion, multi-level selection theory needs further elaboration to fit the case of distant strangers in need.

Gene-Culture Coevolution

Searching for a supplement to remedy the shortcoming of the group-selection theory, the theory of gene-culture coevolution suggested by Boyd and Richerson (2005) seems promising. Gene-culture coevolution seeks to explain modern large scale societies in which individuals interact with non-kin and strangers. In order to explain cooperation in large groups, the theories of reciprocal altruism and kin-selection are recognized to be adequate theories of our (Pleistocene) ancient social instincts. These are hard-wired into the genes and determine our basic capacities for cooperation in small groups. However, as humans gather in large groups, tribal instincts are developed on top of the ancient social instincts. The tribal instincts are sensitive to symbols such as language and dress code that demarcate groups from each other. The tribal instincts are parochial as they hamper cooperation with strangers from outside the group. These instincts are a product of cultural evolution and they distinguish humans from non-human animals as they exemplify a unique human capacity (ibid. p. 197, 235). Human beings have shown to be extremely adaptive,

[1] Importantly, selection among groups (or nations) would still be part of this world order. I owe this interpretation to Gorm Nissen.

thanks to the ability of evolving culturally. Boyd and Richerson argue that climate changes could have invoked culture as a means to adapt faster than what was feasible by genetic evolution. Hence, as an adaptation to living in larger groups, humans acquired cultural capacities such as an extended capacity for altruism. However, the capacity for culture comes at a price. The plasticity exhibited by humans means that they are good at adapting through mimicking successful individuals. Mimicking spares the individual the complicated calculating. The downside of mimicking is that bad ideas are sometimes copied; ideas that might have been successful in one type of environment are less well adapted to another. The example of altruism shows this. An acquired ability to cooperate with strangers and meet them with an altruistic and open mind might have been a successful strategy in tribal societies in which salient symbols and features marked friend from foe. However, in modern societies not differentiated in marked tribes, the altruistic tendency to openness towards strangers leaves the individual vulnerable to exploitation. Boyd and Richerson admit that excessive altruism might be a mal-adaptation (Boyd & Richerson, 2005, p. 150). The question is of course, whether the case of helping distant strangers in need is a case of excessive altruism. Unfortunately, Boyd and Richerson do not engage with this particular issue, though they provide ample evidence for the fact that humans have an extended capacity for altruism towards strangers. Moreover, humans have a capacity to evolve rapidly by invoking culture when faced with environmental changes. Hence, the culture of globalization, of managing global issues such as climate changes, migration and global interdependency, might be issues dealt with by cultural adaptation. Similarly, the case of distant strangers in need conceived as the issue of global poverty might as well be a case to which humans are currently adapting. This means that the theory of gene-culture coevolution might provide a framework for understanding how humans can surpass their inherited social instincts and adapt to the circumstance of global interdependency. Admittedly, this is a charitable interpretation, and it is more likely that gene-culture coevolution, though providing a general framework for altruism towards strangers, does not provide much argument for the feeling that we should spend our resources to help distant strangers in need if these strangers are not recognized as members of 'our' group. Hence, it is debatable whether gene-culture coevolution provides a clear argument in favour of global altruism.

Global Altruism – A Big Mistake?

Phrasing the case of distant strangers in need as 'global altruism' stresses the utopian features of the case, and therefore evolutionary theory should, in fact, not be blamed for shortcomings in explaining it. This gives rise to the so-called 'big-mistake hypothesis' (Boyd & Richerson, 2005, p. 150), according to which modern humans are confronted with challenges to which they respond with their innate ancient/tribal psychology. This implies a mismatch (Kurzban, 2003, p. 119) between modern and ancestral environments. Just as humans have a disposition to indulge in fat and sugary food, they also have the disposition to act altruistically towards distant strangers. These propensities are rational from the point of view of humans of an ancient past, but they are simply a big mistake when applied to the modern circumstance of an abundance of sweet and fat food, as well as, an abundance of distant strangers in dire need. The big-mistake hypothesis presupposes some sort of a discontinuous break between ancient and modern times. Boyd and Richerson tend to agree that such a break did indeed take place and that the cause of the break was cultural evolution (the evolution of human language, symbols and extensive cooperative abilities) (Boyd & Richerson, 2005, p. 150). However, they do not believe that cultural mal-adaptation is the sole product of a mismatch of ancient instincts and a modern environment. They see the excessive and vulnerable propensity towards altruism, in as far as it is a mal-adaptation, as the necessary by-product of a cultural adaptation to cooperating with strangers in large and complex societies. Hence, there is a trade-off between adapting to modern society and the cost of vulnerability which cultural evolution seeks to balance. In that sense Boyd and Richerson provide the framework for an optimistic explanation of 'global altruism'. It may be mal-adaptive but it is the price we must pay to have the ability to live among strangers in modern societies. The downside is of course, that global altruism is indeed a mal-adaptation and in that sense a big mistake. Again, gene-culture coevolution is ambiguous about whether the feeling that distant strangers should be helped has any solid footing in the evolutionary explanation of morality.

Veneer Theory

Assuming that the case of distant strangers in need is fit for scientific inquiry and that the above discussion is a fair treatment of 'global altruism' judged by the standards of evolutionary theory, it leaves the case ambiguously unexplained and unjustified. One possible reply to this would be to invoke the so-called Veneer Theory (VT), according to which morality is "a cultural overlay, a thin veneer hiding an otherwise selfish and brutish nature" (De Waal 2006, p.7). VT may 'save' distant strangers in need, so to speak, since what is required to command and justify obligations towards distant strangers is ultimately a moral or a religious theory that transgresses the shortcomings in human nature. At least, this is one possible solution, if one is confident with the tenets of VT. In the book *Primates and Philosophers – How Morality Evolved* (2006), Frans de Waal discusses this issue in his Tanner Lectures. In this book he is commented on by moral philosopher Peter Singer who has taken the issue of obligations towards distant strangers into consideration in several of his earlier writings (see Singer 1971). The discussion between De Waal and Singer is instructive in this connection. De Waal commits himself to the theory of kin-selection and reciprocal altruism as the most credible theories of human morality (De Waal, p. 16). He is less convinced by the theory of group-selection since migration has made human groups insufficiently variable to elicit substantial selective pressure. This implies that De Waal is less confident that we can expand the circle of altruism to also include distant strangers, since they cannot stand in any real reciprocal relation to us. De Waal exemplifies such an extension of our sentiments to regard the entire human race by referring to The Geneva Convention of 1949 (an example of a human rights document) and states: "But we all know how fragile an effort this is" (ibid. p. 53). De Waal therefore does not deny that we may stand in a moral relation to distant strangers, but he contends this to be a conflicted relation between local commitment and impartial obligations (ibid. p. 164). Moreover, he argues that a circle of commitment depends on the amount of resources available – in times of abundance it is possible to care for distant others, otherwise not (ibid.). So, De Waal confirms the result from the above inquiry, that the case of distant others may not find much support from within evolutionary theories, since such theories are committed to the constrained and hierarchical order of gene-individual-kin-group. Obligations towards humanity or towards distant strangers in need find only indirect explanation in evolutionary theory. As Peter Singer is a strong advocate of global justice, but at the same time rejects VT, he seeks to answer to De Waal's restricted scope of justice. Singer claims that the human ability to reason, to be an 'impartial spectator', is a specific, inherent part of human nature. He thinks of the morally evolved sentiments as partly shared with the moral emotions of non-human primates, but significantly, humans are the only animals with an extensive capacity for reason and impartiality (ibid. 145). Reason is not some peculiar super-natural aspect of humans, but a capacity that humans evolved with. Here Singer is close to a view of the unique human capacity for abstract reasoning that is also found in the Aristotelian virtue-ethical tradition. According to this tradition the capacity for reason is potentially a part of the innate nature of humans; this capacity is, however, only fully realized through upbringing and education (McDowell, 1995). It is not clear whether the Aristotelian, and in this respect Singer's, view of a reconciliation between morality as reason and human nature, is not simply a version of supernatural VT. De Waal is happy to allow for impartiality and reason as long as altruistic commitments are confined to the circle of evolutionarily stable relations. Hence, he does not deny that we do have intuitions of caring for distant strangers, only that such commitments will have to be weighed against evolutionary and morally prior loyalties towards ourselves, our family and community (De Waal, p. 165). So, the disagreement between Singer and De Waal is not whether humans have a capacity for impartial reasoning, but rather what the fact that humans have the capacity means to our conception of morality, and in particular to our obligations towards distant strangers. Singer defends extensive cosmopolitan obligations on the basis of impartiality, whereas De Waal points out that evolution has made us care the most for our closest affiliations and that we are also morally obliged to act in accordance with this fact of our psychology (ibid.).

This debate between Singer and De Waal illustrates that the case of distant strangers is not easily accounted for in evolutionary theory if one is looking for an account that will not compromise equal concern for distant others. As equal concern for distant others is stated in the idea that each human being has an equal moral standing regardless of national belonging, as is found in human rights documents, there is an unavoidable tension with the accounts given in evolutionary theory.

If this tension is formulated as a dilemma, we are faced with the hard choice of either sticking to evolutionary theory, thus giving up the idea of global altruism, or evoking some sort of VT to 'save' global altruism. The dilemma is an uncomfortable one since it would be favourable, if the scientific explanation of global altruism were fully compatible with felt moral commitments to distant others. However, the tension between the two approaches to global altruism need not be at the level of a dilemma as evolutionary theory does indeed provide an account for why we tend to extend our sympathy to distant strangers. This tendency is elicited by our innate capacity for cooperating in large groups. Hence, distant strangers are future co-operators and future members of the group. Even though this drift of thinking might defuse the tension between felt moral commitments and evolutionary theory it still, however, renders the moral status of distant strangers uncertain. And, if it is considered morally required to provide distant strangers with a status of equal moral concern, the tension with the evolutionary account is high and the dilemma persists.

Faced with the dilemma it is possible to choose one of the options. However, it is equally possible to hope for future developments within evolutionary theory that will further defuse the tension. Finally, perhaps the only politically feasible option is to adopt an attitude of neutrality about 'foundational' issues of morality (Rawls, 1996), though this is not entirely satisfactory to someone who prefers coherence between his or her moral intuitions and the best scientific account available.

References

Boyd, R. & Richerson, P.J. (2005). *Not by Genes alone – How Culture Transformed Human Evolution.* USA: University of Chicago Press.

De Waal, F. (2006). *Primates and Philosophers – How Morality Evolved.* Princeton: Princeton University Press.

Fehr. E. & Gächter, S. (2002). Altruistic Punishment in Humans. *Nature, Vol.* 415, 137.

Gilady, L. (2003). A Contest of Beneficence: Explaining Altruism in International Relations. *Paper Presented at ISA,* Portland, Oregon, 26th. Feb.- 1st. Mar., 2003.

Gintis, H. & Smith, E.A. & Bowles, S. (2001). Costly Signalling and Cooperation. *Journal of Theorethical Biology,* 213, 104.

Hamilton, W.D. (1964). The Genetical Evolution of Social Behaviour. *Journal of Theoretical Biology, 7 (1)*, (p.1-52).

Kurzban, R. (2003). Biological Foundations of Reciprocity. In: Ostrom, E. & Walker, J. (eds.), *Trust and Reciprocity – Interdisciplinary Lessons from Experimental Research.* USA: Russell Sage.

McDowell, J. (1995). Two Sorts of Naturalism. In: Hursthouse, R. et al. (eds.), *Virtues and Reasons – Philippa Foot and Moral Theory.* USA: Oxford University Press.

Nowak, M.A. & Sigmund, K. (1998). Evolution of Indirect Reciprocity by Image Scoring. *Nature, Vol. 393,* 11. *June.*

Pogge, T. (2001). Priorities of Global Justice. In: Pogge, T. (ed.), *Global Justice.* Great Britain: Blackwell Publishers.

Rawls, J. (1996). *Political Liberalism.* New York: Columbia University Press.

Singer, P. (1971). Famine, Affluence, and Morality. In: Beitz, C. et al. (eds.), *International Ethics. A Philosophy & Public Affairs Reader.* New Jersey: Princeton University Press.

Sober, E. & Wilson, D.S. (1998). *Unto Others – The Evolution and Psychology of Unselfish Behaviour.* Harvard: Harvard University Press.

Trivers, R. (1971). The Evolution of Reciprocal Altruism. *The Quarterly Review of Biology, Vol. 46,* 1, (p. 35-57).

Wilson, E.O. & Wilson, D.S. (2007). Survival of the Selfless – How does Altruism manage to Thrive in the Ruthless Cut and Thrust of Natural Selection? *New Scientist*, 3. Nov. 2007, (p. 42-46).

Zahavi, A. & Zahavi, A. (1997). *The Handicap Principle: A Missing Piece of Darwin's Puzzle*. New York: Oxford University Press.

CHAPTER THIRTEEN

MORAL OBJECTIVITY AND ALTRUISTIC ADAPTATIONS

GORM NISSEN

Abstract: In this paper I investigate what conclusions to draw in philosophical ethics from the fact that morality is an adaptation. What can we plausibly conclude concerning the meaning of moral terms, when we take this fact seriously? Some philosophers and scientists seem to think that the conclusion to draw from an evolutionary understanding of altruism is that we must be anti-realists regarding the meaning of moral terms because evolution shows us that there is no such thing as an objective morality. Morality is an illusion created by our genes. I beg to differ: There is within evolutionary naturalism plenty of room for a notion of objective morality that is strong enough to support moral realism, and the fact that morality is an adaptation does nothing by itself to rule out moral objectivity and moral realism. Rather, my argument will be that the fact that morality is an adaptation should count in favour of moral realism.

Altruistic Adaptations

Examples of the phenomenon known as evolutionary altruism are plenty in nature. One of the common prima facie examples of such behaviour is the honeybee and its sacrifice, when losing its life to sting an intruder. Other common examples are bird sentinels that issue warning cries when a predator approaches. A trait or behaviour by an organism is altruistic in this evolutionary sense, if that particular behaviour increases the fitness of another organism, while lowering the fitness of the organism with the trait in question, and fitness is defined in terms of survival and reproduction relative to other individuals. This altruistic behaviour then is not what you would expect from the process of natural selection, where the traits or behaviours to be found should be traits that increase the fitness of organisms rather than lowering their fitness. So the crucial question is: Why is altruistic behaviour not weeded out by natural selection?

The first models used to answer this question and explain altruism involved appealing to groups as the level of selection, which Darwin did when explaining the moral behaviour of man:

> There can be no doubt that a tribe including many members who, from possessing in a high degree the spirit of patriotism, fidelity, obedience, courage, and sympathy, were always ready to aid one another, and to sacrifice themselves for the common good, would be victorious over most other tribes; and this would be natural selection. (Darwin 1981 [1871], 166).

Where individual selection favours traits that increase individual fitness, group selection favours traits that increase group fitness, so if the force of group selection is strong enough, it will overcome the force of individual selection and thus, making possible the evolution of altruism.

Since Darwin many biologists relied uncritically on group selection as an explanatory tool. This extended use of group selection as an explanation of evolutionary altruism, however, was sharply criticised by G.C. Williams in *Adaptation and Natural Selection* (1966). Instead he proposed alternative models, drawing mainly on the work of W. Hamilton (1964) and in particular Hamilton's concept of inclusive fitness, developed as a way of seeing evolution from the gene's "point of view". Another important early contribution in this line of thought was J. Maynard Smith's (1964) distinction between group selection and kin selection. Richard Dawkins' *The Selfish Gene* (2006[1976]) follows in the above tradition of inclusive fitness/kin selection theory, where the models are based on the gene's "point of view".

In the same period of time, however, there was a continuous effort to make models of group selection work. Lewontin and Dunn (1960) published a paper on the evolutionary dynamics of the t-allele in the house mouse (which was acknowledged by Williams (1966) as a possible example of group selection), and later Lewontin published an influential paper entitled "The Units of Selection" (1970), in which he defended group selection. In the 70's more work on group selection appeared. A table listing more of the work on group selection from 1970 - 1994 can be found in Wilson and Sober (1994), and many of the conclusions drawn from this tradition are summarized in Sober and Wilson (1998). Today, it seems, a good case can be made for the claim that both of the two above-mentioned research traditions, i.e. genic selection and group selection, can be seen as two inter-translatable mathematical frameworks, where researchers favour one approach over the other for either heuristic (Kerr and Godfrey-Smith 2002), semantic or sometimes perhaps even political reasons. (Segerstråle 2000)

Furthermore, the role of cultural selection in facilitating altruistic behaviour has been explored. Most notably Boyd and Richerson (1985) have proposed that culture plays a considerable role in facilitating many cases of human altruism, through the process of cultural group selection. The implication is that culture makes group selection a much more potent force by making selfish behaviour more costly through reward and punishment. The model however, is still strictly naturalistic and implies that culture has biological implications, even though it highlights the role of culture as a causal force. Let me note here that this point should be of special interest to some of the hysterical critics of socio-biology, who directly accuse this science of genetic determinism and other evils: An evolutionary explanation is not exclusively a genetic explanation but always involves environmental factors, and in the case of humans, at least, sometimes these environmental factors are cultural factors.

In summary: The idea of morality as an adaptation created by natural selection was proposed by Darwin (1981) [1871] and has since found support in the writings of a growing number of evolutionary biologists, anthropologists, sociologists, psychologists, economists and philosophers, where results from kin selection theory, evolutionary game theory, and multi-level selection theory from the 1960's and onwards have confirmed Darwin's proposal. (Barkow et al. 1992, Boehm 1999, Dawkins 2006 [1976], de Waal 1996, Fehr & Gächter 2002, Gintis et al. 2003, Haidt 2007, Hamilton 1964, Maynard Smith 1964, Price 1970, Richerson & Boyd 2005, Sober & Wilson 1998, Trivers 1971, Wilson & Wilson 2007). In other words: That altruistic adaptations exist is a well-established scientific fact.

The Naturalistic Fallacy and Error-Theory

What are we to make of all this and what implications, if any, does it have on questions of objectivity and realism in metaethics? Well, pondering the scientific facts a number of philosophers and scientists have claimed that there can be no such thing as objective morality, because evolution shows us that "morality is a collective illusion foisted on us by our genes", as Michael Ruse (1986, 253) forcefully phrases the point. The conclusion being that if you are a naturalist and a Darwinist the only way to go is to become an anti-realist when it comes to the meaning of moral terms, because there are no objective facts for these terms to refer to: There are only the dictates of the hypothalamus.

To fully understand Michael Ruse's position, as well as my own, I think it will be helpful with some philosophical exegesis. Ruse's theory, and especially his insistence that morality is an illusion, is in fact a way of avoiding the problem known as the *naturalistic fallacy* (Ruse 1986, 86-90, 256). It has become a familiar insight in philosophy that someone commits the naturalistic fallacy when they make an illegitimate move from a descriptive domain to a normative domain. However, it seems that some terminological clarification is required here because of the fact that the term "naturalistic fallacy" has been used ambiguously as shorthand for two distinct theses in the philosophical literature.

One of these is the thesis by G. E. Moore, who coined the term in *Principia Ethica* (1903) and consequently, "naturalistic fallacy" ought to refer to the thesis, based on Moore's famous *Open Question Argument*, stating that any attempt to define the term "good" as some natural property N will be fallacious because for any such attempt it will always be possible, without conceptual confusion for someone familiar with the meaning of the terms "good" and "N", to ask of any X that is "N" if it is also good, i.e. it is always an open question, according to Moore, if anything which is "N" is also good. Anyone attempting to make an analytical definition of "good" in terms of a natural property will therefore commit the naturalistic fallacy, because the terms "good" and "N", if analytically equivalent, would leave no room for open questions of the sort described (compare "A bachelor is an unmarried man").

The second thesis derives from a famous passage by Hume:

> "In every system of morality, which I have hitherto met with, I have always remark'd, that the author proceeds for some time in the ordinary way of reasoning, and establishes the being of a God, or makes observations concerning human affairs; when of a sudden I am surpriz'd to find, that instead of the usual copulations of propositions, is, and is not, I meet with no proposition that is not connected with an ought, or an ought not. This change is imperceptible; but is, however, of the last consequence. For as this ought, or ought not, expresses some new relation or affirmation, `tis necessary that it shou'd be observ'd and explain'd; and at the same time that a reason should be given, for what seems altogether inconceivable, how this new relation can be a deduction from others, which are entirely different from it" (Hume 1978[1740], 469)

Hume's thesis is simply that you cannot deduce an "ought" from an "is", and the term "naturalistic fallacy" is very frequently used to refer to someone who breaks this principle. By now it should be clear that the two theses referred to by the term "naturalistic fallacy" are distinct. One can break Moore's principle without breaking Hume's and vice versa.

The first of these two theses, that is Moore's, seems to be of little relevance for naturalistically minded philosophers or scientists in general, since they do not normally claim to be engaged in making analytical definitions of any kind, rather one of the central tenets for any such philosopher is captured nicely by the title of a Daniel Dennett paper: "Out of the Armchair and into the Field" (Dennett 1988). Any naturalistic definition of a term will be empirical and can therefore never be viewed as an analytical definition. Modern attempts to revive Moore's "Open Question Argument" actually seem to be drawing on Hume's insight by claiming that the problem for the naturalist is to explain why there is a conceptual connection between judging something "good" and being motivated accordingly, when there is no such connection between judging something to be "N" and being motivated. (See e.g. Darwall et. al. 1997, and Miller 2003). It seems therefore that the thesis of most relevance for the naturalist is Hume's and this may explain the tendency to use Moore's term for Hume's thesis, since this apparently is the most troubling problem for the naturalist: How can descriptive statements generate prescriptive statements? How can facts be action-guiding?

A simple solution to this vexing problem is to insist that *the facts* just *cannot* be action-guiding? One form such an insistence could take is simply to claim that the normativity or action-guiding character of moral judgements comes from their close connection to the emotional side of our nature. *Moral judgements do not describe anything*. In spite of appearances, the primary function of moral judgements is not to describe any facts but they should instead be understood as expressions of emotion, sentiment or norms, intended to show how certain acts are valued. This view is known as *expressivism* and has been propounded and developed by several philosophers, most notably by Ayer (1936), Stevenson (1937), Hare (1952) and more recently by Blackburn (1984, 1998) and Gibbard (1990). The primary notion here being that moral discourse has a non-cognitive nature, i.e. that it has no truth-conditions. The *surface structure* of moral discourse may seem cognitive, but its *deep structure* is non-cognitive. Hence the generic term non-cognitivism for all of these positions.

Another form of the insistence that facts are not action-guiding can be found in the claim that we should keep up appearances, so to speak. The surface-structure of our moral discourse should not be seen as misleading. Rather, moral judgements do indeed purport to describe facts, and do have truth-conditions, also in terms of their deep structure. *The facts described just do not exist*. Our moral judgements are all systematically and uniformly false. This solution was proposed by Mackie (1977) as a so-called *error-theory* for moral judgements, and it is this solution that Ruse defends in his writings on evolution and ethics, even though he sometimes misleadingly classifies himself as a non-cognitivist (e.g. Ruse 1995, 257).

The background of Ruse's error-theory therefore can be found in Mackie's original error-theory. It was primarily defended by two sceptical arguments. One of these was simply the observation that morality seems to be culturally relative. The other argument, which was Mackie's main argument, is known as "The argument from queerness" and comes in two steps. First there is a conceptual claim that moral judgements have truth-conditions and purport to describe certain facts of value that are objectively and categorically prescriptive. This is the reason why the error theorist is not directly classified as a non-cognitivist, even though Mackie's and Ruse's positive accounts are very similar to the above-mentioned non-cognitivist positions. The second claim is ontological and simply asserts that no such facts exist because they would be unnatural or "queer":

> If there were objective values they would be entities or qualities or relations of a very strange sort, utterly different from anything else in the universe (Mackie 1977, 38).

And,

> An objective good would be sought by anyone who was acquainted with it, not because of any contingent fact that this person, or every person, is so constituted that he desires this end, but just because the end has to-be-pursuedness somehow built into it. Similarly, if there were objective principles of right and wrong, any wrong (possible) course of action would have not-to-be-doneness somehow built into it. (Mackie 1977, 40)

Again the quotes exemplify that the problem for Mackie is to understand how facts can be action-guiding and objective at the same time. His solution is the simple one of denying that anything can be that way. Our moral judgements are massively and universally fallacious. The reason for our continued use of false moral judgements is, according to Mackie, simply that we objectify our values, following some simple "patterns of objectification", and Ruse concurs. (Mackie 1977, 42; Ruse 1986, 253).

What Ruse contributes to Mackie's argument is simply a level of scientific sophistication because of his general insight into and interest in evolutionary explanations: We have a full explanation of why we have the moral beliefs and sentiments we do from evolutionary theory along the lines alluded to in the introduction of this essay. No further justification of our morality is needed and moral objectivity will therefore be redundant. But why not then just accept a straightforward non-cognitive account? The reason is that the belief in an objective morality, i.e. the belief in objectively and categorically prescriptive facts, is somehow adaptive, which Mackie apparently also held (e.g.1977, 43). Belief in an objective morality is, according to Ruse, somehow fitness-enhancing *even though it is false*, it is, in other words, a useful fiction:

> We do not claim that people are either unthinking genetic robots or that they co-operate only when the expected genetic returns can be calculated in advance. Rather, human beings function better if they are deceived by their genes into thinking that there is a disinterested objective morality binding upon them, which all should obey. (Ruse and Wilson 1986, 179)
> The Darwinian argues that morality simply does not work (from a biological perspective), unless we believe that it is objective. Darwinian theory shows that, in fact, morality is a function of (subjective) feelings; but it shows also that we have (and must have) the illusion of objectivity. …If morality did not have this air of externality or objectivity, it would not be morality and (from a biological perspective) would fail to do what it is intended to do. …In a sense, therefore, morality is a collective illusion foisted upon us by our genes. (Ruse 1986, 253).

First of all, let me note that the question of whether peoples' concepts of morality and what is required of them really involves reference to something objective, i.e. that they *actually* believe in an objective morality is a question that we can only answer by historical and empirical studies. My idiosyncratic impression is that many people, at least in Denmark, have no such beliefs, even though people here in general are very inclined to act in what is normally considered morally respectable ways. (So maybe we are slowly moving toward self-destruction) Furthermore, at least one empirical study conducted by Shaun Nichols seems to undermine this claim in that he found that around 40 percent of respondents in his study seemed to be non-objectivists about morality (Nichols 2004). But for the sake of argument let us grant Ruse the point.

With evolution naturally intended to include different environmental factors, social processes, enculturation, etc., the position of Ruse can be summarized as follows: No ethical statements can be true, because our beliefs in ethical statements can be explained exclusively by evolution.

The Genetic Fallacy and Dependence Relations

The error-theoretic position of Ruse has been pinned down schematically by the philosopher Elliott Sober (1994, 104), who interprets Ruse and Wilson (1986) in the following way:

> (G) We believe the ethical statements we do because of our evolution and because of facts about our socialization.
>
> No ethical statements are true (Sober 1994, 104)

This type of argumentation is known as a *genetic* or *genealogical* argument, because it draws conclusions about the truth of a certain belief on the basis of the *genesis* of that belief. The standard view is that this type of argument commits the so-called genetic fallacy, because it involves a failure to distinguish between the context of discovery and the context of justification.

Sober, however, reminds us that not all genetic arguments are false, only deductive genetic arguments are all false, because we can never make a strict deduction about the truth or falsity of a belief, i.e. its context of justification, based on knowledge of how it was obtained, i.e. its context of discovery. Consider, he says, how the chemist Kekulé, when working to determine the structure of Benzene, came up with the idea of the benzene ring:

> "After a long day at the lab, Kekulé found himself gazing wearily at a fire. He hallucinated a pair of whirling snakes, which grabbed each other's tails and formed a circle. Kekulé, in a flash of creative insight came up with the idea of the benzene ring" (Sober 1994, 105).

We cannot *deduce* whether the structure of benzene as a ring, is a true or false hypothesis because of the way the hypothesis was generated. For that we need experiments and other ways of obtaining independent evidence. We can, however, say something about the probability that some belief is true, that is we can *infer* something about the truth of a statement by genetic arguments, because there is more to inference than strict deduction.

Sober considers two ways of obtaining information about the number of students in a philosophy class. One person, Ben, decides that there are 78 students in the room by drawing a number at random from an urn. The other person, Cathy, decides that there are 34 students by counting the people in the room. Now, Sober's point is that Ben's belief about the number of students has a very low probability of being true, whereas

Cathy's belief has a very high probability of being true. Naturally, it could be that Cathy is very poor at counting and that Ben is extremely lucky, but this only means that we *cannot deduce* anything about the truth of their beliefs. It does not mean that we cannot *infer* anything about their respective probabilities of being true.

The reason that we assign different probabilities to the two ways of obtaining information about the number of students in the room has everything to do with the ways their respective beliefs were formed. Ben's method of obtaining information, i.e. drawing a random number, has nothing to do with the belief held, and there is clearly an independence relation, between method and belief held. Cathy's method conversely has very much to do with the fact of the matter, i.e. the number of students actually present and so there is a dependence relation between method and belief held in this case. The general conclusion to draw from Sober's example is that a belief is probably untrue if it is obtained by methods that are entirely independent of the facts of the matter, i.e. the truth conditions of the belief. And conversely there is a high probability that a belief is true if it is obtained by methods that are dependent on the facts of the matter.

This leads Sober to conclude that Ruse and Wilson could make a good genetic argument out of a bad one, i.e. Ruse and Wilson's original and flawed genetic argument G by adding the extra premise, which states that the processes determining our moral belief are independent of their truth. This makes us able to extract the following non-deductive and plausible genetic argument, (Sober 1994, 107) here named GP to indicate that it is a *genetic* argument about the *probability* that a belief is true, and therefore not a deductive argument:

> (GP) We believe the ethical statements we do because of our evolution and because of facts about our socialization.
>
> The processes determining our moral beliefs are entirely independent of which moral statements are true or false.
>
> The ethical beliefs we currently have are probably untrue.

With Sober I think the conclusion to draw is that, Ruse has no effective argument for the error-theory of moral judgements, unless he can provide us with a strong defence of the second premise, and he has so far done no such thing. Ruse, however, feels otherwise. In replying to Sober's criticism (Ruse 1995, 271), he says that he believes something like the missing premise is true and that he has, in fact, argued for its truth by an analogy. His analogy is that when bereaved people obtain beliefs about their lost ones through an Ouija board, we explain their beliefs by saying e.g. that they deceive themselves because of their need for psychological relief from some kind of anxiety (Ruse 1986, 257). Even if the deceased were actually in heaven playing the harp, or perhaps nowadays playing poker with St. Peter, we would still say that the psychological explanation offers "more than adequate causes for the belief" (Ruse 1995, 271).

This is the same point that Ruse intends to apply to ethical beliefs, we can only scientifically explain why we have these beliefs, we cannot justify their truth by science, because we could have had another set of moral beliefs, had our evolution and socialization been different, e.g. "we could have believed, systematically, that one ought to hate one's neighbour" and that this fact "does indeed point to the independence of moral beliefs from moral truths." (Ruse 1995, 269-270).

Objectivity and Adaptations

The above move by Ruse, however, seems like a very strange one for a naturalistically minded philosopher to make; does Ruse really believe that moral truths are given? That e.g. "Killing is wrong" is a moral truth and that because we would not necessarily have had this belief, as apparently many people do and do not, it is implied that our beliefs have no connection with these moral truths. Naturally, Ruse would reply that there are no such moral Truths with a capital T, but only some kind of internal truths that are true by virtue of the morality game (Ruse 1986, 257). Apparently, we can only gain moral objectivity or moral realism by referring to properties that are external to biological and psychological facts about human beings. But as Sober points out (1994, 102) no such thing follows. The moral realist can point out that the morally relevant facts are precisely these facts about our biology, sociality and psychology. And that is all we need for realism and objectivity.

Rottschaefer and Martinsen (1990) make a similar point in their extensive critique of Ruse's position. They extract from Ruse (1986) two distinct criteria for objectivity and subjectivity of values. On Ruse's account, values are objective (1) if they exist independently of humans in some non-physical way or (2) if they are imposed on us externally. However, values are subjective, (3) if there is no right and wrong without human nature or (4) if human nature is the only source of morality. As they point out there is no logical problem from these criteria for Ruse to maintain that values are subjective in sense 3 and objective in sense 2, i.e. one could simply hold that the human genetic make-up is subjective because it is internal to human nature but that the social and environmental selective forces are not. As they phrase it:

> Only if morality is constituted by human nature, could one argue that it is not objective in either sense of the term. But to claim that morality is constituted by human nature contradicts Ruse's account of the moral sentiments. Human nature is a necessary, but not a sufficient, condition for the existence of moral sentiments since on the Darwinian account human nature itself depends on both genetic and selective environmental factors. (Rottschaefer and Martinsen 1990, 158-9)

Here they claim Ruse could reply by appealing to colour terms, and the view that colour properties are not in any straightforward sense objective, because the property of being "red" relies partly on some causal properties of objects and partly on normal perceivers and normal perceptual conditions. Redness thus, is not a property of the objects. Similarly, he could claim that the same applies to moral goodness, because it is also constituted partly by properties of the environment and partly by the genetic make-up. However, Rottschaefer and Martinsen argue that in the case of colour, the objectivity of naïve realism has been replaced by scientific realism and thus, it is possible for the realist to argue that there are both objective sources for and objective referents of the perception of colours, namely the micro-physical surface structures and the electromagnetic radiation. And they claim that the robust Darwinian can make the same move with regard to moral properties:

> A robust Darwinian has a general characterization of these properties which derives directly from evolutionary theory. They are the properties of the natural and social environment to which the person is adapted and in terms of which one can claim that the person is fit. As such they are relational properties. (Rottschaefer and Martinsen 1990, 160).

Ruse (1995, 290) replies to this by reminding us of his views about the possible alternative trajectory for our evolved morality, which implies that it could have been quite different, but that this does not hold for colour experience: "I can imagine it being true that one ought to hate and try to cheat neighbours. I cannot imagine that the sky be red and the grass blue." This is reminiscent of his reply to Sober, and again the suggestion is baffling. It may be true that the current physical state of the world and our evolved perceptual apparatus give us our current colour experience as it is, but had our evolutionary trajectory been markedly different, then so would our colour experiences. Still we would be able to tell the story about that different set of micro-physical surface structures, the radiation, the perceptual apparatus, and the normal conditions that were responsible for the alternative set of colour experiences. Ruse gives us no reason to think otherwise, and has so far not supplied us with a good argument for his subjectivist position.

There seems therefore to be a road to moral objectivity that is not blocked by Ruse's version of the error-theory. This objectivity, however, is not species independent, and hence would probably not be accepted by Ruse as such, e.g.: "No abstract moral principles exist outside the particular nature of the individual species" (Ruse and Wilson 1986, 179). Rottschaefer and Martinsen (1990) try to build this kind of alternative realist theory that is not species independent by maintaining that morality is objective in sense 2 above, i.e. that moral beliefs are imposed on us externally through the environmental and social properties of the environment to which the person is adapted. Elaborating on this, they propose that we, instead of doing the "end-run round the is/ought barrier"(Ruse 1986, 256) by claiming that objectivity is an illusion, rather should invoke the notion of supervenience and emergence when accounting for moral properties and their objective nature. We simply avoid the naturalistic fallacy by claiming that the moral properties are not reducible to natural facts (Rottschaefer and Martinsen 1990, 161). Much can be said in favour of this approach (e.g. Boyd 1988, Sturgeon 1986), but because we are left with an unexplained notion of supervenience, many questions will still be left unanswered by this type of approach.

An alternative but complimentary approach to that of Rottschaefer and Martinsen is taken by Harms (2000). He invokes what he calls "the current best theory of meaning", referring here to the "biosemantics" or "teleosemantics" of Ruth Millikan (1984, 1993). I fully agree with Harms in his praise of Millikan but, unfortunately, the strength of her theory is still not widely appreciated by philosophers.

On the teleosemantic account the "truth of a signal just is the fact that the signal now stands in the kind of relation to the world which explains the historical contribution to fitness of the signalling system. There are true and false bee dances, true and false monkey warning cries, and if there is the kind of adaptive history we began by supposing, then there are (objectively) true and false moral statements/imperatives." (Harms 2000, 701). Obviously, this is somewhat simplified but the full detailed version of teleosemantics and its implications for more complicated forms of language is far beyond the scope of this paper.

Like Rottschaefer and Martinsen, Harms suggests that species independence is not required for objectivity and explains why this is and should be so. Because of our adaptive histories, we as human beings have numerous different adaptations, some of which we share with other species and some of which are unique for our species. Adaptations have functions and so are capable of malfunctioning. When an adaptation performs properly it does what that particular subsystem was designed by natural selection to do, i.e. in the ways that historically contributed to the survival and reproduction of the system of which it is part. This means that for any such adaptation there will be an *objective* set of standards for proper functioning, and hence there will be a possibility of failure to perform according to function. (Harms 2000, 702)

Harms points to the example of the signalling system used by vervet monkeys in Kenya, as described by Cheney and Seyfarth (1990). Part of the signalling system deals with approaching predators, and because there are three kinds of predators, the vervets predictably have evolved three different signals with three subsequent avoidance strategies: Leopard = run up a tree, Snake = stand up and look around on the ground, Eagles = look up into the sky. It is the coordination between signal, predator and avoidance behaviour that generate the truth conditions of an emitted signal. So if a particular screech is the usual signal sent when a leopard approaches, and that historically resulted in the vervets running up a tree, then this screech is true just when a leopard is actually approaching, and false if it is not (Harms 2000, 704).

One particularly interesting feature of the approach is that it makes no difference to the question of truth whether you construe the screech as an indicative or an imperative statement, i.e. "There is a leopard" or "Run up a tree!" There is no difference between the two in terms of their capability of being true. Truth is prior to indicative content on a teleosemantic account of signalling systems, and hence does not depend on the possibility of assigning descriptive content to a signal. "What it does depend on is whether the signal fulfils or does not fulfil its historical role in the function for which the issuing system was selected" (Harms 2000, 705). This has the interesting consequence that we can be objectivist realists about morality, whether or not we decide to be cognitivists or non-cognitivists with regard to moral language. If it is true, as I believe, that the main reason for the non-cognitivist to claim that moral expressions are not truth apt is because of the widespread assumption that indicative content is required for truth, then we can be objectivist realists while still retaining the core of the non-cognitivist position, which is the insistence that morality is somehow action-guiding.

Harms also offers an example where the only species involved in the sending of signals are vervets. Evidently, this is to avoid the objection that his example involves more than one species, which is not the case for most of our thoughts of morality, even though the interests of other species recently have become a more widespread concern. He considers a case where signals are sent when a vervet accidentally attacks a member of its own troop, or when mothers pick up the wrong infant. Naturally, the teleosemantic story is the same as before: if the signalling system is one of adaptive signals and responses, then there will be objective truth conditions for the respective screeches and hisses. So the general conclusion is the same as before. Even though our human moral language is highly complex, the main points of teleosemantics still hold for human morality, i.e. our moral imperatives and justification claims, as long as it is considered to be a system of adaptive signals, because for any such system there will be objective truth conditions:

> Notice that this does not require a *separate* realm of moral facts inaccessible to scientific enquiry. All that it requires is that morality is an adaptation, and that the way that the moral hardware has paid its bills is by getting certain sorts of signals sent and responses made in particular kinds of circumstances. The signal is true if those circumstances obtain, false if not. My statement/imperative "you shouldn't do that" is morally true if the issuance of this fulfils at least my end of the function of the moral regulatory system. My claim that killing an intruder is justified because he posed a threat to my family is true just in case the language of justification is fulfilling *its* distinctive function. (Harms 2000, 705).

This basic insight is, to put it plainly, all that we need to vindicate the realist objectivist position in metaethics. Furthermore, it is a strictly naturalist position fully compatible with Darwinism, and a position that is not dependent on the claim that the moral structures, social structures, psychological structures, biological structures are supervenient on the natural world, i.e. reduction is not precluded. In fact, I believe the teleosemantic account is also particularly strong in response to Sober's above-mentioned demand for clarity about whether a dependence or independence relation between the moral beliefs and the truth making features obtain, when we make genetic arguments about the probability of a certain statement or class of statements being true or false. The teleosemantic account first of all has a very straightforward answer to how these truth making features are constituted, and hence whether a dependence relation obtains between belief and truth making features. The truth making features are the ones that historically contributed to the survival and reproduction of the ancestors of the adaptive system of which the belief/imperative in question is a part. And Ruse has offered us nothing to show that there *always* is an independence relation between our moral beliefs and the facts of our adaptive history, apart from an obscure analogy concerning messages derived from an Ouija board.

So the only thing to conclude about Ruse's argument that Darwinism shows us how morality is a collective illusion is: Nothing could be further from the truth. The fact that altruistic adaptations exist is precisely what paves the way for moral objectivity and moral realism, albeit not a species independent objectivity and realism. There is a *fact* of the matter in the adaptive history of humans. If human beings are adapted in a way that constrains our behaviour in social interactions then this is a fact we can be wrong or right about, when uttering our moral convictions. Similarly, as there are objective rules for the correct issuance of warning signals in the vervet monkeys, so are there rules that apply to our moral utterances, as well.

Conclusion

In this paper I have investigated what conclusions to draw in philosophical ethics from the fact that morality is an adaptation. We have seen that the attempt by Ruse and others to argue for the claim that we must be anti-realists regarding the meaning of moral terms fails. They claim that an evolutionary understanding of altruism shows us that there is no such thing as objective morality and that in this sense, morality is an illusion created by our genes. This argument is fallacious. First of all, the argument is not successful, because it is a bad genetic argument, and when Ruse tries to improve it we are really offered nothing but a bad analogy. Secondly, it seems that the position is largely due to some confusions concerning what it takes for something to be objective in a robust sense. Consequently, I have argued that within evolutionary naturalism there is plenty of room for a notion of objective morality strong enough to support moral realism, and therefore have shown that if morality is an adaptation, as the evidence strongly suggests, then we should indeed accept moral objectivity and realism in metaethics.

Acknowledgments

Thanks to Kasper Lippert-Rasmussen and Ryan Mckay for valuable comments. Also thank you to those who commented at my presentation at the Human Mind – Human Kind conference.

References

Ayer, A.J. (1936). *Language, Truth and Logic*. London: Gollancz.

Barkow, J.H., Cosmides, L. & Tooby, J., Eds. (1992). *The Adapted Mind.* New York: Oxford University Press.

Blackburn, S. (1984). *Spreading the Word.* Oxford: Clarendon Press.

—. (1998). *Ruling Passions*. Oxford: Clarendon Press.

Boehm, C. (1999): *Hierarchy in the Forest: The Evolution of Egalitarian Behavior.* Cambridge, MA: Harvard University Press.

Boyd, R. (1988). How to be a moral realist. In Sayre-McCord (1988).

Boyd, R. & Richerson, P J. (1985). *Culture and the Evolutionary Process.* Chicago: University of Chicago Press.

Cheney, D. L. & Seyfarath, R. M. (1990). *How Monkeys See the World.* Chicago: University of Chicago Press.

Darwall, S., Gibbard, A. & Railton, P. (eds.) (1997). *Moral Discourse and Practice*. New York: Oxford University Press.

Darwin, C. (1981) [1871]. *The Descent of Man and Selection in Relation to Sex.* Princeton: Princeton University Press.

Dawkins, R. (2006) [1976]. *The Selfish Gene*, 30th ann. ed. Oxford: Oxford University Press.

Dennett, D.C. (1988). Out of the armchair and into the field. *Poetics Today* 9, 205-221.

de Waal, F. B. M. (1996). *Good Natured: The Origins of Right and Wrong in Humans and Other Animals.* Cambridge, MA: Harvard University Press.

Fehr, E. & Gächter, S. (2002). Altruistic punishment in humans. *Nature* 415,137-140.

Gibbard, A. (1990). *Wise Choices, Apt Feelings*. Cambridge, MA: Harvard University Press.

Gintis, H., Bowles, S., Boyd, R. & Fehr, E. (2003). Explaining altruistic behavior in humans. *Evolution and Human Behavior* 24, 153-172.

Haidt, J. (2007). The new synthesis in moral psychology. *Science* 316, 998-1002.

Hamilton, W.D. (1964). The genetical evolution of social behaviour I & II. *Journal of Theoretical Biology* 7, 1-52.

Hare, R.M. (1952). *The Language of Morals*. Oxford: Clarendon Press.

Harms, W.F. (2000). Adaptation and moral realism. *Biology and Philosophy* 15, 699-712.

Hume, D. (1978) [1740]. *A Treatise of Human Nature*. Oxford: Clarendon Press.

Kerr, B. & Godfrey-Smith, P. (2002). Individualist and multi-level perspectives on selection in structured populations, *Biology and Philosophy* 17, 477–517.

Lewontin, R.C. (1970). The units of selection. *Annual Review of Ecology and Systematics* 1, 1-18.

Lewontin, R.C. & Dunn, L.C.(1960). The evolutionary dynamics of a polymorpism in the house mouse, *Genetics* 45, 705-722.

Mackie, J. L. (1977). *Ethics: Inventing Right and Wrong*. New York: Penguin Books.

Maynard Smith, J. (1964). Group selection and kin selection. *Nature* 201, 1145-1147.

Miller, A. (2003). *An Introduction to Contemporary Metaethics*. Cambridge: Polity Press.

Millikan, R. G. (1984). *Language, Thought, and Other Biological Categories: New Foundations for Realism.* Cambridge, MA: MIT Press.

Moore, G.E. (1956)[1903]. *Principia Ethica.* Cambridge: Cambridge University Press.

Nichols, S. (2004). After objectivity: An empirical study of moral judgment. *Philosophical Psychology* 17, 3–26.

Price, G. R. (1970). Selection and covariance. *Nature* 227, 520-521.

Richerson, P. J. & Boyd, R. (2005). *Not By Genes Alone. How Culture Transformed Human Evolution.* Chicago: University of Chicago Press.

Rottschaefer, W.A. & Martinsen, D. (1990). Really taking Darwin seriously. *Biology and Philosophy* 5, 149-173.

Ruse, M. (1986). *Taking Darwin Seriously*. Oxford: Blackwell.

—. (1995). *Evolutionary Naturalism*. London: Routledge.

Ruse, M. & Wilson, E.O. (1986). Moral philosophy as applied science. *Philosophy* 61, 173-192.

Sayre-McCord, G. ed. (1988). *Essays on Moral Realism*. Ithaca, N.Y.: Cornell University Press.

Segerstråle, U. (2000). *Defenders of the Truth: The Battle for Science in the Sociobiology Debate and Beyond.* Oxford: Oxford University Press.

Sober, E. (1994). Prospects for an evolutionary ethics. In Sober, E.: *From a Biological Point of View*. Cambridge: Cambridge University Press.

Sober, E. & Wilson, D. S. (1998). *Unto Others: The Evolution and Psychology of Unselfish Behaviour*. Cambridge: Harvard University Press.

Stevenson, C.L. (1937). The emotive meaning of ethical terms. *Mind* 46, 14–31.

Sturgeon, N.L. (1988) [1984]. Moral explanations. In Sayre-McCord (1988).

Trivers, R.L. (1971). The evolution of reciprocal altruism. *Quarterly Review of Biology* 46, 35-57.

Williams, G.C. (1966). *Adaptation and Natural Selection*. Princeton: Princeton University Press.

Wilson, D.S. & Sober, E. (1994). Reintroducing group selection to the human behavioral sciences. *Behavioral and Brain Sciences* 17, 585-608.

Wilson, D.S. & Wilson, E.O. (2007). Rethinking the theoretical foundation of sociobiology. *Quarterly Review of Biology* 82, 327-348.

CHAPTER FOURTEEN

THE EVOLUTION OF MARTYRDOM? CONSIDERATIONS IN THE STUDY OF RELIGIOUS SELF-SACRIFICE

ANDREAS LIEBEROTH

Abstract: Evolutionary principles such as inclusive fitness have been suggested to explain religious self-sacrifice. Evolutionary arguments are explored and discussed, and it is concluded that the most promising theories point to general purpose cognitive features influenced by situational factors. Cognitive phenomena such as will, imagination, social sensibility and the mind's embeddedness in culture are all necessary, if not sufficient, for the planning and execution of self- sacrificial acts. This implies that there can be no single 'trigger condition' or 'self-sacrifice gene'; only a cluster of cognitive and psychological functions working with and against each other in emotion, planning and action.

You did indeed yearn for death before you met him. Now you have seen him with your own eyes, (and you flinch!)
—The Holy Qur'an, Sura 3: 143

He said, I locked you in this body, I meant it as a kind of trial
You can use it for a weapon, or to make some woman smile
—Leonard Cohen "Lover lover lover".

Introduction

One of the most staggering paradoxes of human existence is the apparent contradiction between evolved systems for self-preservation and humans who are willing to sacrifice their own lives for kin or a cause.

This chapter discusses evolutionary explanations for other-benefiting suicide, their strengths and obvious shortcomings, and attempts to point out insights from the cognitive sciences that may contribute to an understanding of suicide terrorism. The chapter maintains the position that evolutionary psychology can elaborate on certain cultured phenomena, and although a one-to-one relationship between genes and complex actions like suicide attacks may be alluring, it makes little scientific sense.

Suicide attacks have occurred as long as humans have waged war to protect what they hold precious. To limit the scope of this exploration, we focus on contemporary cases in the Muslim Middle East. Here, martyrdom is typically pronounced post-mortem over diligent followers of the faith, but not until the 20th century did we reach the technological capability to make single acts of self-sacrifice militarily significant. The role of the martyr has always been symbolic and largely remains so today, but communications, politics and ways of life have changed in the 21st century, and new martyrologies have found root in public imagination (Cook, 2007). The destruction caused by the many human bombings in Palestine may have little strategic significance in the war fought against Israeli interests, but the signal-value of the willingness to become a martyr for the cause, the ability to get to anyone anywhere (e.g. the assassination of Benazir Bhutto in 2007) through disregard for one's own safety, and the fear that ripples through society in the wake of terrorist-attacks, hold immense psychological value; both in regard to rallying people to the cause, and crippling the enemy's infrastructure and security. In this sense, terrorism has no time and place, like Napoleonic battlefields once did (Baudrillard, 1983). Instead it strikes without warning, and ripples beyond the instance of destruction to create waves of sentiment and expand existing cultural narratives. Thus, suicide attacks end the life of the attacker, but promote the fitness of ideas.

The will to self-destruction only makes sense as a psychological phenomenon if we accept that human existence is fraught with inputs, motives, associations, emotions, fears and appetites tugging at us from different directions. To steer a clear path beyond simple hedonistic indulgence, we need a certain amount of self-control, goal representation and step-by-step planning. Especially if a cause of action might result in risk and hardship. Therefore, recent studies of will (Schulkin, 2007; Elster, 2000, 2005) may aid us in understanding suicide terrorism.

This chapter starts out by presenting cultural and social background factors for religious sacrifice, martyrdom and Middle Eastern militancy, and then moves on to discuss the evolution of social sensibilities, will and the human mind as a battleground between conflicting motives. On the way, different theories of inclusive fitness and institutional manipulation are presented, and discussed in relation to higher order human consciousness.

But first, a quick look at some seemingly parallel behaviours in other corners of biological reality.

Spite, Sacrifice and Altruism in Nature

Genetic theories of altruism, spite, youthful aggression and inclusive fitness have on occasion been suggested as explanations for human warfare and the will to sacrifice (Blackwell, 2005; Blackman, 2004; Atran & Norenzayan, 2004). Divisions of labour, where some specimens forego sustenance and reproduction, fall well within the Dawkinsian paradigm of the "selfish gene". The cells of our organisms are either reproductive "germ" or sterile "soma", but all readily perish in the interest of the complex whole. Such altruism can further be combined with spite (the evolutionary principle of eliminating competition), in cellular "suicide bombings", where "microbal martyrs" release bacteriotocins in death, killing sensitive competitors, but sparing their own relatives (Blackman, 2004) Genetic codes for evolved self-inhibition tantamount to suicide can now also be identified in certain forms of algae, where Nedelcu & Michod (2006) recently found the gene regA to suppress cell growth and thereby suspends reproduction in the algae Volvox Carterii. In a cousin to Volvox, the gene Crsc13 also curbs cell-growth to suspend temporarily useless activities, where scarce resources might be applied better elsewhere. The genes are deemed altruistic since they make carriers express "self sacrificial" features only under conditions, where others might benefit. Had the algae's life been brighter, more nourishing, or in other ways happier, expression of the given trait might not have occurred. In other words, we may be well under way to discovering the genetic basis for genuinely altruistic features, which compels organisms to delay gratification or forego fitness for the long-term good of the collective. Genetic predictors for socially oriented personality traits are even now being discovered in humans (i.e. Ridley, 2003).

In the world of social insects both inherent cooperation and uncompromising spite is exceptionally well documented. A significant portion of a hive matures to be sterile workers and soldiers, serving the interest of their queen who carries and reproduces their genes. Even more interesting, these genetic serfs can also serve as a sort of "gene police", disposing of the occasional worker who goes rogue and starts reproducing like a queen. In argentine ants, soldiers further kill a proportion of male members of the hive, keeping the population majority on the female side, and in red fire-ants, queens with slight genetic deviations are killed outright, keeping the gene pool stable. The same phenomenon can be found in parasitic wasps, where soldiers swiftly kill off the competition within the succulent bodies of the larva, where they are hatched (Blackman, 2004). The fascinating thing about this is that killers prefer non-relative victims, and then categorically move over to killing closer kin, starting with cousins, brothers, then sisters, and preferably sparing clonal sisters whose genetic blueprint they share 100%. Killers they are, but victims can be chosen.

Spite and behaviours dangerous to the individual, however, are not confined to insects and microorganisms. Predators sometimes attack dangerous prey for food, small mammals will attack ferociously when cornered, and battles for status and mating within the flock can have lethal outcomes. Non-alpha males stake their hide in a grab for power, or for a quick tousle with one of the leader's harem (Diamond, 1992) and in hyena, fratricide is a time-honoured tradition among cubs, from the outset eliminating not just unlucky competitors, but carriers of the same genes. Although socially refined and acquired, higher primates occasionally develop outright warfare between tribes, and even against other species such as threatening predators; reciprocal support and investment in the collective are all important to the survival of individuals (Ibid; Dunbar, 1996). In complex systems, some parts must always be sacrificed.

Motivations

Human self-sacrifice is generally wrapped in layers of cultural narrative, institutions, motives and values. It is not unusual for individuals to express wishes to give up everything for their ideals, but creating the right circumstances and carrying through with such an act is rare, and generally reliant on social institutions for practical and emotional support. The wish and its underlying motives do not equal will and opportunity.

Before the will to martyr oneself come certain motivating factors, but sadly, the two phenomena are often confused. Many themes can be discerned in the animal-behaviours mentioned above, but others, such as the idea of eternal life in paradise, are uniquely human. Psychological and psychiatric profiling has been attempted to identify potential terrorists, but most experts agree that the hatching of bombers is usually a product of personal circumstances, desires for social movement and well-planned recruitment, making situational factors much stronger predictors than personal traits, even though some individuals may be more susceptible than others. The "lone wolf killer" is not generally an applicable template, and although trauma and deprivation may be motivating factors (Fields, 2004), most studies indicate that psychopathology is not a typical trait in suicide bombers; quite the contrary, since unbalanced individuals are usually unfit as strategic operatives (Stern, 2003; Atran, 2003; Cook, 2007). The same meta-studies indicate that desires to end one's own life fostered by depression or existential distress are rarely the primary motivation in suicide terrorism.

Although socioeconomic analysis alone cannot supply frames for understanding the allure of becoming a *mujahid* (Muslim holy warrior), there *are* recurring characteristics in young people enlisted for suicide attacks. Since similar phenomena have been seen in other times and places, such as the throngs of young European men flocking to The Holy Land during the crusades, Gunnar Heinsohn (2005) points to the recent demographic "youth bulge" in the Islamic world. Candidates are usually young men recruited from areas suffering from relative economic or political distress. However, random poor Muslims do not fit the bill. Rural workers usually have their hands full providing for their families, and have no time or incentive to worry about political issues, much less abandon their obligations at home to become a militant. On the contrary, recruits tend to have more years of schooling than average[1] (Atran, 2003; Stern, 2003), and live in urban areas where young males face difficulties securing a place in society (Heinsohn, 2005), are exposed to social inequalities and denied job-opportunities. These commonalities in suicide bombers may, however, also be a result of deliberate strategies on the part of recruiters, who routinely target "fighting age males" in urban mosques, or set up charities and madrassah-schools in poorer areas.

Thus, privation coupled with some education are not predictors of engaging in suicide terrorism, but mediating factors for a recruitment-process that ideally starts with "mental preparation" at 10-11 years of age, as Jessica Stern was told by coordinators in one organization (2003). Along this line, alienated Muslims seeking community in mosques around the world seem to be prime candidates for strategic recruitment, owing to language-skills, cultural literacy and situatedness in a potential target-nation. Recruiting and organisational manipulation is now a main focus for terrorism-studies and government operations alike, based on well known social psychological studies conducted by pioneers like Solomon Asch, Stanley Milgram, Phillip Zimbardo, Muzafer Sherif, and even Robert Jay Lifton (Atran, 2003; Stern, 2003; Zillmer, 2006). As we shall see, however, an evolutionary sub-theory may be added to the mix, if not to explain, then to explore the genesis of altruistic action and the human

[1] In countries like Pakistan, however, education usually entails madrassah-schooling, which is often sponsored by fundamentalist organisations, such as Harkat-ul-Mujaihidin (Stern, 2003).

capacity for overcoming pain, fear of death or compassion. The question remains; does this further understanding, or is it simply another attempt by popular evolutionary theory to claw its way into new avenues of the human sciences?

Martyrdom

Muslim martyrdom is not historically or theologically linked to distinctly self-destructive behaviour. Rather, the honorary martyr and associated heavenly rewards are historically attributed to adherents dying a worthy death, which can mean many things including dying of love, and in Shiite discourses often being slain by other Muslims. The Qur'anic term sh-h-d, *shahid,* can mean both witness and martyr, but is traditionally used in the former sense. The term *mujahid,* derived from *jihad* (struggle or endeavour), has increasingly been used to describe holy warriors, and is worn proudly by members of contemporary jihadii-organisations (Cook, 2007; Stern, 2003). Many theological disputes have centred on the validity of suicide bombers as martyrs, since suicide is explicitly forbidden by Islam, and motivations for the martyrium must be pure; hunger for glory or other rewards must not enter into the mind of the prospective martyr; *"the sword does not wash away hypocrisy"* (Ibn al-Mubarak, in Cook, 2007, p. 36). Only God can say who is received in heaven as a *shahid*, but the possibilities were widened considerably in the time after the prophet, where one well-known tradition stated seven categories of martyrdom apart from death in battle, including drowning and dying in childbirth, while several other groups like students and civil servants were later added, indicating the theology of a more settled society (Cook, 2007). Furthermore, only few verses of the Qur'an can be interpreted to single out martyrs for special reward in heaven, while later *hadith*-literature elaborates that theme somewhat, e.g. the famous tradition about the 72 houris and right to intercede on the behalf of 70 relatives. It is therefore important to notice, that 21st century martyrdom is continually being (re)constructed by jihadii organisations embedding their political-religious struggle into the Muslim tradition of Holy War, stemming back to tribal battles in the Prophet's lifetime. Thus, contesting dogmas may be at play in individual and shared imaginations (Augé, 1999) at any one time, even if theological correctness generally prevails.

Aspiring to become a *shahid* requires an alignment between the individual's life-story and the ongoing cultural narrative, up to and including his death. Although personal life-stories are edited along the way, they can refer to long-term storylines, and may mimic cultural fictions, be they religious or recreational. Thus, the human experience of time as "storied", which arises from semantics and autonoetic memories, coupled with strong shared values and images, may be an important part of the individual's will to self-sacrifice. Imagination as an evolved cognitive capability (Lieberoth, 2007) allows humans to navigate storylines owing to culturally acquired points of reference, such as the grand eschatological narratives of the book-religions.

Sacrifice

The concept of sacrifice is known to most world religions as a means of engaging the Gods in exchange, or garnering social status by offering up valuable resources. The willingness to give up oneself for God is usually associated with Middle Eastern terrorist-attacks and cultic mass-suicides, and often regarded as near-pathological expressions of dangerous fanaticism. This is, however, only one extreme end of a scale where we might conceivably also place alms-giving, monastic chastity, doing religious work instead of furthering personal ends, and being persecuted or discriminated against for one's beliefs. Sacrifice seems to draw from the domain of gifts, which has been avidly described in 20th century anthropology, but also differs in many ways. The archetypal image of savages disembowelling enemies before pagan idols whilst rubbing their tummies remains anchored in popular imagination, but apart from certain polemic writings, the history of world religions holds very few indices of human sacrifice. Sacrificial practices usually amount to gift-giving, slaughtering the finest livestock in honour of a religious festival (note the etymological relation to "feast"), and other ceremonial destructions or redistributions of more or less symbolic resources. The sacrificial act is typically a part of a larger ritual system, directing acts towards an otherworldly reality, as well as serving more practical and symbolic purposes in society.

Sacrifices can be symbolic offerings of little actual value, such as candles or a bit of rice, but in any case they involve elaborate exhibition and ritual action. Central to the rationale behind this are communicative displays of giving and participation. Atran and Norenzayan hypothesize that such displays of commitment could be pivotal to group coherence, once our forefathers developed advanced metacognitive abilities for deceit and cheater detection (2004). Thus, the point of sacrifices is not *what* is sacrificed or in essence *to whom*, but the display of general *willingness* to give: "*The result of such convergence is to perpetuate a stable community of cooperators who sacrifice for the group in the short run, but benefit from it in the long run*" (Atran & Norenzayan, 2004, p. 13).

The ideal of offering up human life is consistent with Atran and Norenzayan's framework, and well known in the biblical legacy. Jesus was dubbed the *Agnus Dei*, lamb of God, to emphasize the crucifixion as the final act of sacrifice on behalf of humanity, and the morale in the story of Abraham and Isaac (or Ibrahim and Ismael) is *willingness* to sacrifice that which is most precious, with explicit reference to ceremonial action and the demand for sacrificial blood by God. "We proudly give our son to God", is a much revered phrase in some communities, where jihadii groups hold great sway and boys with little future abound; the poor give their children to the jihad¸ while the rich supply money (Stern, 2003). More often, however, stories of human lives given in the name of god are linked to individual heroic narratives, in which the protagonist or martyr gives his life as part of an important struggle, like Samson pulling down the roof over the philistines or Christian martyrs murdered in defiance of the Roman pagans. Thus, the notion of sacrifice is not necessarily bound up on a ritual context, but can instead be constructed around fictions, and used to bolster the willingness to personal sacrifice in communal struggles. Most sacrificial acts the world over, however, remain symbolic in nature, and usually serve social and practical purposes such as reciprocity, the demarcation of status, payment for services rendered by religious specialists, or as a codified distribution of resources, in addition to ceremonial displays of group coherence and loyalty with reference to omniscient divine recipients.

Sociality, Sensibility and Treacherous Genes

As seen at the start of this chapter, some organisms appear to have been sold out by their genes at birth; more or less destined to give up their life or reproductive opportunities to further the fitness of their biological collective (Blackman, 2004; Nedelcu & Michod, 2006; Wong, 2000). Comparable behaviours are traceable in mammals and even primates and humans, although this mostly translates into higher willingness to risk-taking (i.e. give out a shrill warning-call on appearance of a predator) on behalf of close kin (Wong, 2000) or gift-giving, rather than decidedly suicidal actions. This biological heritage has been linked to the 'kin selection' and 'selfish gene'-paradigms, and thence to modern-day suicide-terrorism (e.g. Dunbar, 1996; Atran, 2003; Blackman, 2004; Blackwell, 2005).

Many theories state that man's ascent to higher order consciousness was the result of a past as Machiavellian animals, and progressive development of more and more complex group dynamics. Beyond the dynamics of animal flocks, humans now also have the means to strengthen communal bonds beyond kin or hunting-party through gossip, narratives and emblems, and we can vividly imagine emotional reactions to our behaviour, thus enforcing social constraints even when we are alone. This omnipresent social womb is a strongly motivating factor, and our abilities for love, cooperation, loyalty, sociality and empathy (Dunbar, 1996; Tomasello et al., 2005), metacognition (Sperber, 2000; Bartch & Hübner, 2005), and self-identity (Edelman & Tononi, 2000), as well as, social outrage and hunger for status (Buss, 2005; Duntley & Buss, 2004) appear to have served us well on the way from simple coalitions to advanced societies. Being human, and therefore beyond blindly obeying impulses of fear, anger or hedonistic gratification, might indeed be what makes truly other-directed actions a behavioural possibility.

The ramifications of the link from selfish genes to self-sacrifice lead us to the extreme end of Hamilton's rule, which dictates that altruistic behaviour evolves if members of cooperative kin-groups *on average* do better than members of groups who are less adept at (or willing to) cooperate (i.e. Dunbar, 1996; Blackman, 2004). Taking a 'selfish gene'-standpoint further suggests that genes which advance altruistic behaviours will have a good chance of surviving in *some* carriers, if others succumb to their own good intentions: It is not the survival of the individual phenotype, but the number of individuals who manage to reproduce, perhaps standing on the corpses of altruistic kinsmen, that counts. In the more economic theory of reciprocal altruism, a pragmatic *quid pro quo*-approach focuses on how animals such as vervets and vampire bats tend to give to members of the group who have on previous occasion given to them (Wong, 2000). This form of *scratch-your-back-if-you-scratch-mine* solidarity can lead to distinctive formation of coalitions, which is a known social phenomenon in primates and, with a few more bells and whistles, humans. Thus, altruism might ultimately serve to garner reciprocal repayment in times of need, generate social benefits through the formation of coalitions, or increase the fitness of distributed genetic material (Wong, 2000; Blackwell, 2005). Many theories on religion suggest that ritualised offerings or, more importantly, displays of willingness to sacrifice signify commitment in roughly the same way as sharing in smaller coalitions, with the added symbolic component of an omniscient divine patient (Atran & Norenzayan, 2004). Thus, action is moved from inherent cognitive systems like the theorised "cheater detection module", into the symbolic niche. This "leeching of culture upon pre-existent evolved dispositions" is at the core of contemporary cognitive theories of religion.

Social coalitions are also at the core of forming personal identities, and notably stigmatising cheaters and out-groups. We seem to have evolved sensibilities about coalition-membership, strangers and cheaters, and most murders appear to be grounded in social slights, or displays of power (Buss, 2005; Duntley & Buss, 2004). This propensity to hate "the other" if he slights us socially or hogs resources, especially found in young males fighting for a place in the social order, is easily recruited into religious and political experience, such as the formation of internally loyal militant cadres (Atran, 2003).

Some science writers have compared inclusive fitness and human self-sacrifice (Blackman, 2004; Blackwell, 2004). After all, if wasps are programmed to kill selflessly on behalf of their kin, cells may explode to take out competitors, and algae might suspend their own growth to allow others to survive (Nedelcu & Michod, 2006), why should this logic not apply to humans? The basic argument, in the words of Blackwell, runs: *"Since suicide attackers are generally young and childless [...], suicide attack effectively eliminates all reproductive success of these individuals"* (Blackwell 2003 p. 5), which at first glance defies biological principles of fitness. However, looking at the specific demographic conditions under which suicide attacks regularly occur[2] one might discern that 1) flocks of siblings are very large compared to western standards, 2) genetic relatives often marry and reproduce, and 3) the hordes "fighting-age males" going on suicide missions often have difficulties finding a bride or place in life due to lack of wealth or appropriate candidates for marriage in the community (Blackwell, 2003; Stern, 2003; Heinsohn, 2005). It is therefore concluded, that inclusive fitness might be the determining factor. Nature might be getting rid of excess biological material and hurting competitors in one stroke. Blackwell thus accepts genetics in the context of Middle Eastern kinship structures as a premise for analyzing suicide terrorism.

Further, cultural factors such as contested resources between siblings, financial rewards to a *shahid*'s near kin, and the relative importance of family status are evoked to explain why the propensity to take on suicide missions would be expressed specifically in young Middle Eastern men. The conclusion drawn following the logic of kin selection is that since "they are all cousins down there", and thus share a *relatively* larger percentage of genes *than* western communities, the biological incentive to

---

[2] Blackwell focuses on Palestine but the argument might easily be extended to Pakistan or Indonesia where conditions are similar, and jihadii-activity extensive.

sacrifice oneself should be greater in countries such as Afghanistan or Palestine. Blackwell's case for this is made by Hamiltonian calculations for populations of insects or animals (e.g. *"assuming families with six children, suicide attackers are expected to have on average thirty-six first cousins, which assuming a genetic coefficient of .125 equates 4.5 times the genetic value of the attacker himself"*, 2005, p. 10), but hardly backed up by empirical proof of the calculus' viability in human cultures, other than through largely anecdotal figures. As Scott Atran suggests, there is no reason to suppose that real kinship still is the decisive factor, since fictive kin created in cultural and institutional settings seems to work even better (Atran, 2003). Thus, systems for kinship-selection have probably become generalized in the course of evolution, and can be influenced by matching certain input conditions for the generation of fierce affection and loyalty.

The Roles of Imagination and Narrative Consciousness

Cultural factors are important to understand how self-sacrifice came to be, since they evolved as a product of human situatedness in time and sociality. Many species have well-evolved apparatuses for social interaction, and this might well have been central to developing the higher order human consciousness (Dunbar, 1996; Tomasello et al., 2005), which is pivotal to thinking above and beyond the present moment, and thus conceptualising time, narrative and self-identity (Edelman & Tononi, 2000; Fludernik, 1996; Lieberoth, 2007).

In other words, cultural narratives of sacrifice require imagination to be invented, shared and brought to life.

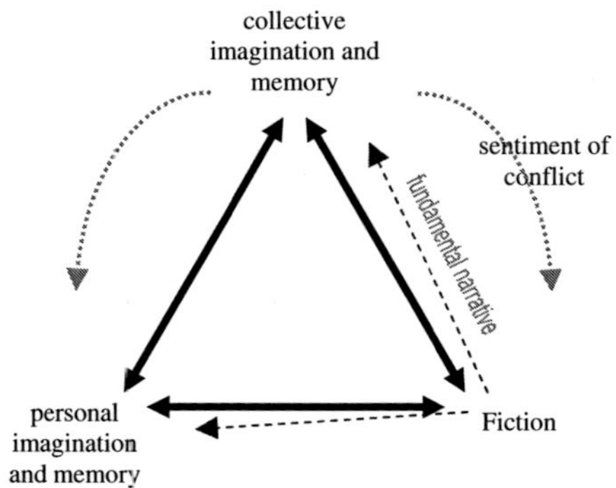

Fig. 14.1 Fiction and imaginations in the fundamentalist turn.
Adapted from Augé 1999.

Alan Leslie calls the ability to think out of the here-and-now *decoupling* (1994), while Gerald Edelman labels our separation from the moment *higher order consciousness* (Edelman & Tononi, 2000). Just like language, individual imaginations are linked to cultural symbols. Post-structuralist anthropologist Marc Augé envisions this relationship as a reciprocal dialogue between the poles of personal imagination, shared imagination and fiction (that is, culturally established symbols, tales, etc. – see fig 14.1). In 'culturally healthy' societies, infusions from the outside are gradually introduced to either of these 'poles', and from there to the others, whereas cultures under symbolic siege may experience that outside fictions and imaginations are imposed with strong force upon the existing equilibrium, which can upset the dynamic flow of adaptation and change. In rising fundamentalism, it is often seen that certain fictions are strategically enforced by dogmatic factions as defence against besieging outside influences. Thus, fundamentalist experience is largely based on motives of conflict, but its fictions need not be. The apparent universality of storied experience (Fludernik, 1996) means, that sentiments of conflict and fundamental fictions are inevitably integrated with personal life experience. Thus, conflict coupled with privation or humiliation may align life narratives to the main storylines of cultural fictions, including schema for resolving conflicts, and the way the heroic narrative must end.

Without decoupling, it would be impossible to generate such personal narratives, constrain future selves, or imagine possible future gains – especially empirically unverifiable ones such as those involving God and heaven. Edelman suggests that human thought was qualitatively separated from animal consciousness by the evolution of language and a coherent metacognitive sense of time and identity (Edelman & Tononi, 2000, see also Bergson, 1911/1998). These factors place [me] in relation to a flow of time [past-present-future], as opposed to most other animals who are confined to the *primary consciousness* of the moment, which humans can easily escape via decoupling, noetic memory and semantics.

Imagination is incidentally known to play a role in most premeditated murders, often involving elaborate fantasy and mental rehearsal of the deed long before homicide actually becomes a realistic possibility in a killer's mind (Hazelwood, 2001; Buss, 2005; Duntley & Buss, 2004). Murder fantasies can further be shared and rehearsed in an institutional setting, as is seen in violent rituals and aggressive social discourses across the globe, which tend to feature prominently in apocalyptic religious movements (Wessinger, 2000). Such "death rehearsals" place the individual and community in direct relation to the grand eschatological narrative, and can make frightening or appalling actions more acceptable and enemies more monstrous by gradual internalization (ibid.).

Also, higher order consciousness grants us a conceptual knowledge of death as a biological or personal termination or transition. The ideas of time, identity and continuity means that although animals may give their lives for their flock or young, only humans can actually grasp the consequences of their own passing, and thus human self-sacrifice must be viewed as qualitatively different from similar behaviours in other species. If suicide-bomber bacteria (Blackman, 2004) had our self-concepts and view of existence as a whole, their actions would probably be quite different. In fact, higher order cognition would appear to be both a hindrance and a stepping stone on the way to self-sacrifice. Higher order cognition allows us to form attitudes and motives directed at our own thoughts and feelings (Bartsch & Hübner, 2005). Thus, the self-conscious human mind can favour some motivations over others, for example taking precautions against future selves, who may stray from a chosen path (Elster, 2000). The generative nature of human society and psychology forces us to look ahead, and envision the world beyond our brief life-span; what will become of our kin, society and memory? But, as we know, higher order consciousness also gives us the power to dream up and institutionalize fantasies about a self separate from the flesh, which might be a soothing or even motivating factor to the religious mind contemplating martyrdom.

Because of the transitional leap to higher cognition, it becomes meaningless to discuss the self-destructive behaviour in ants, and even mammals defending their young, in direct comparison to premeditated human self-sacrifice. Narrative, imagination and self-reflection change the game completely.

Will

Will is central to self-sacrifice, but hard to pinpoint cognitively or neurologically, as it emerges in a wider context including motor-action and appetitive states, and more complex phenomena such as sustainable goal-representation, metacognition, and the ability to imagine possible futures.

In his recent book "Effort" (2007), Jay Schulkin suggests dopaminergetic systems as a prime candidate for a neurophysical correlate to will, and incidentally ends with the musing: *"But amid the great triumphs of human expression are the diverse ways in which we have trampled on one another, sustained the wrong goals, and directed the will towards the less worthy at the expense of others"* (2007: 119). Shulkin thus leaves where this study picks up – namely the carefully orchestrated self-destructive and aggressive actions made possible by higher order consciousness.

Without an evolved human will, sustained behaviours would be next to impossible, and fear would effectively deter martyr missions. Schulkin hypothesises that the prefrontal cortex, the most heavily intertwined dopaminergetic region of the neocortex, is instrumental in sustaining and attributing emotional value to decoupled goal-representations. According to Schulkin's behavioural neuroscience approach, all cognitive systems evolved as part of the organisation of action, which basically makes them tools for homeostatic regulation. Will emerged as part of more complex sets of thought and actions, and thus as a product of basic motor- and cognitive functions. However, evolution of self-control did not start with needs to constrain the myriad of possible futures offered by higher order consciousness. Freeze-reflexes are seen in many diverse species, and predators are well known for their ability to stalk prey and wait for the right moment to strike. Patience pays. The basal gaglia, in tight conjunction with the pfc, are important in the sequential organisation of action and thought, and especially the striatum appears to have a function in prediction of rewards and response to frustration, both apparently involving dopamine. The core of this appears to be simple initiation and termination of movement, but in the evolved will, anticipation and frustration of reward is the key to create more complex sequences of action, and suspending immediate appetites. As Schulkin puts it, *"evolution favoured not only our ability to run, but to sit, wait and think."* (2007, p. 63).

Will working in a context of sociality and higher order cognition can be illustrated through the concept of hyperbolic discounting (Elster, 2000), which describes avoiding 'cheap pops', in order to ensure a certain level of reward or gratification at a later point (see figure 14.2). Leaving your pension in place until a designated retirement date, for instance, guarantees interests and financial security in your old age, while withdrawing it in the meantime to pay for a beautiful wedding reception, would incur a penalty, resulting in both a comparably reduced financial payoff and a lack of income later in life. Varying emotions and motives may conflict, but some goals override other whims, emerging in prolonged action sequences or inhibition of certain acts. To avoid oversimplification, it is also pivotal to notice how the will is often anchored in external constraints, be they social and/or practical. The brain itself is not the only thing that must be understood, to realise how we save up money or muster the will to die.

Fig. 14.2 Hyperbolic discounting

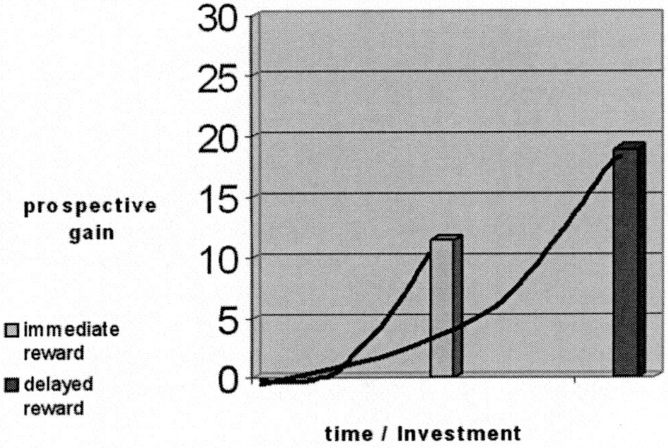

The evolution of will thus reflects the needs for self-binding that come with the greater range of choices offered by higher order cognitive states, and is thus compatible with Edelman's almost Bergsonian theory of our rise to consciousness (Edelman & Tononi, 2000). In other words, will is necessary to sustain behavioural coherence in the human condition, and to keep order in *"the diverse conflicting motives and appetites and desires that compete for cognitive and behavioural expression"* (Schulkin, 2007, p. 97), including aversions and appetites, such as fear, pain and desires for food, sex and eternal glory.

Schulkin's and Elster's perspectives on will provide a framework for understanding the ability to go beyond the basic "in the moment" impulses, such as running scared or recoiling from pain. The term self-control implies self-awareness, and the human ability for metacognition (that is, the ability to direct conscious awareness towards thoughts, motives and emotions) allows checking one's own thoughts, effectively creating meta-motives and even meta-emotions (Bartsch & Hübner, 2005), e.g. being ashamed of fear or wanting to believe in Life Eternal. There is still some debate as to whether a qualitative hierarchy of emotions exists, but at least in some cases anger and related "negative" emotions appear to be able to override fear, which in turn easily overrides appetitive states. Furthermore, distancing consciousness from the self through various meditative techniques and distractions such as prayer or focus on ultimate motives may be able to induce a dissociated state, which keeps acute emotion or self-awareness at bay (Gambetta, 2005).

One could be tempted to think of this in terms of a struggle between 'hot' and 'cold' cognition, but that would be misleading. Emotion and more rational or semantic cognition are so heavily intertwined, that it makes no real sense to separate the two except in a flawed phenomenological sense, stemming more from introspection and philosophical tradition than the emerging scientific picture of cognitive functioning. Just like genes, all physiologically and neurochemically distinct entities in the brain are agile; they are hugely interconnected, and are apparently able to serve many different purposes depending on the order of the day, constantly influencing each other in subtle ways. Thus, our emergent understanding of the brain's complex nature thankfully quells the urge to reduce cognitive explanations to the neat workings of finite "systems". We may constrain ourselves or excite conscious volition, but this always happens in complex interplay between an emotional backdrop, more or less conscious reflection and constraining environmental factors.

Emotions and hedonistic desires can be hard to battle, which is plainly the case in suicide-bombings, where a host of social and psychological constrains is introduced to help the aspiring martyrs steer a clear path. As William James says in a much overlooked passage: *"Reason, per se, can inhibit no impulses; the only thing that can neutralize an impulse is an impulse the other way. Reason may, however, make an inference which will excite the imagination so as to set loose an impulse the other way"* (James, 1890, cit in Ridley, 2003, p. 39). Fear, pain and even compassion for victims form strong deterrents, which may be avoided at any time by simply walking away. As it happens, terror-organisations are masters of introducing highly emotional constraints (Elster, 2005) and psychological leverage like threats, solidarity and questions of honour (Atran, 2003; Gambetta, 2005). The construction of the situation thus appears to create a conflict between fear A (the immediate fear of dying, never seeing family again, etc.) and fear B (the fear of losing all respect, being labelled a hypocrite in the eyes of (fictive) kin, failing to vindicate fallen comrades, dying from being shot instead of doing the right thing, etc.). In other words, the anticipation of what others might think if he should fail is added to the individual's emotional mixture. By adding a decoupled dimension to our social mind, Maurice Halbwachs' insight that *"in reality we are never alone"* (Middleton & Brown, 2005, p. 54) finds new resonance. Social sensibilities and higher order consciousness thus seem to add important evolutionary contributions to the neural working of the will.

Discussion

The puzzles presented here are by no means sufficient to understand suicide missions. On the contrary, the most fruitful approaches so far have centred on well documented social psychological mechanisms (for a review, see Atran, 2003; Stern, 2003; or Zillmer, 2006). This chapter has only presented some evolutionary and cognitive precursors for those.

In accepting kin selection as a premise for self-destructive actions in the Middle East, strictly bio-behavioural approaches fail to account for similar phenomena in parts of the world with different demography, like the risks taken by western revolutionaries or collective suicides in movements such as *Heaven's Gate* or *The People's Temple*. Arguments like those made by Blackwell (2005) and Buss (2005) seem to reduce human reality to a question of genetic populations, and frame human thought and action as mere cogs in mechanical biology, covered by a thin varnish of culture or personal choice (McKinnon, 2005), which might even imply that cultures propagating sacrifice are more psychologically primitive. It is evident that evolved neurocognitive features must play a role in martyr missions, but reducing them to programmed behaviours triggered by certain inputs alone is plainly absurd. Atran (2003; Atran & Norenzayan, 2004) manages to divorce the actions of terrorists from a banal notion of mechanistic pressures to increase the fitness of genes, and instead focuses on the way fictive kin "rides along" these "default settings" of our cognition and emotions, occasionally aided by clever organizers. Atran's framework has explanatory power far beyond purely neo-Dawinistic approaches, since it 1) makes its predictions within the testable scope of social psychology, 2) avoids attribution-errors by keeping explanations at the situational level, 3) affords culture and situation parts in the process, and 4) claims no actualised functionality for the phenomena described. In other words, while Blackwell implicitly argues that suicide terrorism serves an eternal genetic purpose, Atran simply holds that certain evolved features of human psychology may be relevant to suicide terrorism, and that such features may be influenced by people or circumstance.

The similarity between human self-sacrifice and examples from the natural world is striking, but several criticisms might also be levelled against this idea. First of all, the minds of humans, wasps and ground squirrels bear little actual resemblance, and thus, the analogy is flawed at a neurobiological level. Secondly, there are significant similarities between human and animal cognition, but also vast differences, including higher-order processing, metacognition and complex cultural systems, to name but a few. Thirdly, altruism in animals seldom amounts to willingly sacrificing one's own life, but rather generosity when one has resources (such as food or mating-opportunities) to spare, or increased willingness to take risks (such as the chance of being pummelled in a fight, or attracting the attention of a predator) on behalf of allies. The differences are thus found in both investment and ability to represent the severity of possible consequences, including death. Fourthly, the human tendency to anthropomorphism, or the "pathetic fallacy", compels us to ascribe human-like motives to animals, which is mostly an unwarranted trick of the mind. Finally, the naturalistic error, or the "is=ought fallacy", rears its ugly head whenever science looks at primitive urges and abilities. Despite evolutionary pressures to act in our genes' favour, we do have a choice, and human conduct around the world tells us that most people have left their evolutionary past safely behind in favour of more current enterprises, such as choosing career over breeding; as Steven Pinker so memorably put it: "*if my genes don't like it, they can go jump in the lake.*" (1997). Self-sacrifice requires something quite beyond the genetic pull of loving one's relatives and antagonising competitors.

It is conceivable that ritualised rehearsal of sacrifice, ongoing declaration of willingness to die for the cause of God, or recital of cultural narratives (recall fig. 14.2.) may serve as a sort of mental training for acts, harnessing will and courage (Elster, 2005; Wessinger, 2000). This principle is well known from all sorts of military schooling, where recruits meet simulated situations many times over, and thus become used to certain stressors before actually going into battle, as well as creating salient feelings of heroism and camaraderie which may help mentally to override fear or pity. Higher order cognition supplies the stage for such shared imaginations, as well as anticipating future benefits and consequences of human actions. Will would be meaningless without it. The wish to commit suicide for the greater good may well stem at least partly from deeply rooted social sensibilities, but the will and skill to carry out suicide attacks did not evolve to serve that end. They are part of the flexible multi-purpose cognitive systems that make up the anatomically modern brain and cultured mind, which basically evolved to ensure the survival of the individual; not his death. If we accept that suicide terrorists are conscious of their own motives, then biological and zoological examples can only serve as appealing metaphors for human actions. In other words, the evolution of higher order cognition and will effectively ended the sovereignty of programmed behaviour as it is seen in bacteria, wasps and lesser mammals. Drives and motives must exist in a constant dialogue with culture, reflection, identity, and long-term planning.

No modern scientist will dispute that individual human beings are products of DNA, passed down through evolution and expressed in the individual phenotype, but it is unlikely that emergent phenomena like suicide terrorism emerge from single genes with specific labels such as the "homicide-gene", the "cheater-detection gene" or the "writing academic books-gene". Rather, the human genetic code makes such capabilities possible through the combination of different physical features, emerging as stable functional clusters over time.

Concluding Remarks

Many scientific minds have explored Darwinian angles with a healthy dose of critical thought, but sadly such ideas are often grabbed and accepted as canon by excitable fans of the evolutionary paradigm. Therefore, not the scientists themselves, but overly enthusiastic readers bear the bulk of the blame for sometimes turning evolutionary theories on human psychology into a messy circus of ideas and facts tossed about in random order.

Many have hinted at the explanatory power of kin selection in relation to suicide terrorism. We strive to understand what seems frightening and alien, and readily impose our social models on the near mechanical reality of birds and bees - and back again. Counter to the idea of some elusive "self destruct mechanism" hidden deep in our evolutionary heritage, or a hereto unknown way for the selfish genes to mess with us, this chapter presents elements in the interplay between our brains, culture and social reality as an approach to suicide terrorism. The understanding of will and higher order cognition promises to add an additional layer of "hows" to the existing "whys".

Our evolutionary past is forever with us, and can be manipulated by circumstances or shrewd people, be they politicians, demagogues or advertisers, but we are not automatons. Therefore, we may look to the past for individual pieces of the puzzle, that is suicide missions, but we must scrutinise the present to put them together.

References

Atran, S. (2003). Genesis of suicide terrorism, *Science*, vol. 299, March 2003, (p. 1534-1539).

Atran, S. & A. Norenzayan (2004). Religion's evolutionary landscape: Counterintuition, commitment, compassion, communion, *Behavioural and Brain Sciences* 2004, 27:6.

Augé, M. (1999). The War of Dreams – Studies in Ethno Fiction. London: Pluto.

Baudrillard, J. (1983). *Les Strategies Fatales*, Paris: Grasset 1983.

Bartsch, A. & S. Hübner (2005). Towards a theory of emotional communication, CLCWeb: Comparative Literature and Culture: A WWWeb Journal, Thematic Issue Media and Communication Studies at the University of Halle-Wittenberg., (ed.), R. Viehoff, Purdue University Press.

Bergson, H. (1911/1999). *Creative Evolution.* Mineloa, NY: Dover.

Blackman, S. (2004). Spite: Evolution finally gets nasty, *The Scientist* 2004, 18(24): 14.

Blackwell, A. D. (2005). *Terrorism, Heroism, and Altruism: Kin Selection and Socio-Religious Cost-Benefit Scaling in Palestinian Suicide Attack*, Presented at the Human Behavior and Evolution Society 2005, Austin, TX. June 1-5, 2005, http://www.uoregon.edu/~ablackwe/.

Buss, D. M. (2005). The Murderer Next Door – Why the Mind is Designed to Kill, London/New York: Penguin.

Cook, D. (2007). *Martyrdom in Islam*, Cambridge: Cambridge University Press.

Damasio, A. (1994). Descartes' Error – Emotion, Reason and the Human Brain, London: Penguin 2005.

Diamond, J. (1992). The Third Chimpanzee – The Evolution and Future of the Human Animal, Harper: London.

Dunbar, R. (1996). Grooming, Gossip and the Evolution of Language, London: Faber & Faber 2004.

Duntley, J. D. & D. M. Buss (2004). The plausibility of adaptations for homicide. In: P. Carruthers, S. Laurence & S. Stich (eds.) *The Structure of the Innate Mind*, New York: Oxford University Press (pre-publication version).

Edelman, G. M. & G. Tononi (2000). *Consciousness: How Matter Becomes Imagination* London: Penguin 2001.

Elster, J. (2000). Ulysses Unbound – Studies in Rationality, Precommitment and Constraints, Cambridge: Cambridge University Press.

—. (2005). Motivations and beliefs in suicide missions. In: D. Gambetta (ed.) *Making Sense of Suicide Missions* Oxford: Oxford University Press, 2006 (p. 233-258).

Fields, R. M. (2004). The psychology and sociology of martyrdom. In: Rona M. Fields *Martyrdom - The Psychology, Theology and Politics of Self-sacrifice*, (p. 23-83), London: Praeger.

Fludernik, M. (1996). Towards a 'natural' narratology, Routledge, London/NY., 1996 (-p. 43-52).

Gambetta, D. (2005). Can we make sense of suicide missions? In: Diego Gambetta (ed.) *Making Sense of Suicide Missions*, Oxford: Oxford University Press.

Hazelwood, R. w./ S. G. Michaud (2001). *Dark Dreams – A Legendary FBI Profiler Examines Homicide and the Criminal Mind*, New York: St. Martin's Paperbacks.

Heinsohn, G. (2005). *Population, Conquest and Terror in the 21st Century*, available at: http://www.geocities.com/funnyguy_35/HeinsohnPopulation.PDF (March 26. 2008).

Leslie, A., (1994).Pretending and believing: issues in the theory of TOMM, *Cognition*, 50, (p. 211-238), 1994.

Lieberoth, A. (2007). Technologies of experience - the embodied mind in play and imagination. In: J. Donnis, M. Gade & L. Thorup (eds.) *Lifelike.* Copenhagen: Knudepunkt 2007.

McKinnon, S. (2005). Neo-liberal Gentics – The Myths and Moral Tales of Evolutionary Psychology. Chicago: Prickly Paradigm Press.

Middleton, D. & S. D. Brown (2005). The Social Psychology of Experience – Studies in Remembering and Forgetting. London: Sage.

Nedelcu, A. M. & Michod. R. E. (2006). *The Evolutionary Origin of an Altruistic Gene*, Mol. Biol. Evol. 23(8), (p.1460–1464).

Neimeyer, R. A. & D. A. Winter (2006). To be or not to be: Personal constructions of the suicidal choice. In: T. E. Ellis (ed.) *Cognition and Suicide – Theory, Research, and Therapy* Washington: American Psychology Association.

Pinker, S. (1997). *How the Mind Works*, London: Penguin 1999, (p. 149-169).

Ridley, M. (2003). *The Agile Gene – How Nature Turns on Nurture*, New York: HarperCollins.

Schulkin, J. (2007). Effort – A Behavioural Neuroscience Perspective on the Will, Mahwa: Laurence Erlbaum Associates.

Sperber, D. (2000). Metarepresentations in an evolutionary perspective. In: Sperber, D. (ed.) *Metarepresentations – A Multidisciplinary Perspective*, Oxford University Press, New York, (p. 117-137).

Stern, J. (2003). Terror in the Name of God – Why Religious Militants Kill, New York: Ecco/ Harper Collins.

Tomasello, M., M. Carpenter, J. Call, T. Behne & H. Moll (2005). Understanding and sharing intentions: The origins of cultural cognition, *Behavioural and Brain Sciences* 2005, 28, (p. 675-691), Cambridge University Press.

Wessinger, C. (2000). How the Millennium Comes Violently – From Jonestown to Heaven's Gate, New York: Seven Bridges Press.

World Science (2006). *Researching trace origin of an "altruism gene"*, World Science special, www.world-science.net/exclusives/600529_altruism.htm

Wong, R. (2000). *Motivation:A Biobehavioural Approach*, Cambridge: Cambridge University Press.

Zillmer. E. A. (2006). The psychology of terrorists – Nazi perpetrators, the Baader-Meinhof gang, war criminals in Bosnia and suicide bombers. In: C. H. Kennedy & E. A. Zillmer (eds.), *Military Psychology-Clinical and Operational Applications.* London / New York: The Guilford Press, (p. 262-280).

CHAPTER FIFTEEN

RELIGIOUS FORCES BEHIND CULTURAL TRADITIONS

SERGIO LEVI

Abstract: Two theories of the nature and function of religious cognition and behaviour have been competing during the last two decades: The naturalness-of-religion thesis and the theory of costly signalling. Their claims and arguments are reviewed and discussed in the first part of the paper (§§ 2 and 3). The second part of the paper (§§ 4 to 6) presents an evolutionary scenario within which the low fidelity of cultural transmission emerges as a major problem for religion, when speech greatly enhanced the analytical capacities of Homo sapiens and its proclivity to reflect on himself. The conclusion suggests that although ritual behaviour is fuelled by the urge to transcend the ordinary ways of doing things, ritual norms are dictated by built-in constraints of our memory systems and remain among the forces that render human traditions evolutionary stable.

Introduction

In the last ten years the longstanding question of the advantages accruing to religious people for their costly activities has been receiving a tentative answer in the form of models suggesting *that* religious cost may be an adaptive feature of a signalling system to certify altruistic behaviour. As Bulbulia (2004, p. 25) puts it: "*The strategy works because it is based on an illusion, not in spite of any illusion*". If so, one wonders how potentially maladaptive illusions are prevented from making roots among the more beneficial illusions fostering group co-operation. One answer suggesting itself is that all religions abound with norms and prescriptions that seem to be designed to prevent the beneficial illusions from decaying into useless distractions or into socially harmful absolutes. The hypothesis I would like to consider is that the relevant norms are more embedded in ritual forms than couched in terms of religious belief.

Participation in costly rituals has been seen to constitute the social significance of religion since the work of Émile Durkheim, yet the evolutionary stability of ritual (and other religious) forms still raises particular problems for social scientists, and with good reason. We are told, for example, that among animals the invention of a new feeding technique is often sufficient to transform the habitat to the point of extinguishing a much exploited source of food, thus forcing the next generation to develop new food habits (Avital & Jablonka, 2000). So how in a world that is rapidly changing after the appearance of *Homo sapiens* can the food preferences (to offer just an example) of a group of hunter-gatherers cease to respond to the external world and start obeying an artificial net of religious rules independent of the actual state of the environment? It takes little reflection to realise that unless some religious norms somehow pre-selected (and thus pre-shaped) fit modifications from unfit ones, the process would be one with high variation but no gradual selection.

By focusing on those aspects of religious behaviour that could only spread across generations by being explicitly transmitted (and socially learned), I will try to inquire (a) how religions could have evolved mechanisms to limit variations and (b) how this almost unique phenomenon of *enforced* fidelity of transmission might have contributed to the stability (or quasi-stability) of human cultures at large.[1] The paper has two parts. The first part (§§ 2 and 3) reviews theories of religious ritual behaviour that either disregard most of the cultural and theological understanding by concentrating on the ordinary cognitive representations of religious concepts and actions (Boyer, 2001; Atran & Norenzayan, 2004; Barrett & Lawson, 2001; McCauley & Lawson, 2002) or stress the costly features of ritual as phenomena that both need an explanation *and* provide insights into the social regulatory function of religion (Bulbulia, 2004; Sosis, 2004). The second part (§§ 4 to 6) addresses the issue of religious transmission and outlines an evolutionary scenario within which the low fidelity of cultural transmission emerges as a major problem for

---

[1] To do so I will make the assumption that not all religious behaviour is inherited through genetic pathways. So those accounts that attempt to explain religious forms by an exclusive appeal to cross-cultural cognitive structures – thus rendering local meanings explanatorily null – cannot be more than half right. Such accounts tend to speak in terms of religious or supernatural *cognition* even when their topic is ritual action and its (lack of) explanation (e.g. Boyer, 2001). To resist this line I will make the further assumption that religious behaviour cannot be reduced to its cognitive representation where this is taken to exclude the involvement of emotion and motivational structures (Damasio, 1994; Carver & Scheier, 1998).

religious groups – especially so since speech (and writing) greatly enhanced the analytical capacities of *Homo sapiens* and their disruptive influence on society. A major purpose of the ensuing, tentative conclusion will be to make a case for bringing the explanation of religious ritual to bear more closely on the understanding of human behaviour and the evolution of cultural traditions.

The Naturalness of Ritual Behaviour

According to Boyer, "In most human groups people have all sorts of rituals but no good explanation of why they should be performed, especially of why they should be performed in the specific way prescribed" (2001, p. 266). One might want to rebut that those people may be simply unconcerned with "explaining" their ritual behaviour. Yet Boyer's claim has a point: why bother with exact performance if you have no idea of *how* rituals are supposed to work? An answer may be that they work pretty much like everything else in the world. So drawing on Lawson and McCauley (1990), some researchers have proposed that religious behaviour (which includes but does not reduce to religious ritual behaviour) may be not so different from non-religious social action, at least in the sense that both require rational capacities and the cognitive equipment to represent an action in terms of its goals.

On one hand, it is undeniable that religious actions are a particular type of actions: "Someone does something to someone or something in order to bring about some non-natural consequence. That is, rituals are actions that are performed to accomplish something that would not normally follow from this specific action" (Barrett & Lawson, 2001, p. 216). On the other hand, the relevant features are almost the same: "Structurally, religious rituals mirror social actions: someone performs some kind of action in order to motivate another's action or change in disposition. It just so happens that the person being motivated to act is a god or other non-natural agent" (p. 217). The alleged similarity, on closer inspection, is more about action representation than action execution: despite their unusual qualities, ritual actions "are cognitively represented as actions… using ordinary resources" (p. 216).

It is significant that in order to prove the existence of that similarity Barrett and Lawson (2001) interviewed a group of "ritually naive individuals" with different cultural backgrounds. In two different tests, these individuals were asked to select which one from among a dozen ritual variations they considered more likely to achieve the intended non-natural consequence. The results seemed to confirm two important predictions: 1) Ritually naive individuals appreciate the importance to represent superhuman agency somewhere within the ritual structure, 2) After connection to superhuman agency, what is judged most important is the presence of an appropriate agent, that is an agent "which is capable of intending" that the wanted consequence should result from his actions.

A host of scholars had assumed that more than the agent's intention, the relevant ingredient of ritual action was the agent's overall power – as this is expressed in *how well* the action is executed. There must be some reason after all if religious interpretations all over the world tend to place so great an emphasis on perfect execution.[2] But, irrespective of all this, the notion that the core of ritual is perfect execution is dubbed "the folk notion of ritual" (Barrett & Lawson, 2001) or "the popular conception of magic" (Barrett, 2000).

These radical (though empirically testable) claims on what is *essential* to ritual (or to ritual representation) are of a piece with an equally trenchant distinction between *theological* and *religious* concepts. While the former are only used in highly abstract theological disputes, the latter are used in reasoning about deities, recalling narratives and predicting the consequences of religious actions. The basic idea behind the "naturalness-of-religion thesis" (Barrett, 2000; see also Boyer, 2001; Lawson & McCauley, 1990) is that expectations about how the relevant supernatural agents will respond to this or that ritual action are most revealing of the tacit intuitions governing religious concepts. Such intuitions endow the deities with much the same properties we are inclined to ascribe to human-like agents. For this reason those intuitions are said to "violate" the more demanding prescriptions of theology (and of theological correctness). At the same time the prescriptions of theology are said to "violate" important intuitive assumptions about the distinctive properties of different categories of living, moving and thinking things. A similar violation is also imputed to religious concepts, though with an important difference: While theological concepts tend to violate various basic-level assumptions (thus being hardly graspable), religious concepts violate only a few feature-level assumptions at a time (thus being highly memorable). For

[2] As Dennett (2006) puts it: "There is artifice in the design and execution of religious practices, as anyone knows who has ever suffered through an ineptly conducted religious ceremony. A stammering and prosaic minister and boring liturgy, shaky singing from the choir, people forgetting when to stand and what to say and do – such a flawed performance can drive away even the best-intentioned congregants. More artfully celebrated occasions can raise the congregation to sublime ecstasy" (p. 153-154). Also in non-religious social contexts perfect execution is often a necessary condition and the prime purpose of action.

this reason the latter were dubbed by Boyer "minimally counterintuitive concepts" (e.g. Boyer, 2001).

What is the relation between religious concepts and religious actions? According to Barrett (2000, p. 92) "people spread religious concepts in the context of shared religious actions"; and these actions are also executed "in response to those concepts". One would expect that religious ritual actions should correlate with theological (not religious) concepts, because there is a clear parallel between ritual exploration and theological construction. Both attempt to transcend the limits imposed by the tacit assumptions constituting most ordinary cognition – including the concepts we use when we need to remember religious narratives or when forming expectations on their basis. On the contrary, "cognitive scientists of religion argue that… ordinary cognition both structures religious practices and underlies the representation (and thus the execution) of religious actions in participants' and observers' minds" (ibid.).

Now there may be some truth to the suggestion that theological correctness is dropped when devoted individuals need to process information about gods and supernatural beings acting in space-time. Problems arise when claims (or even findings) on how supernatural concepts are represented are uncritically assumed to support the further claims that this very same ordinary cognition (1) structures religious practices and (2) underlies the representation *and thus* (3) the execution of religious actions. The third step (from action representation to action execution) would require more argument to appear less questionable. But on careful consideration each of these claims could be questioned in the light of the notion that our emotion and memory systems can be more efficient than pure cognition in guiding the construction of ritual behaviour. (More on this presently).

Religion and Social Co-Operation

The attempt to downplay religious behaviour to a mere execution of quite ordinary action representations resulted in a theory that explains ritual as a noisy by-product of ordinary cognitive functions. A different tactic is tried by those who maintain there must be something more to ritual forms than merely useless noise.

The notion that ritual (and religion) may help non-kin individuals to live together and co-operate has a far-reaching tradition and a panoply of variations. Classical theories of religious- and moral-based co-operation (see Cronk, 1994 for a review) tend to emphasize (instead of downplaying) the costly features of ritual behaviour as what explains (via the creation of trust) the benefit accruing to those who join at the expense of those who do not. Thus Jared Diamond said that "religion helps solve the problem of how unrelated individuals are to live together without killing each other – by providing them with a bond not based on kinship" (1997, p. 278). But the very idea that *the* purpose of religion is to encourage group co-operation in the face of enemies is cited by Dennett among the ready-made opinions any conscientious scientist should be eager to question, lest he should not succumb to what he calls "premature curiosity satisfaction" (Dennett, 2006, p. 103).

In fact empirical research based on self-reports has produced mixed results and provided only partial support to the pro-social account of religion. More compelling (and unambiguous) results are now emerging from studies using a priming technique to activate implicit concepts in unaware subjects. Thus Pichon et al. (2007) showed that priming religious concepts can activate genuinely pro-social intentions in unaware subjects – and argued that in this case generosity does not reflect egoistic concern for a positive self-image. Using a similar technique Shariff and Norenzayan (2007) found that subjects primed with religious concepts tend to behave less selfishly than control subjects in anonymous versions of the dictator game. A plausible explanation of the pro-social effects of religious concepts should invoke the operation of different mechanisms, from automatic behavioural priming to the activation of reputational concerns.[3]

On a more theoretical level, a recent attempt to articulate fully the theory that religious costs may be efficient signals of altruistic intentions was undertaken by Joseph Bulbulia. And when he says: "I argue that a central function of supernatural cognition is to facilitate efficient solutions to otherwise difficult or intractable co-ordination problems" (Bulbulia, 2004, p. 19), I think that a distance from the "naturalness-of-religion thesis" is being declared, because the notion of "supernatural cognition" he uses obviously includes theological concepts, irrational rituals and a panoply of costly behaviours his approach seeks to explain.

Bulbulia's (2004) fundamental suggestion is that religion may motivate altruistic behaviour by reliably certifying altruistic intentions. As he puts it: "Religion differentiates genuine altruists from impostors by imposing specific costs on altruists that only they are willing to pay. Cost is an

---

[3] Shariff and Norenzayan (2007), in light of others' studies on the pro-social effects of cues of being watched, suggest "that activation of God concepts… matches the input conditions of an agency detector and, as a result, triggers [the] hyperactive tendency to infer the presence of an intentional watcher. This sense of being watched then activates reputational concerns, undermines the anonymity of the situation, and, as a result, curbs selfish behaviour" (p. 807-808).

adaptive feature of this signalling system: *if supernatural cognition were not expensive it would not have evolved*" (2004, p. 21). This might help explain how religious emotion and behaviour reliably mark the presence of specific motivations for altruistic behaviour. But the radical claim is that religious belief (belief in supernatural causation) not only signals but motivates commitment to altruistic exchange. How? In interactions resembling the prisoner's dilemma defection is the single Nash equilibrium, a situation in which no player can do better by changing strategy if others do not change. While co-operation would be strictly efficient (being the strategy whose interaction with itself brings the greatest payoffs) it is an irrational option. As Bulbulia puts it: "Defection dominates strictly efficient moves, hence the dilemma" (2004, p. 23).

Now there are two ways in which co-operation can be encouraged. One is enforcement; it works by altering the punishment structure for games that invite defection, thus rendering co-operation the rational option (Nash equilibrium). Alternatively, players can be led to choose co-operation by misunderstanding the payoff matrix. For example, a false assignment of reward value may also result from a belief in supernatural causation, from a belief in gods who demand altruistic actions. Such illusions tend to alter the payoff matrix in a way that may favour co-operation.[4] Belief in such gods (whether or not they really exist) is then sufficient to reliably motivate altruistic behaviour: "*The strategy works because it is based on an illusion, not in spite of any illusion*" (25). There is no evidence that a nonreligious moral system could not do equally well in affirming the notion that co-operative behaviour will always enhance the fitness of co-operative individuals;[5] though I suspect that a major problem would still involve the institution of a code to signal reliably the intention to co-operate and a method to detect potential defectors.

It must be said that strategies of costly communication are the normal case among non human primates; but animal costly displays are fit to the inherent need of getting the intended recipient to attend and recognise the signal (Zahavi, 1987; Krebs & Dawkins, 1984). In this kind of ritualisation cost is never associated (as it seems to be the case in religious ritual) with

---

[4] "Belief in gods capable of altering individual fortune promotes efficient play by prompting the motivational structure to produce *strategically* co-operative behaviours. The sacrifice of the defection payoff is understood as a kind of investment, the god acting to insure desirable outcomes through supernatural causation" (Bulbulia, 2004, p. 25).

[5] Indeed another finding from Shariff and Norenzayan (2007) is that implicit activation of nonreligious moral concepts tends to restrain selfishness as much as religious priming can do.

no need to grab the recipient's attention. In the case of religious behaviour we have to explain a strange combination: Costly behaviour (to be found with variations across cultures) whose alleged purpose (for the gods to supply the rewards) departs so radically from the likely beneficial effects identified by social scientists (promoting group co-operation).

In summary, what we have considered so far is a theory that maintains that the cognitive representation of ritual forms is all there is to rituals but sheds no light on the reproductive advantage ritual competence may confer on religious individuals; and a theory that stresses the pro-social effects of supernatural cognition (what could explain its persistence) but gives no explanation of *why* the alleged purpose of ritual behaviour should depart so radically from its actual effects on social co-operation. (In other words, whether or not religion is a good solution to a co-ordination problem, it was never meant to provide such a solution in the first place.) How do these theories deal with the problem of cultural stability? The second part of my paper will attempt to make these two approaches converge on a tentative answer. To do so I will discuss in broad outlines an evolutionary scenario on the origin of art, religion and language from joint-deception activities and pretend-play (Knight, 1999). Further insights will be borrowed from other approaches with a view to outlining a general account of the evolution of religion. The issue of religious transmission (and the quasi-stability of human traditions) will be the subject of a tentative conclusion.

Religion and Ritual before Speech

Insofar as religious ritual (and religious behaviour at large) is the cornerstone of religion, we should expect its transmission to depend, at least in part, on social learning. Many think that in order to be learnable and transmissible religious actions have to possess intrinsic qualities that render them appealing and easily memorable (Boyer, 2001). And, as Avital and Jablonka (2000) explained, behaviours that have a practical function (and are learned from more than one teacher) are more likely to stabilise across generations. Ritual forms that are transmitted through social learning can vary nonetheless according to the capacities, needs and other habits of the young, or by simple mis-copying. So if maladaptive variants of learned techniques are usually selected out (through Darwinian natural selection) by crucial applications that reveal their inadequacy in a relatively short time span, in the case of religious behaviour it is less clear how the maladaptive variants are selected out.

In a number of works Chris Knight has made the point that a defining feature of human group behaviour is coalitionary deceptive signalling (Knight et al., 1995; Knight, 1991; 1999). While many other species are capable of deceiving their prey and predators, only humans exhibit co-operative faking behaviour, and systematically resort to joint-deception against enemies.

The idea is that this capacity for joint-deception gradually led to the construction of a domain of counter-reality as the by-product of a disposition to conspire against wild enemies and to form mixed-sex, kin-based coalitions to exploit outgroup mating males to get food. At a certain point in modern human evolution (between 120,000 and 60,000 years ago),[6] human females started to consociate and gradually learned to signal *collective denials* to any male that would not provide food for their offspring. The signal then evolved into a symbolic system of sexual taboos and religious prohibitions and prescriptions (Knight et al., 1995).

When human groups grew too numerous, inventing and preserving bonds that could foster co-operation and discourage fratricidal war became a priority. The most efficient patterns of group pretending behaviour begun to be socially transmitted and in due course made roots in the genetic inheritance system.[7] The social demand for new forms of ritualised behaviour led to the establishment of an articulated system of collectively-devised, coalition-enabling forms of signalling, which eventually evolved into a system of conventional symbols (Knight, 1999). Only such collective activities (ranging from ritualised signalling to religious rituals)

---

[6] Other theories date this phenomenon considerably earlier but most agree on its explanation. For example, Immerman (2003) calls our attention to two human traits that go against predictions from primate homologues. Human fathers travel to procure food and return to the domicile to provision the mother and the young; and they systematically play with their children. The evolution of these behaviours, concludes Immerman following H.G. Fisher, depends "on a successful reproductive strategy of our female ancestors… that enables them to exploit a novel resource for predictable sustenance for themselves and their offspring" (p. 146).

[7] "The requirement that behaviour has to be displayed in order to be transmitted diminishes the potential persistence of socially transmitted adaptations… [However] If there is long-term selection in a particular direction… some of the control of the behaviour will probably be transferred to the genetic inheritance system. This would happen through straightforward Darwinian selection: there would be a gradual selection of the combinations of genes which lead to increasingly more rapid learning, depending on fewer and fewer trials. The result would be an increase in the relative importance of the innate component of the behaviour" (Avital & Jablonka, 2000, p. 136).

could produce and preserve a domain of counter-reality that upon repetition lets ideas of counterintuitive agents make a firm hold on the minds of participants (Boyer, 2001).[8]

In this view, the issue of cultural preservation takes priority over issues of cognitive simplicity of counterintuitive concepts and of religious costs as hard-to-fake signals. Indeed, if we jump to more recent times we stumble upon the panoply of cultures where the views one professes and therefore the words and other low-cost signals one uses are not shrugged off as potentially deceitful and *therefore* irrelevant data; quite the contrary. It might be replied that costly signals and displays are always accompanied by collective, conspiratorial performances (rituals) that over time re-create a context where low-cost signals (words) can be used with a confident attitude. But then why failing to comply with minor doctrinal details should be perceived in so many religions as the hallmark of all wickedness and a more pernicious threat than failing to join? And why those who believe in slightly modified versions of the official creed should be excommunicated instead of being allowed to hold their beliefs *within* the congregation?

It takes very little reflection to realise that the likely consequence of enforcing an individual to abjure will often be an insincere conversion to dogma. Religious insistence on doctrinal orthodoxy appears therefore to be under-explained by the theory of the costly commitment. Two interpretations are possible: (1) The theory is still valid when applied to emotion and action as reliable signs of commitment. Since profession of belief is an easy-to-fake signal, and "a signal that can be imitated is worthless as a signal" (Bulbulia, 2004, p. 25), avowals are not much valued as signs of commitment as they are treated as signals and causes of further, potential subversion – on the part of the subject or whatever co-religionist he might be able to proselytise. (2) The imposition of doctrinal conformity is an adaptation of the mechanisms of cultural transmission that enhances the fidelity of transmission by fixing constraints to select out the potentially maladaptive variations.

---

[8] If Barrett's (2000) HADD helps explain where the moral intangibles came from, the notion of ritual explorations may explain how such intangibles can turn into demanding moral watchers. Early ritual participants were struck by the power expressed by their joint actions, and started to concentrate on their formal and *aesthetic* features. The disposition to suppose that other people (including ancestors and gods) might be watching and judging what one is doing is a first step towards developing a mediated sense of *how well* one is doing in the achievement of a given goal (Rappaport, 1999).

Following the latter, a number of researchers came to suggest that such socially constructed, moral intangibles, besides being minimally counterintuitive and highly memorable entities, may be also *designed* (by a sort of bootstrapping mechanism) to protect the very counter-reality they are made of – in the sense that only those supernatural agents thrive that are fit to prevent the breakdown of the underlying tacit agreements. How? For example, by fostering the joint symbolic activities constituting the premises of ritual; and by inviting repetition (also perfect repetition) in a number of ways that ultimately depend on two strategies: (a) rewarding the participation in the construction of the counter-reality; (b) enhancing the sense of frustration and discontent with reality.[9]

Religion and Ritual after Speech

Whatever its ultimate biological origin, the emergence of speech from pretend-play (Knight, 1999) gave a powerful boost to both cognition and reflection. This process would often lead to unconstrained variation and in the end to scepticism: the enemies of social harmony are now within the coalition. Even more than wild predators and outgroup warriors, over-reflecting (young) individuals would pose a major threat to the very persistence of the counter-reality on which the whole society rests. As Dennett (2006, p. 115-116) vividly puts it:

> The human capacity for reflection yields an ability to notice and evaluate patterns in our own behaviour… Once people start 'catching on', a system

[9] Within the field of affective neuroscience the view that valence (the affective quality marking experience) should be construed as formed by *two* bipolar dimensions was proposed by Carver and Scheier (1998). The idea is that "certain feedback systems monitor and regulate the effectiveness with which people move toward incentives and move away from threats. These particular feedback systems are assumed to compare a signal corresponding to rate of progress against a reference rate. The 'error signal' of these loops is manifest subjectively as affect… The idea that there are distinct approach and avoidance behavioral systems (and thus distinct approach and avoidance behaviors) suggests a basis for the existence of differences in affect qualities. That is, perhaps affects differ as a function of which system – approach versus avoidance – is managing the behavior and thus generating the affect. Doing well at moving *toward an incentive* is not the same experience as doing well at moving *away from a threat*… The idea that eagerness, excitement, elation and so on should relate to an approach process is fairly intuitive. It is also fairly intuitive that fear, anxiety, and so on should relate to an avoidance process" (Carver, 2001, p. 347).

> that has 'worked' for generations can implode overnight. Traditions can erode more swiftly than stone walls and slate roofs, and preventive maintenance of an institution's creeds and practices can become a full-time occupation for professionals.

The most prominent effect of this mounting scepticism, according to Dennett, was the domestication of religion: "the wild (self-sustaining) memes of folk religion became thoroughly domesticated. They acquired stewards" (2006, p. 170).[10]

Once speech and language have set in, the priority is to find every way to disable the critical faculties these new activities bolster. Religious ritual behaviour becomes a major instrument whereby people have their problem-solving capacities discredited as reliable means, and in this way disabled. Lawson and McCauley (1990) had assumed that both religious and non-religious people possess the cognitive equipment to reduce ritual behaviour to its basic conceptual elements (agent, action, patient, instrument, together with their attendant qualities). And we saw that Barrett and Lawson (2001) confirmed that even "ritually naive individuals" tend to converge in crediting the agent, not the action, with the achievement of the wanted non-natural consequences. This is how people "justify in their minds" those consequences and may well be *what* some ritual prescriptions are there to prevent: that all the credit (praise or blame) should fall on the agent as a person as opposed to the agent as a role.

In this context counter-reality started to provide an alternative way of thinking whenever the appeal to ordinary cognitive resources (fit to solve real-time problems) was prohibited. The new role played by moral intangibles is made possible by the social, intentional construction of situations where bearing on one's ordinary cognitive capacities is deemed useless, morally inconsiderate, a social impediment.[11] The young are

---

[10] The evolutionary framework Dennett outlines fails to unify the domestication story with an account (à la Boyer) of what makes a ritual memorable. Having insisted (Dennett, 2006, p. 156) that "the rituals that persist are those that are *self*-perpetuating, whether or not anybody devotes serious effort to the goal of maintaining them", Dennett sidesteps the intermediate forms of social self-maintenance as if he doubted that a ritual system can evolve mechanisms that ensure its stability much before the emergence of a class of foresighted religious ministers.

[11] As Richard Sosis reminds us, "Rappaport explained… that although religious and secular rituals can both promote cooperation, religious rituals ironically generate greater belief and commitment because they sanctify unfalsifiable statements that are beyond the possibility of examination". He comments that:

trained to cast their needs and urges in terms of what the gods would want them to do (or omit). Here the relevant mechanism may be a complex emotional device whereby people are led to join for the sake of whatever reward they hope (or expect) to receive from the gods. The rationale for disguising such pro-social needs behind the prescriptions of some gods may be that this will help the underlying mechanisms to work unattended.[12]

Concluding Remarks

A ritual performance is composed of a series of meaningful, symbolic actions, each of which has good execution as its inherent goal. Since a ritual execution is deemed good according to qualitative criteria linked to god's approval, not on some causal effect, the accomplishment of a long-term goal is always contingent on the completion of a series of short-terms goals that are inherent in the ritual structure. Moreover, although ritual participants are strongly discouraged from applying their problem-solving categories to what they are doing, most ritual actions *are* accomplishments: they have an agent acting on a patient to bring about a non-natural consequence (Lawson & McCauley, 1990; Barrett & Lawson, 2001).

Any full-fledged explanation of a set of ritual forms must involve the articulation of the grammatical norms that regulate and intertwine different acts and participants and qualities into a coherent system. And this is what the naturalness-of-religion theory elegantly fulfils. However, its assumption that cognitive optimality could be all there is to religious transmission shows that it thinks of memory as a general, all-purpose cognitive device, whereas there are important differences between implicit and explicit memory, and (with respect to explicit, long-term memory) between semantic and episodic memory.

"Since statements containing supernatural elements, such as "Jesus is the son of God", cannot be proved or disproved, believers verify them 'emotionally'" (Sosis, 2004, p. 172).

[12] It is only now and then that the underlying mechanisms can emerge without undermining their effectiveness, as the author of the following statement on the practice of witness surely knows: "Without in any way compelling us to believe, the witness calls for a free assent that involves personal respect and trust. To reject the message is to withhold confidence in the witness. To accept it is a trusting submission to the witness's authority. To the extent that we believe, we renounce our autonomy and willingly depend on the judgement of others" (Avery Cardinal Dulles, 2004, p. 22; quoted in Dennett, 2006, p. 364).

Bearing on these different types of memory, the anthropologist Harvey Whitehouse has elaborated on a theory of religious and ritual transmission hinging around two different "modes of religiosity": doctrinal and imagistic (Whitehouse, 2004). A doctrinal mode of religiosity is one in which ritual actions and teachings tend to be highly routinised; this leads to the activation of two types of memory systems: implicit memory for ritual actions (rehearsed like behavioural scripts) and long-term semantic memory for religious doctrine (which allows the transmission of complicated bodies of knowledge). In order to prevent the rise of a "tedium effect" (and the fall of motivation) the faithful are induced to believe in otherworldly sanctions and rewards by the use of rhetoric and logically integrated theology. The imagistic mode of religiosity is characterised by highly arousing and infrequently rehearsed rituals (for example, initiation rites) which tend to trigger vivid and enduring episodic memories (called "flashbulb memories"). If the high levels of arousal provide a motivating force for the transmission of the ritual within the community, infrequent repetition hinders the establishment of doctrinal knowledge, and fosters spontaneous exegetical reflection (SER) which is often experienced as personal revelation (Whitehouse, 2004).

The explanatory and heuristic powers of this model spring from the light it sheds on the evolution of cultural systems at large. Instead of the all-too-obvious idea of a transition from illiterate groups of free believers to the more complex congregations monopolised by literate guilds basing their truth claims on texts, Whitehouse (2004) depicts a different story. He hypothesises that "the presence of doctrinal orthodoxies favours the subsequent development of writing systems, rather than being caused by them". As he explains (p. 80):

> Through elevated arousal, cognitive shocks, and the creation of consequential events, the rituals of the imagistic mode set off trains of exegetical thinking that are enduring and (over time) capable of generating highly elaborate semantic knowledge… These are the general conditions, I would suggest, in which the great philosophers of Aboriginal Australia, Amazonia, Africa, and Melanesia… come into existence… But another kind of philosopher was also born with the advent of the doctrinal mode of religiosity. What made this new breed of religious experts different was that their knowledge could be transmitted *verbally*, via highly repetitive regimes of teaching and reminding.

Once the very existence of human groups begun to depend on the stability of the local religious systems, a breakdown in the efficiency of the ritual machinery would mean the disruption of society, the fall of the group *as* a group. Sudden breakdowns in the ritual efficiency must have been the frequent result of unconstrained variation, and a major cause of turmoil. Lacking any rational understanding of the aetiology of such crises would lead most often to the disruption of the ritually defined coalition. Even the elders, finding a total mystery how the trick could produce the benefits on which they all depended, had no criteria to identify and countenance the harmless variations on the part of the young; the only safe instruction would be that no modification should be tolerated. High fidelity copying became a necessity, an intrinsically valuable good; and the general principle when it came to safeguarding the ritual transmission was, do not let the runaway sexual selection (encouraged by ritual promiscuity) overwhelm the notion that the *core* of ritual is a constant effort to please God – not the other exchange partners. Far from being the mere by-product of an ordinary endowment of religious concepts, theology as an essentially normative enterprise should be viewed as a viable (and deliberately inaccessible) solution to the need of limiting the variation of ritual systems whose beneficial qualities are as mysterious as they are easily lost.

References

Atran, S. & Norenzayan, A. (2004). Religion's evolutionary landscape: Counterintuition, commitment, compassion, communion. *Behavioural and Brain Sciences* 27, (p. 713-770).

Avital, E. & Jablonka, E. (2000). *Animal Traditions: Behavioural Inheritance in Evolution*. Cambridge: Cambridge University Press.

Barrett, J.L. (2000). Exploring the natural foundations of religion. *Trends in Cognitive Science* 4, (p. 29-34); reprinted in D.J. Slone (ed.), *Religion and Cognition*. London: Equinox 2006, (p. 86-98).

Barrett, J.L. & Lawson, E.T. (2001). Ritual intuitions: Cognitive contributions to judgments of ritual efficacy. *Journal of Cognition and Culture* 1, (p. 183-201); reprinted in D.J. Slone (ed.), *Religion and Cognition*. London: Equinox 2006, (p. 215-230).

Boyer, P. (2001). Religion Explained: The Human Instincts that Fashion Gods, Spirits and Ancestors. London: Random House.

Bulbulia, J. (2004). Religious costs as adaptations that signal altruistic intention. *Evolution and Cognition* 10, (p. 19-42).

Carver, C.S. (2001). Affect and the functional bases of behaviour: on the dimensional structure of affective experience. *Personality and Social Psychology Review* 5, (p. 345-356).

Carver, C.S. & Scheier, M.F. (1998). *On the Self-regulation of Behaviour*. New York: Cambridge University Press.

Cronk, L. (1994). Evolutionary theories of morality and the manipulative use of signals. *Zygon* 29, (p. 81-101).

Damasio, A. (1994). Descartes' Error: Emotion, Reason, and the Human Brain. New York: Grosset/Putnam.

Dennett, D.C. (2006). Breaking the Spell: Religion as a Natural Phenomenon. New York: Viking.

Diamond, J. (1997). Guns, Germs and Steel: The Fates of Human Societies. New York: Norton.

Immerman, R.S. (2003). Perspectives on human attachment (pair bonding): Eve's unique legacy of a canine analogue. *Evolutionary Psychology* 1, (p. 138-154).

Knight, C. (1991). Blood Relations: Menstruation and the Origins of Culture. London: Yale University Press.

—. (1999). Sex and language as pretend-play. In: R. Dunbar, C. Knight & C. Power (eds.) *The Evolution of Culture*. Edinburgh: Edinburgh University Press, (p. 228-247).

Knight, C., Power, C. & Watts, I. (1995). The human symbolic revolution: A Darwinian account. *Cambridge Archaeological Journal* 5, (p. 75-114).

Krebs, J.R. & Dawkins, R. (1984). Animal signals: Mind-reading and manipulation. In J.R. Krebs and N.B. Davies (eds.) *Behavioural Ecology: An Evolutionary Approach*. Oxford: Blackwell, (p. 380-402).

Lawson, E.T. & McCauley, R.N. (1990). *Rethinking Religion: Connecting Cognition and Culture*. Cambridge: Cambridge University Press.

McCauley, R.N. & Lawson, E.T. (2002). *Bringing Ritual to Mind: Psychological Foundations of Cultural Forms*. Cambridge: Cambridge University Press.

Pichon, I., Boccato, G. & Saroglou, V. (2007). Nonconscious influences of religion on prosociality: A priming study. *European Journal of Social Psychology* 37, (p. 1032-1045).

Rappaport, R.A. (1999). *Ritual and Religion in the Making of Humanity*. Cambridge: Cambridge University Press.

Shariff, A.F. & Norenzayan, A. (2007). God is watching you: Priming God concepts increases prosocial behaviour in an anonymous economic game. *Psychological Science* 18, (p. 803-809).

Sosis, R. (2004). The adaptive value of religious ritual. *American Scientist* 92, (p. 166-172).

Whitehouse, H. (2004). Modes of Religiosity: A Cognitive Theory of Religious Transmission. Walnut Creek: AltaMira Press.

Zahavi, A. (1987). The theory of signal selection and some of its implications. In U.P. Delfino (ed.) *International Symposium on Biological Evolution*. Bari: Adriatica Editrice, (p. 305-327).

CHAPTER SIXTEEN

EVOLUTIONARY POLITICAL SCIENCE – INTEGRATING EVOLUTIONARY PSYCHOLOGY AND THE SOCIAL SCIENCES

MICHAEL BANG PETERSEN AND LEIF EDWARD OTTESEN KENNAIR

Abstract: Evolutionary psychology is a mainstream approach in psychology. Yet, political and social science has only very recently implemented evolutionary arguments. In this chapter, we seek to contribute to this process by establishing the relevance of evolutionary psychology to political science. First, by introducing evolutionary psychology as a general theoretical approach. Second, by outlining avenues of evolutionary psychological research (on cooperation, punishment and coalitions), which can enhance our understanding of modern political phenomena. Finally, by describing in detail how evolutionary psychology might provide a novel framework for the study of a core political science field, public opinion. At the same time, we outline how modern politics provide an important case for evolutionary psychologists to explore how our evolved psychological programmes interact with novel circumstances. We conclude that conceptual integration between evolutionary psychology and political and social science might contribute fruitfully to all disciplines.

"When it comes to human behavior, evolutionary theory is more than a metaphor for societal change, and biology is more than an interesting but irrelevant discipline"

—Hibbing & Smith, 2007, p. 13

Politics, Political Science and Evolutionary Theory

Politics relates to the processes that give rise to the distribution of power and resources in society (Easton, 1953; Lasswell, 1990). Individuals carry out these processes, determine the rules that govern them and are in the end affected by them. Without an empirically validated model of the nature and architecture of individuals' motivations and cognitions, it would seem difficult to reach an adequate understanding of political life. This requires researchers to seriously examine findings in the psychological disciplines, i.e. disciplines with motivation and cognition as their empirical object, when theorizing about political behaviour. As is increasingly recognised in the psychological literature, the construction of any valid actor model requires an evolutionary and natural science approach to understanding human nature (Kennair, 2002). However, such approaches are not readily available within the very discipline that has politics as its object of study, political science.

This is remarkable. Theorizing about human nature constituted a core element of much classical political thinking. Thus, political philosophers from Plato and onwards were intensely preoccupied with outlining the political nature of man (Poulsen, 2004). Now, more than ever, there is compelling reason to pick up the classical line of inquiry. First, recent evolutionary approaches have presented and successfully tested a broad range of theories which imply that humans are biologically adapted to solve a range of psychological and social problems fundamental to modern politics, e.g. public goods provisioning, resource division, inter-group conflict, antisocial co-members and negotiation of hierarchies (Alford & Hibbing, 2004). Seemingly, man is indeed a *Zoon Politikon* – a political animal. This opens for the introduction of probably the strongest hypotheses-generator ever built – the theory of natural selection (cf. Dawkins, 1986) – into the study of core political phenomenon. The integration of these findings into political science should yield an immediate gain in explanatory leverage.

This chapter will first introduce modern evolutionary thinking in the form of evolutionary psychology. As will be explained, this approach focuses our attention on how biological evolution by selection has shaped our psychological architecture (rather than our behaviour *per se*). Thereby evolutionary psychology theory provides the tools necessary for dissecting how our species' evolutionary past influences modern day behaviour. Next, we will look at how and why evolutionary psychology might be fruitful within political science, and then turn to why evolutionary psychologists themselves ought to address topics that traditionally belong to political science. Finally, we describe the study of public opinion as one possible research agenda around which political science and evolutionary psychology could successfully converge. This chapter uses political science as an example of social sciences that might profit from adopting evolutionary psychology as one of its approaches. The arguments ought to be relevant to all other social sciences.

What is Evolutionary Psychology?

Evolutionary psychology is a specific *approach* to the study of psychology (for overviews, see Barkow, Tooby & Cosmides, 1992; Buss, 2004, 2005; Gaulin & McBurney, 2001; Kennair, 2005; Pinker, 1997, 2002). It is defined by two clear properties. First, a specific model of mind compatible with modern neuroscience (Duchaine, Cosmides & Tooby, 2001; Gazzaniga, 2000), and second, a specific research programme designed to discover the evolved human mental architecture (Buss, 2004; Tooby & Cosmides, 1990; Tooby & DeVore, 1987).

The evolutionary psychological model of the mind emerges from the unification of cognitive science and evolutionary biology (Tooby & Cosmides, 1992). The message from cognitive science is that the mind is comprised by myriads of domain-specific computational systems, i.e. information-processing systems designed to process a narrow set of cues pertaining to a particular problem (for an introduction, see e.g. Bless et al., 2004). The message from evolutionary biology is that any species-typical reasoning system is the product of evolutionary processes. Thus, evolution by natural selection is the only known explanation of complex design in biological tissue (such as the brain) (Dawkins, 1986). Evolutionary psychology utilises both insights by using our knowledge of natural selection processes and the structure of our ancestral environment to generate hypotheses about the structure of the mind's computational systems. Thus, the evolutionary psychology model of mind is one where the brain and therefore mind consists of an array of psychological adaptations, which evolved to solve specific adaptive problems in our species' past. An adaptive problem is a stable, species-specific problem facing our ancestors, whose solution affected (however distally) their reproductive success (i.e., their inclusive fitness).

The human mind is a mosaic of information processing circuits that were selected by biological evolution to extract specific kinds of contextual information, process this information in a specific way, and generate specific behavioural responses. Thus, to understand the design of our (present day *Homo sapiens sapiens*) psychology, we need to know the environment to which it has been adapted. The current archaeological and genetic evidence suggests that the part of the Homo lineage from which our species eventually evolved, lived in North East Africa as hunter/gatherers until anatomically and psychologically modern humans migrated and colonised the rest of the planet within the last 90 thousand years (Boyd & Silk, 2003). Biological evolution is a slow process. To build complex psychological adaptations selection pressures need to be stable across numerous generations. Since migrating out of Africa, rapid cultural evolution has constantly changed our species' environment (the innovation of agriculture, the nation states, the Internet, etc.) and hence the pressures operating on us. This implies that the most substantial parts of our psychological architecture are adapted to the life of our Pleistocene Stone Age ancestors (Tooby & Cosmides, 2005; see Kelly [1995] for an introduction to these life conditions). Research documents that social life in these ancestral hunter-gatherer groups involved many of the same issues that are fundamental to modern politics, such as group-living, public goods provisioning, foraging, resource division, within and between group social exchange of resources and objects, intergroup warfare, antisocial co-members, the use of symbolic markings and negotiation of hierarchies (see e.g. Buss, 2005; regarding the relation to modern politics, see Alford & Hibbing, 2004). However, an important difference between then and now concerns the scale of the social interaction. The hunter-gatherer bands, which apart from the family constituted our primary social group, were presumably no larger than twenty-five persons, and overarching tribes, which comprised a number of bands, contained presumably no more than a couple of hundred individuals (Dunbar, 1993; Kelly, 1995). Thus, social interaction was a small-scale phenomenon.

Our knowledge of the structure of the ancestral environment allows us to judge the ex ante plausibility of hypotheses on diverse phenomena. For example: A common hypotheses in psychology has been that humans naturally code information about other persons' "race" (cf. Cosmides, Tooby & Kurzban, 2003). In an evolutionary perspective, the existence of such psychological machinery would require the ancestral existence of several physically different groups interacting through evolutionary history – and that perceiving this information and acting on it would be relevant in some specific fitness enhancing manner. This is not plausible given current knowledge. Human evolution happened in small groups of hominins, living in reasonable proximity of each other – and with a great amount of genetic relatedness between the groups due to females migrating (willingly and unwillingly) between the groups. Presumably, there was no interaction between ancestral groups of humans where skin colour could provide any relevant ecological information. For example: Recent research suggests that genes for white skin colour are very recent mutations, occurring between 12 and 6 thousand years ago (Gibbons, 2007). On these grounds, Kurzban, Tooby & Cosmides (2001) predicted that skin colour or "race" is not coded naturally, but is coded as an arbitrary social marker for group (coalitional) membership. In support of this prediction, their detailed empirical study shows that subjects disregard information about "race" when presented with a situation where other and more valid cues of coalition are present. Thus, evolutionary psychology utilises knowledge of the selection pressures operating in our ancestral environment to form testable, predictive hypotheses about human behaviour.

Compared to earlier approaches, evolutionary psychology advances the study of the evolved underpinning of complex human behaviour in two ways. First, by directing our focus to the psychological level. Through the domain-specific computational mechanisms – built by evolutionary processes and tailored to solve particular adaptive problems – our biological past influences our present-day reactions (Tooby & Cosmides, 1990a). Innate psychological mechanisms regulating behaviour are the evolved solutions to adaptive problems in our species' past. These are what evolved – not behaviour itself (Cosmides & Tooby, 1987). In this respect, it is important to note that these mental programmes evolved due to being adaptive in our species' past. Thus, it is within the causal structure of the ancestral world that our computational systems are evolutionary functional. They might not be adaptive here and now. But the contexts in which they will function and in which they will misfire should be predictable and understandable (see Nesse & Williams, 1996). Second, the focus on mechanisms reshapes our understanding of nature and nurture. Thus, contrary to some earlier socio-biological theories, which primarily made claims regarding universal essences, natural goodness or aggressiveness for example, evolutionary psychology is not as much about essences as it is about functional relationships between elicitors and responses. Due to this focus, evolutionary psychology is as much a theory about variation as a theory about universality. Universal mechanisms can thus give rise to manifest differences if their received input varies (Tooby & Cosmides, 1990b, 1992). Traditionally, individual-level and group-level (i.e., cultural) differences are explained by reference to socialization: Socialization processes transmit the different cultural customs and individual values from one generation or individual to the next. However, given the framework of evolutionary psychology, it is perfectly plausible that such differences arise in other ways. If we imagine a phenomenon such as complex cooperation (cf. below), the adaptive response is not uniform across all conditions. The design of the computational systems that regulate our response to the phenomenon would take the following form: In relation to A (the phenomenon), do X if B but do Y if C. Under conditions B our response would be X; under conditions C our response would be Y. In this connection, the existence of universal mechanisms will not be manifestly expressed. The constant (our innate psychological programmes) would lead to variation, because this constant would induce different individuals or groups situated in different environments to react differently. Environment B would elicit behaviour X and environment C would elicit behaviour Y. But, and this is important, this variation would occur in a predictable pattern. The intertwining of our evolved architecture and the given environment thus creates evoked differences across individuals and cultures. Ultimately, we can base hypotheses on the conditions generating these differences on knowledge of the selective pressures behind our psychological architecture. Thus, the human flexibility in this perspective is not reached by having few instincts, but by having many, and, dependent on the specific environmental conditions in which one is situated, some of these instincts will be made manifest and others will be kept latent resulting in the observed differences.

Human nature is far from random and chaotic. Rather, human psychological architecture is a collection of species-specific and context-specific adaptations for perceiving and processing information that evolved by biological evolution. By invoking knowledge of recurrent adaptive problems facing our ancestors, important features of this architecture are predictable (cf. below). There is scientific promise and optimism in this –an investigation as to whether an evolutionary perspective can deliver what it promises ought to be worthwhile pursuing.

Why is Evolutionary Psychology Important to Political Scientists?

The aim here is to introduce political scientists to basic arguments from evolutionary psychology on the structure of human social psychology and describe how these explorations illuminate questions in political science on the foundations of human motivation.

In the last section of this chapter – after considering why political science is important to evolutionary psychologists – we go into more detail with specific regards to public opinion and show how this could be an important point for conceptual integration between political science and evolutionary psychology, contributing fruitfully to both.

It is a challenging endeavour to undertake evolutionary analyses of complex adaptive problems and the psychological adaptations designed through selection to solve them; political scientists might be reluctant to undertake the task. However, the range of findings from genetics, ethology, anthropology and psychology support that any correct model of human nature needs to be informed by evolutionary thinking (Barkow, Cosmides & Tooby, 1992; Pinker, 2002). If political scientists are concerned about whether their models of decision-makers, voters and street-level bureaucrats are in fact good approximations to external reality, this implies that they do need to consider evolutionary theory.

Opposed to this view, the economist Friedman (1953) argues that scientific theories are instrumental constructs, which need not be 'true' in any normal sense of the term. In this perspective, good theories are simply good predictive tools. In some sense, such instrumental philosophies of science pose a reasonable challenge to evolutionary-informed political science, which can even be sustained by more realistic philosophies of science. This could be phrased as 'don't tell it, show it'. If political man is in fact endowed with an evolutionarily shaped nature, this should enable evolutionary psychological theories to out-compete non-evolutionary theories empirically.

As an illustration, we turn towards Friedman's own field, economics. Traditional economic analysis is based on the model of *Homo Economicus* – a model endowed with self-interested preferences, a rational mind and an incapability to commit to cooperative social solutions. Yet, as economists informed by evolutionary research have argued, such a model is, evolutionarily speaking, highly implausible as it cannot harvest the immense evolutionary benefits related to cooperation (see e.g. Frank, 1988). As will be developed in more detail below, early advances in evolutionary psychology predicted that humans evolved to cooperate conditionally by reciprocating cooperation and retaliating defection (Trivers, 1971). In line with this, a large amount of experimental work on economic decision-making in game theoretic dilemmas (e.g. the well-known Prisoners' Dilemma) has shown that cold self-interest is far from the only (or necessarily most important) ingredient in economic decisions. They are soaked in moral emotions and notions of reciprocal justice. First, experiments reveal that subjects cooperate conditionally. Subjects tend to reciprocate cooperation even when defection is the rational choice (Dawes & Thaler, 1988; Ostrom, 1998), but the defection of others quickly undermine otherwise stable cooperative solutions (see e.g., Fehr & Gächter, 2000). Recently, these studies have been supplemented by studies using methods from cognitive neuroscience, which have uncovered specialized circuits in the human brain for engaging in reciprocal cooperation (for reviews, see Zak, 2007; Petersen, Roepstorff & Serritzlew, 2008). Second, the defection of others does not only motivate us to withhold cooperation, but also to seek out and punish these individuals. Thus, experiments reveal that subjects are willing to punish defectors despite lack of rational incentive and at even large costs to themselves (Cameron, 1999; Fehr & Fischbacher, 2004). Apparently, this behaviour is related to neural activity in brain regions that produce pleasure, which indicate that we literally enjoy when justice is served (de Quervain et al., 2004). Finally, it has spawned significant interest that we punish and cooperate even in one-shot games (where the subjects only interact once) between anonymous subjects, i.e. in situations with absolutely no rational incentive for educating the other or for considering the other's welfare. Yet, as such one-shot interactions presumably were not frequent ancestral occurrences, one plausible interpretation backed by empirical evidence is that this behaviour results from the operations of evolved cognitive programmes, which represent the single-shot interaction as repeated and process a range of subtle cues outside conscious awareness (Hagen & Hammerstein, 2006; Haley & Fessler, 2005; for other interpretations, see Fehr & Henrich, 2003). So, despite our conscious awareness of the single-shot structure of the game, we react *as if* it is repeated, because our psychology is designed for that kind of interaction. While the first two observations challenge the classical idea of selfish preferences, this latter observation significantly questions the traditional economic notion of rationality.

In summary, by showing how evolved notions of fairness enter into economic decisions through the activation of distinct neural circuits, these observations testify to the predictive power of evolutionary theorising in economics. Hard facts have compelled researchers in behavioural economics to ground their theories explicitly in the current knowledge of evolutionary biology, neuroscience and ancestral adaptive problems. As a consequence, the traditional actor model of *Homo Economicus* in behavioural economics is being replaced with *Homo sapiens sapiens*. The question now of course is: Can we expect similar advances in political science from explicitly integrating evolutionary psychology?

Below, we outline three different yet interlinked avenues of evolutionary psychology research. This research shows that our evolved architecture includes mechanisms for facilitating cooperation, dealing with antisocial members, delimiting the boundaries of cooperation through group formation and for facilitating hostility or cooperation at the group level. The investigation of phenomena such as cooperation, punishment and conflict has from the very beginning been central to political science and these phenomena are at the core of politics. These lines of research in evolutionary psychology are thus directly applicable to political science problems and conceptual integration should yield immediate explanatory leverage.

The Evolved Psychology of Cooperation

The current evidence suggests that our ancestors have engaged in social exchange for several million years (Cosmides & Tooby, 1992). An important factor in the evolution of these capacities might relate to climatic changes in the Pleistocene period, when the African savannah was formed. With its diversity in animal life, the formation of the savannah enabled a shift in human diet towards larger intake of protein (Kaplan, Hill, Lancaster & Hurtado, 2000). In other words, a niche for evolving more fine-tuned abilities to hunt was formed. Presumably, hunting is one important context for the evolution of neurocomputational programmes to facilitate cooperation (Cosmides & Tooby, 1992; de Waal, 1996; another important context could be aid in combat, see Tooby & Cosmides, 1988; Trivers, 1971). First, this is because hunting as a foraging strategy is influenced by a number of factors outside the hunter's control (the prey can hide, run away, fight back etc.) (Cosmides & Tooby, 1992; Kaplan & Hill, 1985). Not just effort and skill, but also luck is vital to hunting success. By implication, our ancestors' supply of meat would to some extent fluctuate randomly. This must have constituted an enormous adaptive problem, first, because meat constituted a main part of the diet, and second, because not only the hunter but also his whole family depended on the meat. Yet, storing excess meat could solve this problem of random variation in foraging success. Obviously, it was not an option to physically store excess meat (as there were no refrigerators), but there were other ways. Thus, resources can be 'stored' in the form of reciprocal obligations. If a lucky hunter shares with those unlucky and if these return the favour, when fortunes are reversed, then random variation in resource access will be nullified. Presumably, there has been a strong selection pressure for exploiting these possibilities. Second, our ancestral environment was home to large prey – the Pleistocene megafauna. When hunting and killing such prey, abilities to engage in collective and coordinated action would increase hunting efficiency and reduce risk. Given the importance of meat in the ancestral diet, such abilities would have been strongly selected for (Kaplan et al., 2000).

Explaining cooperation has been pivotal in evolutionary theorising ever since Darwin, and today a range of findings based on diverse empirical paradigms (including experiments, cross-cultural research and neuro-imaging) strongly suggests that our abilities to cooperate have deep biological roots (see below). Yet, the evolution of these abilities contains an intriguing theoretical problem. The reason is that cooperation involves an altruistic component: When A helps B, A experiences costs. This is in direct opposition to the fundamental mechanism in evolution, natural selection, which selects for those designs that out-reproduce other designs. As implied by the above discussions, the solution to this paradox relates to reciprocity (Trivers, 1971; Axelrod, 1984): In the long run, A can increase his fitness by helping B, if A is sure B will return the favour, and if the benefit of B's help is larger than the cost A himself experienced by helping B. It is, however, important to notice that B would do even better if helped by A, but did not return the favour. A would, admittedly, be considerably worse off as a consequence.[1]

The potential benefits of reciprocity and the potential dangers of free-riders have created a double selective pressure which has shaped important parts of man's social mind: We want to harvest the fruits of cooperation, but keep a keen eye on the cost, in other words; we like to cooperate, but hate being cheated (Alford & Hibbing, 2004; Poulsen, 2004). Extant evolutionary research has, first, uncovered highly specialized mechanisms at the cognitive (Cosmides & Tooby, 1992; 2005) and the neurological level (Ermer et al., 2006; Stone et al., 2002; Zak, 2007), which facilitate and regulate cooperation, and second, validated this political incentive structure across cultures (Henrich et al., 2004) and species (de Waal, 1996).

Our evolved psychology of cooperation could explain a range of puzzling observations on why people behave as they do. Why do people cooperate, vote, pay taxes, recycle, obey the law, support welfare, etc.? Self-interest is a powerful predictor of human behaviour in general but it seems to fail in explaining many of these political behaviours (see e.g. Green & Shapiro, 1994; Ostrom, 1998). Recently, researchers in political science have alluded to so-called social capital in the form of trust and have shown how people behave pro-socially, if they trust others to do the same (see e.g. Putnam, 1993). But there is still no clear consensus on what social capital is and why it matters (cf. Sønderskov, 2008). By stressing how we evolved to engage in small-scale conditional cooperation, evolutionary psychology seems to provide a rigorous theoretical

[1] While the language used here relates to dyadic (two person) exchanges, the analysis can be applied to true collective n-person exchange. A current debate in evolutionary literature relates to whether this requires so-called group selection (a rare, but in principle possible selection mechanism) or whether ordinary selection at the individual level is enough (Tooby, Cosmides & Price, 2006; Boyd, Gintis, Bowles & Richerson, 2003).

foundation in this respect (Ostrom & Walker, 2003). This interaction between modern politics and the evolved psychology of cooperation is increasingly being explored. For example: Hibbing & Alford (2004) investigate how this psychology induces citizens to be more inclined to accept certain political outcomes; Scholz & Lubell (1998) show how this psychology interacts with individuals' level of trust in deciding taxpaying; Orbell et al. (2004) apply the evolutionary logic to the problem of cooperation in political theory; and McDermott, Fowler & Smirnov (2008) explore how related evolved circuits (for foraging) shape our reactions to gains and losses, which illuminate (among other things) the structure of party politics.

The Evolved Psychology of Punishment and Reparation

The selection pressures that structure our species-typical psychology of cooperation also give rise to a distinct evolved psychology of punishment and reparation. Given that we have adapted to systems of social reciprocity, capacities to detect and react upon cheating have been important to our survival (Trivers, 1971; Cosmides & Tooby, 1992). Cross-cultural evidence suggests that we are indeed highly adept at detecting cheating (Cosmides & Tooby, 1992, 2005; Sugiyama, Tooby & Cosmides, 2002), and studies in neuroscience indicate that specialized neural regions facilitate this (Stone et al, 2002; Ermer et al, 2006).

When it comes to reactions to cheating, it is possible to specify two contradictory selection pressures (Trivers, 1971; Petersen, Sell, Tooby & Cosmides, 2008). On the one hand, protection is obviously a central concern, and we have presumably evolved a psychological system of moral aggression to ensure this protection. By triggering a punitive reaction, this aggression will change the decision rules of the cheater by showing that one is able to impose costs (Sell, 2005), removing the extra benefits the cheater has accumulated because of his cheating (Price, Cosmides & Tooby, 2002) or in extreme cases directly eliminate the cheater (see Daly & Wilson, 1988; Duntley & Buss, 2005). Further, aggression will signal a reliable threat to other potential cheaters thereby facilitating cooperation more generally (Frank, 1988; Fehr & Gächter, 2000). Ethological, anthropological, economic and neurological evidence indicate that we and other animals actually do have a well-developed instinct for punishment and are willing to incur costs in order to have others punished (Brosnan & de Waal, 2003; Clutton-Brock & Parker, 1995; de Quervain et al., 2004; Fehr & Gächter, 2000; Cameron, 1999). One important feature of this protective system is that the more one invests in a social relationship, and the more one stands to lose in the case of cheating, the more strongly should the moral aggression be triggered (Price, Cosmides & Tooby, 2002). This has been confirmed in diverse experiments: The more one contributes to the public good, the more one aggresses against cheaters (see for example Fehr & Gächter, 2000; Price, 2005).

On the other hand, it is important to notice that in an evolutionary perspective, social relations exist because they create reproductive benefits (Petersen, Sell, Tooby & Cosmides, 2008; de Waal, 1996; Tooby & Cosmides, 1996). If the cheated party reacts by cutting off all future acts of aid, the cheater has paid dearly, and it is in his interest to avoid this. Similarly, because of possible future benefits, it might also be in the cheated party's own interest to avoid a breakdown of the reciprocal relation – despite the cheating. This selective pressure has plausibly given rise to specific computational programmes enabling the cheated party to facilitate the restoration of the broken social bond, if the bond is assessed as sufficiently valuable (de Waal, 1996; Petersen, Sell, Tooby & Cosmides, 2008). In other words, reparation might be an evolved strategy. Consistent with this perspective, reconciliation has often been observed in animal societies (see e.g., de Waal, 1996), it is a universal feature of human societies (Braithwaite, 1998), and the study of the neurological underpinnings of the behaviour is progressing (Farrow et al., 2001). Just as one goal of punitive strategies is to change the decisions-rules of the malefactor, the cognitive and emotional programmes evolved to facilitate reconciliation should motivate the cheated party to engage in behaviour, which will up-regulate the cheater's perception of the value of the prior and the possible future relationship. Specifically, adaptive reparative strategies should target the malefactor's intrinsic valuation of the victim and enable (when successful) the victim to harvest the value of the malefactor more efficiently than would be the case if fear of punishment was his/her sole motivation for behaving pro-socially in the future (Petersen, Sell, Tooby & Cosmides, 2008; on behavioural consequences of fear of punishment versus need for reparation, see Caprara et al., 2001). From these selection pressures, evolutionary psychology produces more detailed predictions about the design of our counter-exploitation programmes. Especially, it has been argued that the choice of strategy (punishment or reparation) will be guided by cues to the value of re-associating with the malefactor (Petersen, Sell, Tooby & Cosmides, 2008).

The existence of an evolved psychology for being vigilant and reacting against exploiters could explain the political importance of criminal justice, why people in even relatively safe societies are intensely preoccupied with crime and why citizens oscillate between viewing restorative and punitive reactions, respectively, as the right reaction to specific crimes depending on the information they receive about the specific criminal (Cullen, Fisher & Applegate, 2000; Roberts, 1992). Again, such investigations of the interaction between modern politics and our evolved psychology are already being explored. Petersen (2007) has shown how our evolved psychology for counter-exploitation shapes and structures public opinion on criminal justice issues in predictable ways, and Smirnov (2007) investigates how it influences possible solutions to the so-called 'commons problem'.

The Evolved Psychology of Coalitions

The selection pressures inherent in the evolution of cooperation also carry important consequences for our capacities to form coalitions (groups). The archaeological and anthropological evidence clearly suggest that already the social world of our Pleistocene ancestors consisted of interacting, competing and nested small-scale social groups or coalitions (Dunbar, 1993; Kelly, 1995; Manson & Wrangham, 1991; Tooby & Cosmides, 1988). Collective action enabled the production of a range of benefits unavailable to solitary organisms such as social insurance against variance in resource production in the form of exchange systems (Cosmides & Tooby, 1992), division of labour and collective vigilance against predators (Kameda & Tamura, 2007), and we have seemingly evolved a range of specifically designed neurocomputational systems for forming and representing coalitions and for navigating in dense environments of interacting coalitions (see e.g. Cosmides, Tooby & Kurzban, 2003; Gil-White, 2001).

While the evolution of these mental capacities has yielded enormous evolutionary benefits, natural selection has had to overcome at least two seemingly insurmountable adaptive problems before these benefits could be reaped. First, there must be a solution to the free-rider problem described above. Second, there must be a solution to the coordination problem, i.e. how to align the mental representations and behaviours of the individuals involved, all performing different functions, working over long periods of time, and having somewhat divergent interests (McElreath, Boyd & Richerson, 2003; Petersen, Delton, Robertson, Tooby & Cosmides, 2008; Tooby, Cosmides & Price, 2006). The existence of these problems explains why collective action and cooperation is nested within relatively stable social groupings. First, given the free-rider problem, mutual trust becomes a fundamental precondition for establishing a viable system of cooperation. Repeated interactions (if successful) between individuals secure the necessary trust (Ostrom, 1998). Second, given the coordination problem, participants need to share representations about each other's interests and goals. This again calls for repeated interactions, which enables a convergence of representations. These selection pressures explain the human tendency to form general-purpose groups, which are organized around sets of shared interests, and which are ready to act in concert when more specific needs arise (Tooby, Cosmides & Price, 2006). Due to the adaptive problems of free-riding and coordination, it has not been adaptive to form groups *de novo* when the need for collective action arises and our coalitional psychology clearly reflects this. The existence of an evolved group psychology presumably explains the intrinsic pleasure universally found in forming stable cooperative alliances around common visible markers such as those of the nation, ethnicity, race or religion. Research has explored and supported this explanation in a variety of disciplines (see e.g. Sosis & Alcorta, 2003; Poulsen, 2004; Kurzban, Tooby & Cosmides, 2001; Gil-White, 2001; Boyd & Richerson, 1987). As illustrated by a recent study, this evolved psychology might also shed light on leadership and followership in modern politics (see also Tooby, Cosmides & Price, 2006). Thus, Smith et al. (2007) demonstrate how the evolved psychology of coalitions induces citizens in modern society to prefer leaders, who acquire office without making any effort to do so.

Groups do not only exist to facilitate within-group cooperation but also to augment the power of individuals in situations of intergroup conflict over resources such as land, food and status (Manson & Wrangham, 1991; Tooby & Cosmides, 1988). Accordingly, it is expected that we have a distinct psychology, which regulates and facilitates coalitional aggression (Kennair, 2007b; Smirnov, Arrow, Kennett & Orbell, 2007; Tooby & Cosmides, 1988). This, first of all, implies that modern forms of out-group hostility such as ethnocentrism might be by-products of evolved psychological programmes for coalitional reasoning. Second, this implies that we have distinct psychological programmes facilitating war, and numerous studies in political sciences have already explored how this evolved psychology shapes modern warfare (see Johnson, 2004; Kennair, 2007b; McDermott et al., 2007; Smirnov et al., 2007; Thayer, 2004). This does not imply that intergroup hostility is inevitable. Just as social exchange between individuals has been adaptive when cooperative surpluses could be harvested, so have cooperative relationships between groups (Kurzban & Neuberg, 2005). This is supported by the anthropological record (Cosmides & Tooby, 1992). Given this, one promise of evolutionary psychology could be to specify the conditions under which meetings between groups and cultures give rise to conflict (Wrangham, 1999; Mesquida & Wiener, 1999) and when peaceful co-residence can be achieved.

Political Science and the Evolved Structure of Human Psychology

When the cues surrounding modern political phenomena resemble cues, which in our species past have been related to recurrent adaptive problems, modern citizens should automatically represent the former (modern politics) through the computational lenses selected in response to the latter (ancestral problems). Political science and evolutionary psychology could be conceptually integrated through the notion that modern day humans live in large-scale political systems but represent politics using evolved categories such as 'cooperative partner', 'cheater' and 'rival coalition'. As these categories and the mechanisms in which they are embedded evolved in small-scale societies, this would imply that we represent modern day politics as if it were small-scale and that we are motivated accordingly. For example: The representation of politics as a small-scale phenomenon might explain the paradox that citizens feel that it is important to vote (i.e. support their coalition) even though their actual influence on election outcome is close to zero.

Evolutionary psychology provides researchers with a distinct set of tools (from evolutionary biology) for generating hypotheses about the structure of the psychological processes that regulate behaviour. By utilizing knowledge of recurrent ancestral adaptive problems and the evolutionary processes operating on them, the promise of this integration is that political scientists will be able to specify elicitors and responses of distinct psychological processes in much more detail, thereby enabling more powerful dissections and predictions of the conditionality of political behaviour. Indeed, there seems to be an increasing tendency for using evolutionary psychology to explore questions raised by political science. As revealed above, studies in political science now appear, which utilize the framework of evolutionary psychology to generate and test novel hypotheses about the public's representations of politics. Another indication of this is presented in figure 1. The figure displays the number of publications from 1990 until 2006 matching the search phrase ["political science" and "evolutionary psychology"] in Google Scholar. While the exact numbers should be interpreted very cautiously (e.g. the search engine allows papers to appear both in the form of working papers and as eventual publications), the steadily increasing number of hits in which both "political science" and "evolutionary psychology" appear, reveals that the two fields are indeed approaching each other (see also Alford & Hibbing [2004] and the entire volume of Hibbing & Smith [2007]).

Figure 1. The dialogue between political science and evolutionary psychology

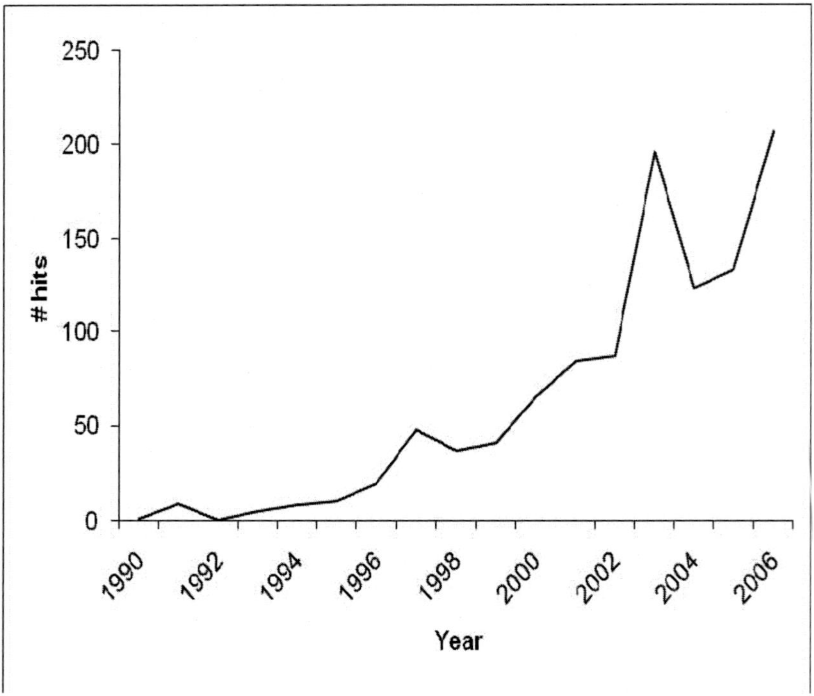

Notes. The figure displays the number of hits in Google Scholar for the search phrase ["political science" AND "evolutionary psychology"] in each year. Search conducted April 14 2008.

Why is Political Science Important to Evolutionary Psychologists?

Evolutionary psychology is not only important to political scientists. There also seems to be a number of incentives, which make it fruitful for evolutionary psychologists themselves to investigate questions traditionally associated with political science. Simply put, studying political scientific problems forces evolutionary psychologists to develop more adequate accounts of one core issue, namely how psychological mechanisms adapted to ancestral environments function in the radically different modern environments.

Ancestral social interaction was small-scale. Our species-typical architecture evolved to regulate behaviour based upon the myriad of cues that would be readily available in social interactions that happened face-to-face (we would thus have direct access to e.g. facial expressions; see Ekman & Rosenberg, 1997), and, secondly, would involve intimate social knowledge of each others' past behaviour, familial and social affiliations and so on (see e.g. Kurzban & Neuberg, 2005).

Although modern politics involves many issues fundamental to social life in ancestral hunter-gatherer groups, modern society provides a radically different context in which these issues have to be solved (Cosmides & Tooby, 2006; Petersen, 2007). Modern politics is – as political scientists have repeatedly pointed out – essentially about non-local phenomena (see e.g. Lippmann, 1922; Zaller, 1992). Today, the politically relevant communities are comprised by millions of people, and the opinions of the public have to be formed e.g. regarding social groups such as welfare recipients, immigrants and criminals with whom most of the population does not have intimate contact. Thus, at first glance, our computational systems seem to lack fully the necessary cues for their activation. Further, this lack of cues is accentuated by the fact that a central goal of modern politics is to formulate general rules (Stone, 1988). Politicians talk about how to solve *the* immigration problem or *the* problem of crime; not just how to deal with specific instances. While we have adapted to carefully modulate our reactions to the specific circumstances at hand, modern politics demands solutions at the most general level without specific reference.

The difference between the modern setting and the setting in which our innate programmes evolved can be illuminated by the anthropologist Dan Sperber's (1996) distinction between the proper and actual domains of psychological systems. The proper domain is the set of information that the system evolved to process, and where it has its adaptive relevance. The actual domain is all the information that is in fact processed by this system. In ancestral time the two domains would be fully overlapping, but cultural evolution creates divergence between the two. In Sperber's vocabulary, the proper domains of a range of politically relevant innate systems are local and concrete information-dense examples, while their actual domains today include modern politics characterized by information-sparse de-contextualised generalisations about non-local phenomena.

The implication is that a full account of how our Stone Age minds process the modern world requires of evolutionary psychologists to specify how computational programmes are provided with adequate stimuli to have made the enlargement of the programmes' domains possible. This includes how our psychology extracts information about unobservable phenomena from secondary sources, assigns truth values to such information, create generalizations from examples to whole categories, develops representations of non-local phenomena, enlarges the domains of innate categories to include novel topics, and allows these innate categories to interact with the representations about the non-local persons, groups and events. To consolidate evolutionary psychology's claim that even behaviour in radically novel and modern environments is guided by ancestral psychological mechanisms, it is vital to explore such questions. Given the conditions discussed above, modern politics constitutes an advantageous arena for this exploration.

Providing testable accounts of how evolved psychological mechanisms represent information and guide decisions in the context of modern politics thus provides an important challenge to evolutionary psychology. Theoretically, a series of important steps has already been taken (see e.g. Boyer, 2001; Cosmides & Tooby, 2000; Sperber, 1996; Thornhill et al., 1997).

An Evolutionary Psychological Framework for the Study of Public Opinion

To illustrate the benefits of disciplinary integration for evolutionary psychologists and political scientists alike, it would be fruitful to perform a more detailed description of the relationship between public opinion research and evolutionary psychology.[2] From the perspective of evolutionary psychology, focusing on public opinion directs our attention towards how old psychology functions in radically new cultural settings. From the perspective of political science, public opinion is a core field. Thus, a key insight of political science is that although political systems are supported by law, they could not function without being viewed as legitimate (Easton, 1957). For example: Citizens pay tax not only because they will be punished by the state if they do not, but also because they feel

[2] This section builds on Petersen (2007), where the arguments have been more thoroughly developed.

morally obliged to contribute to the community. The foundation of this diffuse support, as Easton (1957) calls it, is the political opinions and beliefs of the public.

Despite the importance of public opinion, political scientists are still trying to tackle basic questions such as how citizens form opinions at all (see e.g. Sniderman & Bullock, 2004). One reason for this is the observation that public opinion is highly inconsistent over time and across situations, which naturally makes it hard to grasp. First, citizens display 'mushy' or ambivalent attitudes, when asked general political questions. For example, citizens often oscillate between supporting different policy proposals even though the different proposals might rest on contradictory political principles (Alvarez & Brehm, 2002; Zaller, 1992). Second, citizens have unstable specific attitudes. Their attitudes in relation to specific welfare recipients, immigrants or criminals are often highly dependent on the targets' specific characteristics (see e.g., Oorschot, 2000; Sniderman, Hagendoorn & Prior, 2004; Stalans, 2002). Basically, public opinion is in flux. While these inconsistencies are well documented, we know surprisingly little about why they arise.

One relatively dominant explanation focuses on the social sources of opinion, arguing that citizens 'learn' to process political information and form opinions by paying attention to the political elites (politicians, the media, commentators, etc.) (Converse, 1964; Sniderman & Bullock, 2004; Zaller, 1992). But because citizens are only minimally engaged in politics (see e.g. Carpini & Keeter, 1997) – the argument continues – they do not pay adequate attention to develop any solid understanding of 'what goes with what' and inconsistencies abound.

At the same time, public opinion researchers are increasingly studying emotions (Marcus et al., 2000; Neumann et al., 2007; Redlawsk, 2006), and some have incorporated methods from cognitive neuroscience (Morris et al, 2003; Westen, 2007). These studies have successfully documented that political attitudes are highly influenced by emotional reactions. This proved importance of emotions in public opinion seems somewhat to contradict the idea of minimal engagement. What still seems to be lacking in the political science literature on emotions, however, is a theoretically grounded understanding of when and why specific emotions are elicited, which in turn can inform us as to why public opinion is so malleable. By providing the necessary tools for understanding emotional systems as evolved information processing systems, evolutionary psychology has an enormous potential for contributing in this respect (McDermott, 2007; Petersen, 2007; Schreiber, 2007).

The basic argument to be presented is: We are born with emotional and computational systems designed to enable us to decide on fundamental political questions. These systems operate based on distinct cues with ancestral relevance. When these cues shift from one decision-making context to another, our emotional responses shift as well. Therefore, inconsistencies might not be the result of empty-headedness but a consequence of the existence of sophisticated and context-sensitive psychological mechanisms, which enables us to form political opinions. Importantly, through careful dissections of relevant adaptive problems, evolutionary psychology should enable us to specify the input conditions of these mechanisms and to predict which shifts in cues and context cause patterns of attitudinal consistencies and inconsistencies. To outline this argument in more detail, we need to distinguish carefully between specific and general political opinions. We will begin the discussion by focusing on the former.

The Generation of Flexible Specific Opinions

In the modern political world, publicly accessible information that indirectly put us in touch with non-local phenomena abound (Zaller, 1992; Kinder, 2003). We are confronted with important political events and groups through television, newspapers, the Internet and so forth. Information stored in these types of media is not different from direct observations in the sense that information in both forms may be considered bundles of cues which our innate computational systems can pick up and process. In certain situations this enables opinion-formation to closely mimic decision-making in small-scale environments. Thus, information-dense descriptions should incline us to represent political phenomena psychologically as local phenomena impending on our welfare. For example: In episodic media stories, citizens are presented with concrete descriptions of the circumstances of specific crimes, of the plights of specific welfare recipients, etc. When specifically forming opinions in relation to such depicted individuals, our computational mechanisms can exploit the wide variety of information presented. These mechanisms evolved to match output closely to circumstances, and the motives they elicit will carefully be modulated by the exact information. By implication, inconsistencies in public opinion should automatically occur from one specific situation to another.

Thereby, the context-sensitivity of specific political opinions might be a by-product of our evolutionary past in small groups. Importantly, our opinions should be especially sensitive to information, which ancestrally has been correlated with increasing or decreasing evolutionary benefits. For example: As argued above, our minds are designed to be sensitive to exploitation from potential cheaters. Accordingly, the psychological systems that regulate the activation of our social emotions should carefully assess cues relating to the motivations and intentions of others (see e.g. Ermer et al., 2006). In line with this, research shows how small changes in information with regard to e.g. the laziness of a welfare recipient or the intentionality of a criminal, cause large shifts in opinion and emotions (see e.g., Darley & Pittmann, 2003; Weiner, 1995). Furthermore, we should expect that specific emotional reactions and opinions are highly consistent across situations, which contain *similar* information with ancestral relevance. Therefore, every time a welfare recipient is depicted as a potential free-rider, opinions towards helping should be consistently negative. And the reverse: When a welfare recipient's plight is framed as due to bad luck, we feel that it is legitimate (even morally obligatory) for the government to help him or her.

The Generation of Ambivalent General Opinions

Specific opinions are formed in information-dense contexts in relation to specific policy proposals or specific individuals (e.g. a specific welfare recipient or criminal). Yet, as argued above, many political discussions are directed toward more general questions. In contrast to specific opinions, general opinions are formed in information-sparse contexts in relation to abstract ideological or principal political questions.

Just as specific opinions, general opinions have been found to be ridden with inconsistencies. Yet, given the evolutionary model of human psychology, we should expect this type of inconsistency to differ markedly from the kind of inconsistencies characterizing specific opinion-formation. In the latter context, inconsistencies arise from shifts in information; in the former context, they arise from lack of information. This might not only give rise to externally observed inconsistency, but to internally felt ambivalence. Thus, to guide action, cognitive and emotional processing systems need information. Given this, conflicting motivations seem likely to arise: (1) when information cannot be obtained and (2) when it is deemed important to reach a decision. General political questions on political problems, which mimic ancestral social problems, seem to provide just this combination of features. Indeed, ambivalence seems to be a central feature of general political attitudes (see e.g., Craig & Martinez, 2005). In contrast, we should expect specific opinions – despite their surrounding inconsistencies – to be firmly expressed.

The Consistency-Generating Role of Stereotypes with Emotional Fit

Inconsistency is not all there is to general opinions. In a computational perspective, consistency requires (1) that the lack of information is compensated (2) in a way consistent across situations (3) such that the input conditions of emotion-programmes that give rise to a singular motivation are unequivocally met. To satisfy the first two requirements, our minds need to contain systems for storing politically relevant information (extracted from personal experience, from the media, etc.) and exploiting it consistently in novel situations. Even though the precise evolutionary origins for this cognitive apparatus are still underdeveloped, stereotypes are one type of cognitive inference structures, which seems to meet these requirements.[3] Stereotypes are generalizations about members of a category, which are used to pass judgments in novel situations and which seem resilient to change (Brown, 1986).

The third requirement does not relate to the apparatus for storing and exploiting representations but to the content of the stored representation. Basically, representations will fit the input conditions of different

[3] One important feature of the cognitive machinery underlying stereotypes might be so-called internal regulatory variables (Tooby & Cosmides, 2005; Tooby et al., 2008). A fundamental adaptive problem has been the temporal distance between the presence of information with adaptive significance and the presence of decision-making contexts in which that information has to be utilized. Delayed reciprocal exchange is an example in point. Here we need to be able to exploit information about previous interactions – did we receive help, was it delivered by the others' own free will and so forth – which by definition is not present when a decision about potential reciprocation has to be taken. To overcome this kind of adaptive problem, our minds comprised myriads of so-called internal regulatory variables. These variables are summary magnitudes calibrated by extracting relevant information from situations with adaptive significance. Subsequently, they guide behaviour in contexts in which the past information is relevant. The existence of internal regulatory variables might explain how and when consistent general opinions emerge. If internal regulatory variables are parts of our minds' design, it is plausible that the mental storage of information is comprised by two processes: One whereby information is stored in consciously accessible memory and one whereby the information recalibrate our internal regulatory variables. This recalibrational process makes information accessible to subsequent processing. The input to these recalibrational processes are presumably extracted from a infinite number of sources: Genetic factors (Alford, Funk & Hibbing, 2005), physiological factors (e.g. strength, see Sell, 2005), parental socialization, personal experiences, the media, etc.

emotional systems to varying degrees (see e.g. Sperber, 1996; Ekman, 2004). We should only expect decoupled representations such as stereotypes to facilitate consistency (and serve as substitutes for the cues contained in direct observations), if they embody cues that satisfy the input conditions of our emotion-programmes in ways fostering singular motivations. Accordingly, the impact of emotions on political opinions and their expressed consistency should be highly conditioned by what can be called *the emotional fit* of the stored representations. For example: As can be expected from literature on the design of our psychology of social exchange, it makes a huge difference whether citizens through the use of stereotypes represent welfare recipients in general as being lazy or as being victims of bad luck (e.g., Larsen, 2006; Iyengar, 1991). According to the argument here, it is such representations, which serve as a bridge between our evolved systems and general political debate. In the domain of criminal justice, this idea has been directly tested. The evidence suggests that stereotypes about criminals, which met the input conditions of relevant evolved programmes (e.g. are either represented as having high or low social value), do indeed facilitate consistent general criminal justice opinions (Petersen, 2007).

The political science literature has long stressed the importance of stereotypes, but the argument presented here goes further. While stereotypes have sometimes been seen as purely cognitive structures, e.g. as working hypotheses freeing up cognitive resources (see e.g., Darley & Gross, 1983), evolutionary psychology suggests that simultaneously with the construction and reconstruction of representations at a consciously accessible level, processes of emotional calibration take place. The stereotypical 'pictures in our heads' (cf. Lippmann, 1922) have profound effects (at least, in part) because they allow our emotional systems to be calibrated to react to non-local political phenomena. Rather than 'just' be one in a series of heuristic devices that can help cognitive misers form an opinion, holding such representations with emotional fit – at least, to the extent the opinion is based on emotion – constitute a necessary precondition for forming general political opinions.

That information might fit our evolved inference systems to a lesser or greater degree also carries implications for how we are expected to process the information that provides the raw material for our stereotypes. In political science, John Zaller (1992) has forcefully argued that we need to discard what he calls the file drawer model of political opinion if we are to understand how citizens make up their minds on political issues. We do not have a fully-formed opinion stored in memory that we retrieve when asked e.g. a survey question. Instead, we construct one on the spot from various inputs. The implication of the argument presented here, however, is that this file drawer model also needs to be discarded in relation to the storage ('internalisation') of information. As Sperber (1996) has argued, we do not take a piece of information as a file and store it in memory in the same form that it was presented to us. Rather, the process of memory storage is an active process whereby we reconstruct a mental representation of the information offered in e.g. a newspaper article. The degree of resemblance between the mental and the public representation will depend upon the ease with which this reconstruction can take place. The most important factor influencing this ease is the extent to which the information fits our innate psychological structure (Sperber, 1996; Boyer, 2001). It is simply easier to reconstruct information about e.g. social relations than about quantum physics because we have hardwired cognitive structures that evolved to process precisely information about social relations while quantum physics seem to contradict a range of innate categories. Thus, because information storage is an active reconstruction process, information with emotional fit will tend to spread more easily (see also Boyd & Richerson, 2005; Tooby & Cosmides, 1992). By implication, the concept of emotional fit might explain the tendency for the higher levels of consistency in attitudes towards moral rather than technical political issues (see e.g., Carmines & Stimson, 1980). Moral issues invoke a vocabulary that fit into our innate psychological categories ('cheater', 'rival coalitions', etc.), making it easier to store the necessary information to regulate our emotions. In a similar vain, the concept might contribute to the study of media effects on public opinion. Chong & Druckman (2007) argue that one of the challenges facing such studies is the identification of factors that make a media frame suggestive. Evolutionary psychology might provide a solution to this puzzle by stressing how information that fits the themes our emotions evolved to process in Pleistocene hunter/gatherer-society, is more easily internalised and elicits stronger emotional reactions.

Functional Deactivations of Stereotypes

According to the arguments presented here, concrete opinions emerge from the flexible operations of evolved psychological programmes, attending to the available cues in thick descriptions of concrete political situations. In contrast, consistent general opinions are expected to emerge from interactions between evolved psychological mechanisms and stereotypes (and other generalizing types of representations), functioning as replacements of concrete observable cues. This will automatically create inconsistencies between the specific and the general opinions. Such inconsistencies are well-documented. Public opinion research shows that the public's perceptions of the deservingness of specific recipients are relatively unaffected by general welfare attitudes (Oorschot, 2000; Petersen et al, 2007) and that punitiveness towards specific criminals is unconstrained by punitiveness in general (Petersen, 2007; Stalans, 2002). Similarly, research on political tolerance shows how people do not hold on to their political principles in specific situations (McClosky & Brill, 1983). Furthermore, related effects have been observed in experimental economics. Thus, behaviour in specific situations in economic games is relatively independent of the general beliefs of the subjects (Camerer & Fehr, 2004).

So why are people's specific opinions not strongly coloured by their stereotypes? From an evolutionary perspective, stereotypes as cognitive tools might exist because they allow us to pass swift judgments under pressure (Macrae et al., 1994). Yet, when it comes to making inferences of a given individual, stereotypes will necessarily give rise to misrepresentations and offer poor guidance compared to more detailed specific information. Accordingly, an adaptive feature of the machinery for deploying stereotypes could be the existence of rules for regulating the scope of this deployment (see also Cosmides & Tooby, 2000). Especially, we should expect them to be 'turned off', when more individualized information is present. Thus, despite the resilience of stereotypes to change and their importance to general opinions, stereotypes should briefly wither in the face of contradictory specific information with adaptive significance (only to blossom again when compensating information is needed; see e.g., Cosmides, Tooby & Kurzban, 2003). The existence of such scope rules could be a primary explanation for the observations of the inconsistencies between general and specific opinions. Furthermore, psychological research shows that consciously accessible stereotypes are structured in a way, which facilitates their deactivation in the face of specific information. Thus, a stereotype is akin to a probability distribution, where some characteristics are believed to be overrepresented among members of certain categories (Brown, 1986). Stereotypes are specifically *not* representations that imply the expectation that *all* category members display certain characteristics. Thus, the very structure of stereotypes implies that specific information can make it irrelevant. This might be an evolved functional design feature of stereotypes as a cognitive tool.

Public Opinion and Evolutionary Psychology

To summarize, this combination of opinion-formation context (specific versus general), information provided within that context (individualized versus stereotyped) and input conditions of evolved emotion programmes, provide us with a fully-fledged explanation of the way in which emotions influence public opinion and how consistency and inconsistency arise at the specific and general level. Of course, in itself this framework is rather weak in terms of explanatory leverage. It needs to be coupled with detailed analyses of the relevant ancestral adaptive problems and the input conditions of relevant psychological systems. It is on this basis that explicit content can be given to the notion of emotional fit (see e.g., Cosmides & Tooby, 2006; Petersen, 2007).

It should also be noted that the above remarks far from exhaust the ways in which evolutionary psychology is relevant to public opinion formation. For example: It is highly possible that we process political parties as ancestral coalitions and we represent our own support for a political party as a matter of coalitional affiliation (Petersen, Delton, Robertson, Tooby & Cosmides, 2008). Such representations can induce citizens to express certain political attitudes to signal coalitional affiliation. Thus, signalling should be done in a highly strategic manner, which e.g. involves that citizens should be willing to change their attitudes in order to show loyalty to and coordinate with a specific party (as is frequently observed, see Gilens & Murakawa, 2002), not just because we have a hard time deciding for ourselves, but also because it helps us accomplish specific coalitional goals.

Conclusion

Evolutionary psychology is relevant to political scientists. A range of findings indicates that human motivation cannot be reduced to products of rationality or socialization. We are political animals and our psychological mechanisms are adaptations evolved to solve a range of problems resembling modern political issues. As a consequence, individual and social models need to be combined with biological findings (evolutionary biology, neuroanatomy, neurophysiology, endocrinology, genetics, etc). Any understanding of human nature and human society and culture needs to be founded on (or at least, consistent with) current knowledge of our biology and history as a species. The life sciences and the human sciences ought not to be kept separate.

We recommend a multidisciplinary approach, which focuses on the relevance of and interplay between different levels of analysis. This approach is not reductionistic in the sense that there is nothing but biochemistry and each discipline may need to include it's own level of analyses (in the case of political science, for example, it is necessary to analyse and specify how information about non-local phenomena become available for our evolved cognitive programmes to process). Nevertheless, the inclusion of other levels of analysis and the findings of other disciplines ought to increase the predictive success of social science theories. Adopting basic principles from a broader set of phenomena and scientific disciplines might help build more accurate theories of social and political behaviour.

In this chapter, we have outlined three different avenues of evolutionary psychological research (on cooperation, counter-exploitation and coalitions, respectively), which should contain important insights for political scientists. These avenues of research could readily be integrated with the proposed framework for the study of public opinion. This integration would facilitate the development of testable accounts of how certain information and evolved emotional and cognitive programmes interact to produce predictable patterns of consistency and inconsistency across different political contexts. Furthermore, as revealed by the discussions, insights from evolutionary psychology pertain to numerous other classical topics in political science such as cooperation, war, ethnocentrism and leadership.

How should a political scientist interested in such conceptual integration proceed? One first step would be to reinterpret observations from political science in light of current psychological knowledge. The quality and type of data used in political science implies that many of the disciplines' theoretical arguments on the structure of individual motivation are empirically underdetermined, and a sensible criterion for selecting between possible alternatives would be to look to those specific disciplines, which focus directly on individual motivation as their chief empirical object (e.g., psychology and cognitive neuroscience). As outlined in this chapter, these disciplines clearly stress the evolved nature of motivation programmes. Such an endeavour of making explanations of political phenomena consistent with well-established facts from other sciences constitutes in itself an important step forward in terms of scientific progress. A second step could be to use evolutionary psychology to explain otherwise unintelligible behaviour such as behaviour produced by psychological biases (Johnson, 2007). Political scientists have traditionally considered biases and 'irrational' behaviour as something produced by cognitive limitations, errors and mistakes. However, evolutionary psychologists consider psychological biases to be context specific problem solving mechanisms (e.g. Haselton & Buss, 2000; Haselton & Nettle 2006), which misfires in present environments due to the mismatch between ancestral environments and modern society. As discussed, one example of this could be our tendency to cooperate with perfect strangers we only meet once. A third step would be to focus directly on ancestral adaptive problems structurally resembling the problems of modern politics, map the psychological inference mechanisms which would solve these problems, define the cues which these inference mechanisms were designed to process, state how modern politics contain these cues, specify how modern politics tap into the slots of the evolved mechanisms and, finally, provide adequate empirical tests of the steps in this chain of arguments. While a fully-fledge evolutionary political science requires researchers to take this third step, smaller less demanding steps towards an integration of political science and evolutionary psychology should also yield explanatory leverage.

In the end, the promise of the integration of political science and evolutionary psychology is that knowledge of the adaptive problems our species evolved to solve in ancestral time, will enable researchers to map and dissect the structure of the psychological mechanisms that regulate modern political behaviour, and to develop hypotheses about political phenomena solidly grounded in findings from the broadest possible range of disciplines (e.g., evolutionary biology, genetics, anthropology, neuroscience). We encourage political scientists to assess whether evolutionary theory makes contributions to political science. This requires researchers to make serious attempts at implementing the theory in research and compare predictive evolutionary hypotheses to other existing hypotheses (Kennair, 2006, 2007a). Researchers remaining naïve about these insights run the risk of continuing outdated discussions and replicating erroneous theories of human nature.

References

Alford, J. R., Funk, C. L., & Hibbing, J. R. (2005). Are political orientations genetically transmitted? *American Political Science Review, 99*, (p.153-167).

Alford, J. R., & Hibbing, J. R. (2004). The origin of politics: An evolutionary theory of political behavior. *Perspectives on Politics, 2,* (p.707-723).

Alvarez, M. & Brehm, J. (2002). *Hard Choices, Easy Answers*, Princeton: Princeton University Press.

Axelrod, R. (1984). *The Evolution of Cooperation*. New York: Basic Books.

Barkow, J. H., Cosmides, L. & Tooby, J. (eds.) (1992). *The Adapted Mind*. New York & Oxford: Oxford University Press.

Bless, H., Fiedler, K. & Strack, F. (2004). *Social Cognition. How Individuals Construct Social Reality*. Hove and New York: Psychology Press.

Boyd, R., Gintis, H., Bowles, S. & Richerson, P. (2003). The evolution of altruistic punishment. *Proceedings of the National Academy of Sciences, 100*, (p.3531-3535).

Boyd, R. & Richerson, P. J. (1987). The evolution of ethnic markers. *Cultural Anthropology, 2*, (p.65–79).

Boyd, R. & Silk, J. B. (2003). *How Humans Evolved*, 3rd ed. New York & London: W. W. Norton & Company.

Boyer, P. (2001). Religion Explained. The Evolutionary Origins of Religious Thought, New York: Basic Books.

Braithwaite, J. (1998). Restorative justice. In: Tonry, M. (ed.), *The Handbook of Crime and Punishment*. (p. 323-344). Oxford: Oxford University Press.

Brosnan, S. & de Waal, F. (2003). Monkeys reject unequal pay. *Nature, 425*, (p.297-299).

Brown, R. (1986). *Social Psychology. 2nd ed.* New York: The Free Press.

Buss, D. M. (2004). Evolutionary Psychology: The New Science of the Mind. 2nd ed. Boston: Allyn & Bacon.

—. (ed.) (2005). *The Handbook of Evolutionary Psychology*. Hoboken, N.J.: John Wiley & Sons.

Camerer, C. F. & Fehr, E. (2004). Measuring social norms and preferences using experimental games: A guide for social scientists. In: Heinrich et al. (eds.), *Foundations of Human Sociality*. New York & Oxford: Oxford University Press.

Cameron, L. (1999). Raising the stakes in the ultimatum game: Experimental evidence from Indonesia. *Economic Inquiry*, 37, (p.47-59).

Carmines, E. & Stimson, J. (1980). The two faces of issue voting. *The American Political Science Review, 74*, (p.78-91).

Caprara, G. V., Barbaranelli, C., Pastorelli, C., Cermak, I. & Rosza, S. (2001). Facing guilt: Role of negative affectivity, need for reparation, and fear of punishment in leading to prosocial behaviour and aggression. *European Journal of Personality, 15,* (p.219-237).

Carpini, M. X. D. & Keeter, S. (1997). *What Americans Know about Politics and Why It Matters*. New Haven: Yale University Press.

Chong, D., & Druckman, J. N. (2007). Framing theory. *Annual Review of Political Science, 10,* (p.103-126).

Clutton-Brock, T. H. & Parker, G. A. (1995). Punishment in animal societies. *Nature, 373*, (p.209-16).

Converse, P. (1964). The nature of mass belief systems in mass public. In: Apter, D. (ed.), *Ideology and Discontent.* (p. 206-262). New York: The Free Press.

Cosmides, L., & Tooby, J. (1987). From evolution to behavior: Evolutionary psychology as the missing link. In: J. Dupré (ed.), *The Latest on the Best: Essays on Evolution and Optimality.* (p. 276-306). Cambridge, MA: The MIT Press.

Cosmides, L. & Tooby, J. (1992). Cognitive adaptations for social exchange. In: Barkow, J. H., Cosmides, L. & Tooby, J. (eds.), *The Adapted Mind.* (p. 163-228). Oxford: Oxford University Press.

Cosmides, L. & Tooby, J. (2000). Consider the source: The evolution of adaptations for decoupling and metarepresentation. (p. 53-115). In: D. Sperber (ed.), *Metarepresentations: A Multidisciplinary Perspective*, New York: Oxford University Press.

Cosmides, L. & Tooby, J. (2005). Neurocognitive adaptations designed for social exchange. (p. 584-627). In: D. M. Buss (ed.), *The Handbook of Evolutionary Psychology*. Hoboken, NJ: Wiley.

Cosmides, L., & Tooby, J. (2006). Evolutionary psychology, moral heuristics, and the law. (p. 182-212). In: G. Gigerenzer & C. Engel (eds.), *Heuristics and the Law*. Cambridge, MA.: MIT Press.

Cosmides, L., Tooby, J. & Kurzban, R. (2003). Perceptions of race. *Trends in Cognitive Sciences, 7,* (p.173-179).

Craig, S. & Martinez, M. (eds.) (2005). *Ambivalence and the Structure of Public Opinion*. New York: Palgrave Macmillan.

Cullen, F. T., Fisher, B. S., & Applegate, B. K. (2000). Public opinion and crime and corrections. *Crime and Justice, 27*, (p.1-79).

Daly, M., & Wilson, M. (1988). *Homicide*. NY: Aldine de Gruyter.

Damasio, A. (2003). *Looking for Spinoza*. London: Vintage.

Darley, J. M. & Gross, P. H. (1983). A hypothesis-confirming bias in labeling effects. *Journal of Personality and Social Psychology, 44*, (p.20-33).

Darley, J. M. & Pittman, T. S. (2003). The psychology of compensatory and retributive justice. *Personality and Social Psychology Review, 7*, (p.324-336).

Dawes, R. M. & Thaler, R. H. (1988). Anomalies: Cooperation. *The Journal of Economic Perspectives, 2*, (p.187-197).

Dawkins, R. (1986). *The Blind Watchmaker*. London: Penguin Books.

de Quervain, D., Fischbacher, U., Treyer, V., Schellhammer, M., Schnyder, U., Buck, A. & Fehr, E. (2004). The neural basis of altruistic punishment. *Science*, 305, (p.1254-1258).

de Waal, F. (1996). Good Natured. The Origins of Right and Wrong in Humans and Other Animals. Cambridge, Mass.: Harvard University Press.

Duchaine, B., Cosmides, L., & Tooby, J. (2001). Evolutionary psychology and the brain. *Current Opinion in Neurobiology, 11*, (p.225-230).

Dunbar, R. (1993). Coevolution of neocortex size, group size and language in humans. *Behavioral and Brain Sciences, 16*, (p.681-735).

Duntley, J. D. & Buss, D. M. (2005). The plausibility of adaptations for homicide. In: Carruthers, P. Laurence, S. & Stich, S. (eds.), *The Structure of the Innate Mind*. (p. 291-304). New York: Oxford University Press.

Easton, D. (1953). The Political System. An Inquiry into the State of Political Science. New York: Alfred A. Knopf.

—. (1957). An approach to the analysis of political systems. *World Politics, 9*, (p.383-400).

Ekman, P. (2004). *Emotions revealed*. London: Weidenfeld & Nicolson.

Ekman, P., & Rosenberg, E. L. (eds.) (1997). What the Face Reveals: Basic and Applied Studies of Spontaneous Expression Using the Facial Action Coding System. New York: Oxford University Press.

Ermer, E., Guerin, S., Cosmides, L., Tooby, J. & Miller, M. (2006). Theory of mind broad and narrow: Reasoning about social exchange engages ToM areas, precautionary reasoning does not. *Social Neuroscience, 1*, (p.196-219).

Farrow, T. F. D., Zheng, Y., Wilkinson, I. D., Spence, S. A., Deakin, J. F. W., Tarrier, N., Griffiths, P. D. & Woodruff, P. W. R. (2001). Investigating the functional anatomy of empathy and forgiveness. *NeuroReport*, *12,* (p.2433-2438).

Fehr, E. & Fischbacher, U. (2004). Third-party punishment and social norms. *Evolution and Human Behavior, 25*, (p.63-87).

Fehr, E. & Gächter, S. (2000). Cooperation and Punishment in Public Goods Experiments, *The American Economic Review,* 90, (p. 980-994).

Fehr, E. & Henrich, J. (2003). Is strong reciprocity a maladaptation? On the evolutionary foundations of human altruism. In: P. Hammerstein (

ed.), *The Genetic and Cultural Evolution of Cooperation.* (p. 55-82). Cambridge, Mass.: MIT Press.

Frank, R. H. (1988). Passions Within Reason: The Strategic Role of the Emotions, New York: W. W. Norton & Company.

Friedman, M. (1953). The methodology of positive economics. (p. 3–43). In: M. Friedman, *Essays in Positive Economics*. Chicago: University of Chicago Press.

Gaulin, S. & McBurney, D. (2001). *Psychology: An Evolutionary Approach*. New Jersey: Prentice-Hall.

Gazzaniga, M. S. (ed.). (2000). *The New Cognitive Neurosciences.* 2nd ed. Cambridge, MA: The MIT Press.

Gibbons, A. (2007). European skin turned pale only recently, gene suggests. *Science, 316*, 364.

Gilens, M. & Murakawa, N. (2002). Elite cues and political decision-making. In: Carpini, M. X. D. & Huddy, L. (eds.) *Political Decision Making, Deliberation and Participation*, JAI Press.

Gil-White, F. J. (2001). Are ethnic groups biological 'species' to the human brain? *Current Anthropology, 42,* (p.515-554).

Green, D. P. & Shapiro, I. (1994). *Pathologies of Rational Choice Theory.* New Haven: Yale University Press.

Hagen, E. H. & Hammerstein, P. (2006). Game theory and human evolution: A critique of some recent interpretations of experimental games. *Theoretical Population Biology, 69*, (p.339-348).

Haley, K. & Fessler, D. (2005). Nobody's watching? Subtle cues affect generosity in an anonymous economic game. *Evolution and Human Behavior*, *26,* (p.245-256).

Haselton, M. & Buss, D. (2000). Error management theory. A new perspective on biases in cross-sex mind reading. *Journal of Personality and Social Psychology, 78*, (p.81-91).

Haselton, M. & Nettle, D. (2006). The paranoid optimist: An integrative evolutionary model of cognitive biases. *Personality and Social Psychology Review, 10*, (p.47-66).

Henrich, J., Boyd, R., Bowles, S., Camerer, C., Fehr E. & Gintis, H. (eds.) (2004). *Foundations of Human Sociality: Economic Experiments and Ethnographic Evidence from Fifteen Small-scale Societies*, Cambridge: Oxford University Press.

Hibbing, J. R. & Alford, J. R. (2004). Accepting authoritative decisions: Humans as wary cooperators. *American Journal of Political Science, 48*, (p.62-76).

Hibbing, J. R., & Smith, K. B. (2007). The biology of political behavior: An introduction. *The ANNALS of the American Academy of Political and Social Science, 614,* (p.6-14).

Iyengar, S. (1991). *Is Anyone Responsible? How Television Frames Political Issues*. Chicago & London: The University of Chicago Press.

Johnson, D. D. P. (2004). *Overconfidence and War: The Havoc and Glory of Positive Illusions.* Cambridge, MA: Harvard University Press.

—. (2007). *Adaptive Politics: Strategic Advantages of Psychological Biases?* Paper presented at *HBES 2007*, Williamsburg, VA, June, 2007.

Kameda, T. & Tamura, R. (2007). To eat or not to be eaten? Collective risk-monitoring in groups. *Journal of Experimental Social Psychology, 43,* (p.168-179).

Kaplan, H. & Hill, K. (1985). Food sharing among Ache foragers: Tests of explanatory hypotheses. *Current Anthropology, 26*, (p.223–245).

Kaplan, H. S., Hill, K. R., Lancaster, J. B. & Hurtado, A.M. (2000). A theory of human life history evolution: Diet, intelligence, and longevity. *Evolutionary Anthropology, 9*, (p.156-185).

Kelly, R. L. (1995). *The Foraging Spectrum: Diversity in Hunter-gatherer Lifeways*. Washington: Smithsonian Institution Press.

Kennair, L. E. O. (2002). Evolutionary psychology: An emerging integrative perspective within the science and practice of psychology. *Human Nature Review, 2,* (p.17–61). http://www.human-nature.com/nibbs/02/ep.html

—. (2005). *Evolutionspsykologi – en indføring i menneskets natur.* København: Akademisk Forlag.

—. (2006). Editorial: Evolutionary political science? Why not! *Politik, 9*, (p.56-60).

—. (2007a). Statsvitenskap og evolusjonspsykologi: På tide å inkludere studiet av menneskets biologiske natur? *Norsk Statsvitenskapelig Tidsskrift, 23*, (p.54-75).

—. (2007b). Følelsenes funksjon og krigerens mentalitet. In: Moldjord, C., Arntzen, A., Firing, K., Solberg, O. A., & Laberg, J. C. (eds.), *Liv og lære i operative miljøer «Tøffe menn gråter!»*. Bergen: Fagbokforlaget.

Kinder, D. R. (2003). Communication and politics in the age of information. In: Sears, D. O., Huddy, L. & Jervis, R. (eds.), *Oxford Handbook of Political Psychology*. (p. 357-391). Oxford: Oxford University Press.

Kurzban, R., & Neuberg, S. (2005). Managing ingroup and outgroup relations. (p. 653-675). In: D. M. Buss (ed.), *The Handbook of Evolutionary Psychology*, Hoboken, NJ.: John Wiley & Sons.

Kurzban, R., Tooby, J. & Cosmides, L. (2001). Can race be erased?: Coalitional computation and social categorization. *Proceedings of the National Academy of Sciences, 98(26),* (p.15387-15392).

Larsen, C. A. (2006). The Institutional Logic of Welfare Attitudes. Ashgate.

Lasswell, H. (1990). *Politics: Who Gets What, When and How?* Goulster, MA: Peter Smith Publisher.

Lippmann, W. (1922). *Public Opinion*. New York: The Free Press.

Macrae, C. N., Milne, S. B., & Bodenhausen, G. V. (1994). Stereotypes as energy-saving devices: A peek inside the cognitive toolbox. *Journal of Personality and Social Psychology, 66,* (p. 37-47).

Manson, J. & Wrangham, R. (1991). Intergroup aggression in chimpanzees and humans. *Current Anthropology, 32*, (p.369-390).

Marcus, G. E., Neuman, W. R., & MacKuen, M. (2000). *Affective Intelligence and Political Judgement*. Chicago: The University of Chicago Press.

McClosky, H. & Brill, A. (1983). *Dimensions of Tolerance*. New York: Russell Sage Foundation.

McDermott, R. (2007). Cognitive neuroscience and politics: Next steps. In: Neumann, R., Marcus, G., Crigler, A., & MacKuen, M. (eds.). *The Affect Effect.* (p. 375-397). Cambridge: Cambrige University Press.

McDermott, R., Fowler, J. H. & Smirnov, O. (2008). On the evolutionary origin of prospect theory preferences. *Journal of Politics, 70,* (p. 335-350).

McElreath, R., Boyd, R. & Richerson, P. J. (2003). Shared norms and the evolution of ethnic markers. *Current Anthropology, 44,* (p.122-129).

Mesquida, C. G., & Wiener, N. I. (1999). Male age composition and severity of conflicts. *Politics and the Life Sciences, 18*, (p.181-189).

Morris, J. P., Squires, N. K., Taber, C. S. & Lodge, M. (2003). Activation of political attitudes: A psychophysiological examination of the hot cognition hypothesis. *Political Psychology, 24*, (p.727-745).

Nesse, R. M. & Williams, G. C. (1996). *Why We Get Sick: The New Science of Darwinian Medicine*. New York: Vintage Books.

Neumann, R., Marcus, G., Crigler, A. & MacKuen, M. (eds.) (2007). *The Affect Effect.* Cambridge: Cambrige University Press.

Oorschot, W. (2000). Who should get what, and why. *Policy and Politics, 28*, (p.33-49).

Orbell, J., Morikawa, T., Hartwig, J., Hanley, J. & Allen, N. (2004). "Machiavellian" Intelligence as a basis for the evolution of cooperative dispositions. *American Political Science Review*, 98, (p.1-15).

Ostrom, E. (1998). A behavioral approach to the rational choice theory of collective action: Presidential address, American Political Science Association. *American Political Science Review, 92*, (p.1-22).

Ostrom, E., & Walker, J. (eds.) (2003). *Trust and Reciprocity. Interdisciplinary Lessons from Experimental Research.* New York: Russell Sage Foundation.

Petersen, M. B. (2007). Straf eller rehabilitering? Evolutionspsykologi, følelser og politisk holdningsdannelse [Punishment or Rehabilitation? Evolutionary Psychology, Emotions, and Public Opinion formation]. Aarhus, Denmark: Politica.

Petersen, M. B., Slothuus, R., Stubager, R. & Togeby, L. (2007). Hvem fortjener velfærd? Danskernes syn på kontanthjælp til unge, ældre og indvandrere [Who Deserves Welfare?]. *Politica, 39*, (p.31-48).

Petersen, M. B., Delton, A., Robertson, T., Tooby, J., Cosmides, L. (2008). Politics in the evolved mind: Political parties and coalitional reasoning. Paper presented at the 66th MPSA Annual National Conference, Chicago, IL, USA, April 3-6, 2008.

Petersen, M. B., Roepstorff, A. & Serritzlew, S. (2008). Social capital in the brain? In: Svendsen, G. T. & Svendsen, G. L. H. (eds.), *Handbook of Social Capital*. Edward Elgar Publishing.

Petersen, M. B., Sell, A., Tooby, J. & Cosmides, L. (2008). Evolutionary psychology and criminal justice: A recalibrational theory of punishment and reconciliation. In: H. Høgh-Olesen (ed.), *Human Mind – Human Kind. Evolutionary Perspectives on Human Morality and Sociality.*

Pinker, S. (1997). *How the Mind Works*. London: Penguin Books.

—. (2002). The Blank Slate: The Modern Denial of Human Nature. London: Penguin/ Allen Lane.

Poulsen, J. J. (2004). *The General Theory of Solidarity*, doctoral thesis. University of Aarhus, Department of Political Science.

Price, M.E. (2005). Punitive sentiment among the Shuar and in industrialized societies: Cross-cultural similarities. *Evolution and Human Behavior, 26*, (p.279-287).

Price, M. E., Cosmides, L. & Tooby, J. (2002). Punitive sentiment as an anti-free rider psychological device. *Evolution and Human Behavior, 23,* (p.203-231).

Putnam, R. D. (1993). *Making Democracy Work.* Princeton: Princeton University Press.

Redlawsk, D. R. (ed.) (2006). *Feeling Politics. Emotion in Political Information Processing*. New York: Palgrave Macmillan.

Roberts, J. V. (1992). Public opinion, crime, and criminal justice. *Crime and Justice, 16*, (p.99-180).

Scholz, J. T. & Lubell, M. (1998). Trust and taxpaying: Testing the heuristic approach to collective action. *American Journal of Political Science*, 42, (p.398-417).

Schreiber, D. (2007). Political cognition as social cognition: Are we all political sophisticates? In: Neumann, R., Marcus, G., Crigler, A., & MacKuen, M. (eds.). *The Affect Effect.* (p. 48-70). Cambridge: Cambridge University Press.

Sell, A. (2005). *Regulating Welfare Tradeoff Ratios: Three Tests of an Evolutionary-computational model of Human Anger.* Ph.D. dissertation, Center for Evolutionary Psychology, University of California, Santa Barbara.

Smirnov, O. (2007). Altruistic punishment in politics and life sciences, *Perspectives on Politics*, *5,* (p.489-501).

Smirnov, O., Arrow, H., Kennett, D. & Orbell, J. (2007). Ancestral war and the evolutionary origins of heroism. *The Journal of Politics*, 69, (p.927-940).

Smith, K. B., Larimer, C. W., Littvay, L., Hibbing, J. R. (2007). Evolutionary theory and political leadership: Why certain people do not trust decision makers. *The Journal of Politics*, *69,* (p.285-299).

Sniderman, P. M. & Bullock, J. (2004). A Consistency Theory of Public Opinion and Political Choice: The Hypothesis of Menu Dependence. In: Saris, W. E. & Sniderman, P. M. (eds.), *Studies in Public Opinion.* (p. 336-357), Princeton: Princeton University Press.

Sniderman, P. M., Hagendoorn, L. & Prior, M. (2004). Predisposing factors and situational triggers: Exclusionary reactions to immigrant minorities. *American Political Science Review, 98*, (p.35-49).

Sosis, R. & Alcorta, C. (2003). Signaling, solidarity, and the sacred: The evolution of religious behavior. *Evolutionary Anthropology, 12*, (p. 265-274).

Sperber, D. (1996). *Explaining Culture: A Naturalistic Approach.* Oxford: Blackwell Publishing.

Stalans, L. J. (2002). Measuring attitudes to sentencing. In: Roberts, J. V. & Hough, M. (eds.), *Changing Attitudes to Punishment* (p. 15-32). Portland: Willan Publishing.

Stone, D. (1988). *Policy Paradox and Political Reason.* New York: HarperCollins Publishers.

Stone, V., Cosmides, L., Tooby, J., Kroll, N. & Knight, R. (2002). Selective Impairment of Reasoning About Social Exchange in a Patient with Bilateral Limbic System Damage. *Proceedings of the National Academy of Sciences, 99*, (p.11531-11536).

Sugiyama, L. S., Tooby, J. & Cosmides, L. (2002). Cross-cultural evidence of cognitive adaptations for social exchange among the Shiwiar of Ecuadorian Amazonia. *Proceedings of the National Academy of Sciences, 99*, (p.11537-11542).

Sønderskov, K. (2007). The environment. Forthcoming in Svendsen, G. T. & Svendsen, G. L. H. (eds.) (2008). *Handbook of Social Capital.* Edward Elgar Publishing.

Thayer, B. A. (2004). Darwin and International Relations: On the Evolutionary Origins of War and Ethnic Conflict. Lexington: University Press of Kentucky.

Thornhill, N., Cosmides, L., Maryanski, A., Meyer, P., Tooby, J., & Turner, J. (1997). Evolutionary theory and human social institutions: Psychological foundations. In: Weingart, P., Richerson, P. J., Mitchell, S. D., & Maasen, S. (eds.), *Human by Nature: Between Biology and the Social Sciences.* (p. 201-252). Mahwah, NJ: Erlbaum.

Tooby, J., & DeVore, E. (1987). The reconstruction of hominid behavioral evolution through strategic modeling. In: W. G. Kinzey (ed.), *The Evolution of Human Behavior: Primate Models.* (p. 183-237). Albany, NY: State University of New York Press.

Tooby, J., & Cosmides, L. (1988). The evolution of war and its cognitive foundations. *Institute for Evolutionary Studies Technical Report,* 88–1.

Tooby, J., & Cosmides, L. (1990a). The past explains the present: Emotional adaptations and the structure of ancestral environments. *Ethology and Sociobiology, 11,* (p.375-424).

Tooby, J., & Cosmides, L. (1990b). On the universality of human nature and the uniqueness of the individual: The role of genetics and adaptation. *Journal of Personality, 58,* (p.17-67).

Tooby, J. & Cosmides, L. (1992). The psychological foundations of culture. In: Barkow, J. H. Cosmides, L. & Tooby, J. (eds.), *The Adapted Mind.* (p. 19-135). Oxford: Oxford University Press.

Tooby, J. & Cosmides, L. (1996). Friendship formation and the bankers' paradox: Other pathways to the evolution of adaptations for altruism. *Proceedings of the British Academy, 88*, (p.119-143).

Tooby, J. & Cosmides, L. (2005). Conceptual foundations of evolutionary psychology. In: Buss, D. M. (ed.), *The Handbook of Evolutionary Psychology.* (p. 5-67). Hoboken: John Wiley & Sons, Inc.

Tooby, J., Cosmides, L. & Price, M. E. (2006). Cognitive adaptations for n-person exchange: The evolutionary roots of organizational behavior. *Managerial and Decision Economics*, 27, (p.103-129).

Tooby, J., Cosmides, L., Sell, A., Lieberman, D. & Sznycer, D. (2008). Internal regulatory variables and the design of human motivation: A computational and evolutionary approach. In: Elliot, A. J. (ed.), *Handbook of Approach and Avoidance Motivation*. Mahwah, NJ: Lawrence Erlbaum Associates.

Trivers, R. L. (1971). The evolution of reciprocal altruism. *The Quarterly Review of Biology, 46,* (p.35-57).

Weiner, B. (1995). *Judgments of Responsibility*. New York & London: The Guilford Press.

Westen, D. (2007). *The Political Brain*. New York: Public Affairs.

Wrangham, R. (1999). Evolution of coalitional killing. *American Journal of Physical Anthropology, 110*, (p.1-30).

Zak, P. J. (2007). The neuroeconomics of trust. (p. 17-33). In: R. Frantz (ed.), *Renaissance in Behavioral Economics*. Florence, KY: Routledge.

Zaller, J. R. (1992). *The Nature and Origins of Mass Opinion*. Cambridge: Cambridge University Press.

SECTION III

HUMAN SEXUALITY AND MATING STRATEGIES

CHAPTER SEVENTEEN

PARENTS, OFFSPRING AND MATE-CHOICE CONFLICTS

JUSTIN H. PARK, SHELLI L. DUBBS, AND ABRAHAM P. BUUNK

Abstract: Evolutionary approaches to human mating have largely ignored the fact that mating decisions are heavily influenced by parents and other kin. This is significant, because parents and offspring may have conflicting opinions regarding the offspring's mating behaviour, including opinions regarding what makes an ideal mate. Specifically, parents may have a relatively stronger preference for children's mates with characteristics suggesting high parental investment and cooperation with the ingroup, whereas children may have a relatively stronger preference for mates with characteristics signalling heritable fitness. We report results from a study among Uruguayan students that replicated previous results (Buunk, Park, & Dubbs, 2008).

Although Darwin gets most of the fame, much of evolutionary psychology began with Trivers. Parental investment theory (Trivers, 1972) and reciprocal altruism theory (Trivers, 1971) spawned programs of research (e.g., sex differences in mating behaviour, cognitive mechanisms for social exchange) that defined evolutionary psychology in its early years; even now, those theories continue to generate and constrain new hypotheses. Trivers' other ideas (e.g., self-deception; Trivers, 2000) are likely to inject fresh insight into phenomena (e.g., positive illusions) that many psychologists have been studying without the benefit of an evolutionary perspective.

In this article, we focus on one of Trivers' theories that has yet to receive more than cursory attention from psychologists, within a domain that has received almost no attention from psychologists. The theory is parent–offspring conflict theory (Trivers, 1974), and the domain is the battleground of mating. By "battleground of mating", we are referring neither to intrasexual competition for mates nor to intersexual conflict between mates—those are relatively well-trodden grounds. Rather, we are referring to the conflict between parents and their children that emerges when those children come of age and begin their reproductive careers.

If indeed parents and their offspring have conflicting opinions regarding the mating behaviour of the offspring, this may require a substantial redrawing of the picture of human mating. This is because among humans, parents and other kin have a powerful impact on individuals' mating behaviour. Therefore, any systematic opinions that parents have concerning how their children should mate are expected to have significant consequences, not only for how individuals "choose" mates, but also for the process of sexual selection itself—specifically, parents' in-law preferences might be at least as important as individuals' mate preferences in the "intersexual selection" of traits. Although few people would deny that family opinions matter when it comes to mating behaviour—and despite the fact that anthropologists and sociologists had long observed that marriage is often arranged (e.g., Goode, 1959; Murstein, 1974; Sprecher & Felmlee, 1992; Westermarck, 1921)—much of the theory and research on human mating within evolutionary psychology had, until recently, proceeded under the assumption that people freely choose mates based on their personal preferences. To be fair, several researchers had acknowledged that kin influence mating decisions (e.g., Gangestad & Simpson, 2000; Haselton & Buss, 2000); the issue of kin influence on mating simply had not been on the forefront of anyone's agenda.

Mating is a Family Business

Things quickly changed in 2007 with a flurry of articles dealing with the issue of parental/kin influence on mating, each article focusing on a specific aspect of the issue and reporting distinct kinds of data. First, Apotolou (2007a, 2007b) noted that because free-choice mating is rather rare across human societies (and likely always has been), human mate choice (and its consequences for sexual selection) may be more accurately understood as a process in which parents choose desirable daughters- and sons-in-law. Apostolou (2007b) reported data from 190 hunting and gathering societies showing that in the vast majority of societies, marriage is arranged by parents and other kin; only in 4% of societies was courtship the primary form of marriage. Of course, the presence of arranged marriage does not rule out the possibility that children may exert their preferences—by influencing the parents' decisions, for example—but it does imply that parents are the primary decision makers.

If parents are doing most of the mate-selection work, do they have specific preferences when selecting in-laws? If so, what are they? Adapting methods that had previously been used to assess individuals' mate preferences (e.g., Buss & Barnes, 1986), Apostolou (2007a) assessed people's in-law preferences, with the expectation that parents should prefer some characteristics more than others and that parents will have specific preferences for daughters-in-law and for sons-in-law. One reason—among others—for these sex-specific preferences is that daughters are an especially valuable reproductive resource (see Trivers, 1972), and parents may utilise this resource in a manner that maximises their own inclusive fitness when selecting in-laws. The results from Apostolou's (2007a) study conducted in the UK showed that people do consider some characteristics to be more important than others when considering potential spouses for their children. For instance, *kind and understanding* was highly desirable, and *religious* was less desirable. Other results revealed differences between preferences for daughters-in-law versus sons-in-law. For example, *good financial prospect* was especially desirable in a son-in-law; *good cook, housekeeper* was especially desirable in a daughter-in-law.

Of course, parents are not the only interested party. Adopting an inclusive fitness perspective more broadly, Faulkner and Schaller (2007) reasoned that the tendency for people to care about their kin's mating relationships can be conceptualised as an instance of nepotism. In their own words,

> If one's total inclusive fitness depends, in part, on the reproductive fitness of kin, one's fitness is maximized when kin members are themselves maximally reproductively fit. One's fitness is potentially reduced whenever kin members get involved in relationships with mates who seem unlikely to help those kin members produce reproductively viable offspring. This implication is greatest for genetically closer kin. (p. 431)

In a series of studies conducted in Canada, Faulkner and Schaller (2007) found that (a) people care more about the mating behaviour of genetically closer kin, (b) people care more about the mating behaviour of female kin, (c) females care more about the mating behaviour of kin and (d) people care more about mating behaviour of kin in long-term relationship contexts. The rationale for expecting people to care more about the mating behaviour of female kin was that—as noted by Apostolou (2007a)—females are reproductively more valuable. The rationale for expecting females to care more about the mating behaviour of kin was that, compared with males, females may benefit more from indirect reproductive effort (i.e., enhancing kin's reproductive success at the cost of one's own) than direct means (i.e., enhancing one's own reproductive success).

In research conducted in the US, Jonason, Izzo and Webster (2007) assessed people's willingness to help kin and nonkin members find long-term mates and short-term mates. Complementing Faulkner and Schaller's (2007) finding, Jonason et al. (2007) found that people were willing to help kin more than nonkin to find long-term mates (in contrast, people were willing to help nonkin more than kin to find short-term mates). Furthermore, consistent with the implications of parental investment theory (Trivers, 1972), people were willing to help female kin more than male kin to find long-term mates, and male kin more than female kin to find short-term mates.

Together, these articles painted a clear picture: Individuals' family members care about their mating behaviour and this fact deserves more attention. Of course, this fact is less interesting if family members' opinions happened to be no different from that of mating individuals themselves. For instance, suppose that a man wants a mate who is youthful, attractive, kind and faithful, and his parents and other relatives want exactly the same type of mate for him. Then, whether he finds a mate through courtship or through arranged marriage makes little difference—he would end up with essentially the same mate. But it seems unlikely that family members have identical preferences as do mating individuals. For one thing, if parents' in-law preferences simply mirrored their offspring's mate preferences, there would have been little reason for parents across cultures to go through the trouble of attempting to control their offspring's mating behaviour—indeed, there would have been little reason for parents to possess any in-law preferences at all. As it turns out, there are good reasons to expect differing opinions between offspring and parents (and other kin), which follow from Trivers' (1974) parent–offspring conflict theory. We have articulated the idea in detail elsewhere (Buunk, Park, & Dubbs, 2008), so only the main points are summarised here.

Parent–Offspring Conflict in Mate Preferences

In mating, there are trade-offs. For instance, individuals may pursue short-term or long-term mating, each associated with specific costs and benefits. For this particular trade-off, males have more to gain (in terms of reproductive fitness) from short-term mating than do females, and research indicates that males are indeed more inclined toward short-term mating (Buss & Schmitt, 1993). Likewise, individuals may have the option of allocating their resources toward mating effort versus parenting effort, each associated with specific costs and benefits. For this trade-off, one's heritable fitness (especially for males) seems to influence how they resolve the trade-off: Higher-quality males tend more toward mating effort (Gangestad & Simpson, 2000).

Such trade-offs are inherent in the choice of mate as well. For instance, when a woman is faced with the choice between a man with traits connoting heritable fitness and a man without such traits, this essentially represents a choice between having children with good genes versus having a partner who is likely to invest in her offspring (Gangestad & Simpson, 2000). The resolution of this trade-off has been found to depend on various factors. For instance, women's preferences vary according to aspects of local ecological context (e.g., Gangestad & Buss, 1993; Penton-Voak, Jacobson, & Trivers, 2004). In addition, women may have evolved a conditional strategy—preferring men with traits connoting genetic quality when the benefits are personally likely to be highest, such as when they are considering short-term relationships or when they are in the fertile phase of the menstrual cycle (e.g., Gangestad, Garver-Apgar, Simpson, & Cousins, 2007; Haselton & Miller, 2006; Penton-Voak et al., 1999).

Now, why might parents and offspring's opinions clash? They might clash, because a specific choice of mate may have different fitness consequences for parents and for offspring. Essentially, parents are expected to evolve preferences for offspring's mates that maximise the fitness of all of their grandchildren. In the trade-off involving an investing mate versus a genetically high-quality mate, parents' own inclusive fitness may be relatively better served if all of their children acquire highly investing mates (who will invest in grandchildren). Whenever individuals opt for a mate with heritable fitness, they may obtain genetic benefits for their offspring, while assuming an elevated risk of having a low-investing partner and having eventually to depend on their parents (and other kin) for support, to the potential detriment of their parents' other grandchildren. From the mating individual's perspective, this may be a risk worth taking, making them more inclined toward mates high in heritable fitness. The upshot is that any conflict that exists between parents and children in mate choice is likely to revolve around mate characteristics that connote heritable fitness versus parental investment—mating individuals are likely to prefer the former characteristics and parents the latter.

Parents' conflicting preferences may also reflect a desire for in-laws who promote ingroup solidarity, even at the expense of genetic quality of grandchildren (R. Trivers, personal communication, July 13, 2007). For instance, where there is arranged marriage, there is often consanguineous marriage (Shaw, 2001), a practice that may promote group solidarity while increasing the odds of inbreeding. Of course, the prospect of inbreeding—which reduces offspring genetic quality—is expected to be a stronger turn-off for offspring. Indeed, parents strongly prefer in-laws from the same ethnic group, religion, and social class (e.g., Murstein, 1974). Additional evidence comes from a comparison of criteria for love-based marriage and arranged marriage: One study found that personality, physical attractiveness and athleticism were especially important considerations for love-based marriage, whereas religion, social class, education, family and caste were especially important considerations for arranged marriage (Sprecher & Chandak, 1992). Also informative is the finding—in a cross-cultural analysis—that free mate choice was correlated highly (.70) with the importance of physical attractiveness (Rosenblatt, 1974).

In sum, parents may have functional reasons for harbouring preferences that diverge from those of their children. Parents may be particularly attentive to the degree to which their children's mates display traits connoting parental investment and cooperation with the ingroup; children may be particularly attentive to traits connoting heritable fitness.

Evidence for Conflicting Mate Preferences

What evidence is there for conflicting mate preferences between parents and offspring? The results of Apostolou (2007a) provided some indirect evidence: When parents indicate their in-law preferences, physical attractiveness appears to be less important than when individuals themselves indicate their mate preferences. More direct evidence came from studies that we conducted (Buunk et al., 2008). Our methodology was designed to overcome an inadequacy associated with simply asking individuals or their parents to indicate their preferences—they are likely to provide highly similar responses (e.g., a woman and her parents may both indicate that they would prefer a man that is attractive rather than ugly, and from the same ethnic group rather than different). Our methodology was designed to closely track the mating trade-offs: Individuals of mating age were presented with a list of traits, formulated to represent the undesirable variant of trait variables (e.g., physically unattractive, different religious beliefs) and were asked to indicate whether this would be *more unacceptable* to themselves or to their parents. To the extent that an undesirable variant of a trait is perceived as more unacceptable to self, this would indicate that possessing the desirable variant of the trait (e.g., physically attractive) is relatively more important for offspring. To the extent that an undesirable variant of a trait is perceived as more unacceptable to parents, this would indicate that possessing the desirable variant of the trait (e.g., same religious beliefs) is relatively more important for parents. Responses around the middle of the scale would indicate that, for these traits, parents and children are in agreement about their importance.

Data gathered across several samples (Americans, Dutch, Kurdish and exchange students from many different countries studying in the Netherlands) provided a fairly clear picture: Most of the undesirable variants of mate characteristics that connote a lack of heritable fitness were considered more unacceptable to the participants themselves, and most of the undesirable variants of mate characteristics that connote parental investment and cooperation with the ingroup were considered more unacceptable to the parents (Buunk et al., 2008). Characteristics that recurred as especially unacceptable to children included lacking a sense of humour, being physically unattractive and having a bad smell; characteristics that recurred as especially unacceptable to parents included being divorced and having a different ethnic background. In other words, traits such as sense of humour, physical attractiveness and good smell are desired especially by offspring; traits such as no prior marriage and similar ethnic background are—at least according to offspring's reports—desired especially by parents.

The objective of the present research was to replicate our previous findings (Buunk et al., 2008) in a sample of Uruguayans. Toward that end, the previously used questionnaire was translated into Spanish and administered to a sample of students in Uruguay.

Method

Participants

Participants were one hundred eighty-four students (152 women, 31 men; mean age = 23.40, *SD* = 2.28) at the University of the Republic (Uruguay). The students were from the Faculty of Medicine, and they completed the questionnaire voluntarily during class time at the request of the professor (all students consented).

Measure

The measure was identical to the 22-item version of the questionnaire in Buunk et al. (2008); it was translated into Spanish by a professional translator. Participants were provided with 22 traits formulated to describe the undesirable end of the trait variables and were presented with the following instruction:

> There are certain traits that you look for in a potential romantic partner. Your parents also prefer certain kinds of traits in your potential partner. However, you and your parents may emphasise different kinds of traits. In some cases, you and your parents may agree. Below are questions intended to measure such conflicts. Imagine a potential romantic partner for yourself, and assign each of the following characteristics to the potential partner. In your opinion, would this be *more unacceptable* to you or your parents?

Each question was posed in the following manner: "If my potential partner was [physically unattractive], this would be..." followed by a 7-point scale on which the possible answers ranged from (1) "much more unacceptable to me," through (4) "equally unacceptable to me and my parents," to (7) "much more unacceptable to my parents." Male and female participants received the same questions, except for one item pertaining to height. For the male participants, this item was presented as "If my potential partner was considerably taller than me" and for the female participants, the item was presented as "If my potential partner was considerably shorter than me."

Nine characteristics were those that (when formulated as the desirable variant) connote heritable fitness, and nine other characteristics were those that (when formulated as the desirable variant) connote likelihood of parental investment and cooperation with the ingroup; there were four additional characteristics that are important in mating decisions, though they did not clearly fall into either category.

Results

Table 1: Mean Levels of Unacceptability of Characteristics to Self versus Parents

Characteristics connoting heritable fitness	Means	Characteristics connoting parental investment and cooperation	Means	Additional characteristics	Means
Physically unattractive	2.73*** (1.64)	Lacks good family background	3.95*** (1.68)	Unfriendly and unkind	3.38 (1.85)
Considerably shorter/taller than self	3.13 (1.71)	Different ethnic background	4.21*** (1.90)	Very different attitudes than self	2.60 (1.63)
Physically unfit	3.23 (1.69)	Different religious beliefs	3.85*** (1.91)	Physical or mental illness	3.42 (1.75)
Fat	2.41*** (1.66)	Lower social class than self	4.00*** (1.74)	Not a virgin	3.78 (1.96)
Bad Smell	2.28*** (1.63)	Divorced	3.94*** (1.94)		
Lacks sense of humour	2.33*** (1.43)	Poor	4.09*** (1.74)		
Lacks artistic abilities	2.97* (1.62)	Not respectful and obedient	3.26 (1.73)		
Lacks creativity	2.63*** (1.52)	Low education	3.50* (1.71)		
Unintelligent	2.93** (1.65)	Does not like children	2.58 (1.70)		
	2.74		3.71		3.29

Note. Lower values indicate greater unacceptability to children, and higher values indicate greater unacceptability to parents; asterisks indicate significant differences from the mean score for all 22 items (3.24) in the predicted direction: * $p \leq .05$, ** $p \leq .01$, *** $p \leq .001$ (one-tailed). Values in parentheses are *SD*s.

The results are presented in Table 1. We analyzed the data as in Buunk et al. (2008). The mean score for all 22 items was 3.24, which we used as the comparison point (this was done to ease cross-cultural comparisons; see Buunk et al., 2008). Thus, for each of the 22 items, scores lower than 3.24 indicated a higher degree of unacceptability to self and scores greater than 3.24 indicated a higher degree of unacceptability to the parents. To examine whether each characteristic was more unacceptable to parents or to children, we conducted directional one-sample *t*-tests that assessed whether each score differed significantly from 3.24 in the predicted direction. As shown in Table 1, 7 out of 9 characteristics that connote heritable fitness (and thus were hypothesised to be more important for offspring) yielded means that were significantly lower than 3.24. On the flip side, 7 out of 9 characteristics that connote parental investment and cooperation with the ingroup (and thus were hypothesised to be more important for parents) yielded means that were significantly greater than 3.24. On the remaining characteristics (in the right-most column of the table), we did not conduct *t*-tests, as we had no specific hypotheses. Across all 22 items, the three mate characteristics that were considered most unacceptable to children were having a bad smell, lacking a sense of humour and being fat; the three characteristics that were considered most unacceptable to parents were having a different ethnic background, being poor and being from a lower social class.

We also computed two mean scores—one representing the mean score for the 9 characteristics that connote heritable fitness (M = 2.74, Cronbach's alpha = .82) and another representing the mean score for the 9 characteristics that connote parental investment and cooperation (M = 3.71, Cronbach's alpha = .85). These two mean scores differed significantly from each other, $t(183) = 12.98$, $p < .001$, providing further support for the hypothesis. For characteristics connoting heritable fitness, women (M = 2.74) and men (M = 2.72) indicated equal levels of unacceptability to self versus parents (p = .93). For characteristics connoting parental investment and cooperation with the ingroup, women (M = 3.64) indicated somewhat greater unacceptability to self (versus parents) than did men (M = 4.06), although this difference was not significant (p = .072). This is in line with previous results (Buunk et al., 2008): In most of the samples in which this measure was employed, there were no sex differences in these means.

On the whole, the results from the Uruguayan sample replicated those obtained in other samples and added further support for the hypothesis that, when it comes to conflicts between parents and offspring regarding mate characteristics, those connoting heritable fitness may be especially important to offspring, and those connoting parental investment and ingroup cooperation may be especially important to parents.

Discussion

It has been known for a long time that parents have played—and continue to play—an important role in human mating. As reviewed above, however, only recently has this fact received any serious attention from psychologists. We hypothesised (Buunk et al., 2008) that parents and their children may often be in conflict with respect to mate preferences. Specifically, we hypothesised that conflicts are especially likely to emerge on mate characteristics that either strongly signal heritable fitness or parental investment and cooperation. In the present study, we used a methodology that was used in our previous studies, and we found that traits connoting a lack of heritable fitness tend to be more unacceptable to children, and traits connoting a lack of parental investment and cooperation with the ingroup tend to be more unacceptable to parents. These results largely replicated our previous findings (Buunk et al., 2008).

Although informative, the present results are limited by the fact that our assessment of parents' preferences was based on the participants' perceptions of how their parents may respond. An important next step is to present a similar questionnaire to parents who have children of mating age in which they are asked to indicate whether specific undesirable characteristics would be more unacceptable to them or to their children.

There are additional questions that emerge from the present research. For instance, in this study, participants likely had a long-term relationship in mind when answering the questionnaire; the results may have differed if we had asked participants to imagine a potential short-term partner. However, this raises additional complexities, because there may be parent–offspring conflict involving relationship type itself: Parents (and other kin) may try to discourage short-term relationships altogether, especially for women (see Jonason et al., 2007). Clearly, there are different levels and types of conflict in the realm of mating that require further research attention.

Although parent–offspring conflict in mate preferences is presumably universal, the degree to which parents actually exert pressure on their children is highly variable across cultures. In fact, the degree of parental influence seems to be associated with the well-known cultural dimension *individualism–collectivism* (Buunk, Park, & Duncan, submitted). This has some interesting implications for the study of culture. Most obviously, kin involvement in mating seems to constitute an important aspect of collectivism. Less obviously, further research on additional variables that influence the degree of kin involvement might shed light on the origins of cultural differences themselves. In other words, given the correlation between collectivism and kin involvement in mating, one speculation is that kin involvement (which emerged for as-yet-unidentified reasons) played a causal role in the emergence of collectivism.

Other intriguing hypotheses emerge from the fact that parents and offspring have conflicting preferences, and that cultures vary in the level of parental influence. Within cultures characterised by high levels of parental control, sexual selection might in fact be more accurately viewed as a dynamic involving parents and potential in-laws—including both intrasexual competition to gain approval from parents-in-law, and "intersexual selection" by parents for specific traits in daughters- and sons-in-law. Given that parents have somewhat different preferences, sexual selection in such cultures may lead to different outcomes than in cultures in which individuals have historically had more freedom in mate selection. Might such a scenario explain some of the existing phenotypic differences observed across cultures today?

Acknowledgements

This work was supported by the Royal Netherlands Academy of Arts and Sciences. We thank Gabriela Bañuls and Silvia Franco for their assistance.

References

Apostolou, M. (2007a). Elements of parental choice: The evolution of parental preferences in relation to in-law selection. *Evolutionary Psychology, 5*, 70-83.

—. (2007b). Sexual selection under parental choice: The role of parents in the evolution of human mating. *Evolution and Human Behavior, 28*, 403–409.

Buss, D. M., & Barnes, M. (1986). Preferences in human mate selection. *Journal of Personality and Social Psychology, 50*, 559–570.

Buss, D. M., & Schmitt, D. P. (1993). Sexual strategies theory: An evolutionary perspective on human mating. *Psychological Review, 100,* 204-232.

Buunk, A. P., Park, J. H., & Dubbs, S. L. (in press). Parent–offspring conflict in mate preferences. *Review of General Psychology.*

Buunk, A. P., Park, J. H., & Duncan, L. A. (submitted). Cultural variation in parental influence on mate choice.

Faulkner, J., & Schaller, M. (2007). Nepotistic nosiness: Inclusive fitness and vigilance of kin members' romantic relationships. *Evolution and Human Behavior, 28*, 430–438.

Gangestad, S. W., & Buss, D. M. (1993). Pathogen prevalence and human mate preferences. *Ethology and Sociobiology, 14,* 89-96.

Gangestad, S. W., & Simpson, J. A. (2000). The evolution of human mating: Trade-offs and strategic pluralism. *Behavioral and Brain Sciences, 23,* 573-644.

Gangestad, S. W., Garve-Apgar, C. E., Simpson, J. A., & Cousins, A. J. (2007). Changes in women's mate preferences across the ovulatory cycle. *Journal of Personality and Social Psychology, 92,* 151–163.

Goode, W. J. (1959). The theoretical importance of love. *American Sociological Review, 24,* 38-47.

Haselton, M. G., & Buss, D. M. (2000). Error management theory: A new perspective on biases in cross-sex mind reading. *Journal of Personality and Social Psychology, 78,* 81-91.

Haselton, M. G., & Miller, G. F. (2006). Women's fertility across the cycle increases the short-term attractiveness of creative intelligence. *Human Nature, 17,* 50-73.

Jonason, P. K., Izzo, P. L., & Webster, G. D. (2007). Helping others to find long-term and short-term mates: A test of inclusive fitness, reciprocal altruism, and parental investment theories. *Evolutionary Psychology, 5*, 716–732.

Miller, G. (2000). *The Mating Mind: How Sexual Choice Shaped the Evolution of Human Nature.* New York: Random House.

Murstein, B. I. (1974). *Love, Sex, and Marriage Through the Ages.* New York: Springer.

Penton-Voak, I. S., Jacobson, A., & Trivers, R. (2004). Population differences in attractiveness judgements of male and female faces: Comparing British and Jamaican samples. *Evolution and Human Behavior, 25,* 355-370.

Penton-Voak, I. S., Perrett, D. I., Castles, D. L., Kobayashi, T., Burt, D. M., Murray, L. K., et al. (1999). Menstrual cycle alters face preference. *Nature, 399,* 741-742.

Rosenblatt, P. C. (1974). Cross-cultural perspective on attraction. In T. Huston (Ed.), *Foundations of Interpersonal Attraction,* (pp. 79-95). New York: Academic Press.

Shaw, A. (2001). Kinship, cultural preferences and immigration: Consanguineous marriage among British Pakistanis. *Journal of the Royal Anthropological Institute, 7*, 315–334.

Shostak, M. (1983). *Nisa: The life and words of a !Kung woman.* New York: Vintage Books.

Sprecher, S., & Chandak, R. (1992). Attitudes about arranged marriages and dating among men and women from India. *Free Inquiry in Creative Sociology, 20,* 59-70.

Sprecher, S., & Felmlee, D. (1992). The influence of parents and friends on the quality and stability of romantic relationships: A three-wave longitudinal investigation. *Journal of Marriage and the Family, 54*, 888–900.

Trivers, R. L. (1971). The evolution of reciprocal altruism. *Quarterly Review of Biology, 46*, 35–57.

—. (1972). Parental investment and sexual selection. In B. Campbell (Ed.) *Sexual selection and the descent of man, 1871-1971,* (pp.136-179). Chicago, Aldine.

—. (1974). Parent–offspring conflict. *American Zoologist, 14,* 249-264.

Trivers, R. (2000). The elements of a scientific theory of self-deception. *Annals of the New York Academy of Sciences, 907*, 114–131.

Westermarck, E. A. (1921). *The History of Human Marriage* (5th ed.). London: Macmillan.

CHAPTER EIGHTEEN

WOULD YOU GO TO BED WITH ME? GENDER DIFFERENCES IN RECEPTIVITY TO SEXUAL INVITATIONS

GERT MARTIN HALD
AND HENRIK HØGH-OLESEN

Abstract: In the following we present the first replication of Clark & Hatfield's (1989) and Clark's (1990) much cited and famous field experiments 'Consent to Sex with a Stranger', investigating men and women's receptivity to sexual invitations. As in the original studies, a total of 96 men and women, 48 of each gender, were approached by a reasonably attractive confederate of the opposite gender who after a standard introduction asked one of the following three questions: 1) 'Would you go on a date with me tonight?'; 2) 'Would you come to my place tonight?'; 3) 'Would you go to bed with me tonight?'. For women, the results showed an inverted relationship between explicitness of the sexual invitation and consenting rates, i.e. the more explicit the sexual invitation the lower the consenting rate. For men, the results showed a parabola-like relationship. That is, men did not respond favourably to sexual invitations of either too implicit ('date' condition) or too explicit ('bed' condition) a nature, but favourably to a sexual invitation fairly explicit in nature, yet still not 'in the box' ('home' condition). In important ways, the results contradict those originally reported by Clark and Hatfield (1989) and Clark (1990).

Introduction

Common stereotypes have it that men and women would react differently to sexual invitations, with men being more receptive than women. These stereotypes are supported by empirical evidence consistently showing gender differences in receptivity to sexual invitations across different methodology and data domains, with men being more receptive than women (Mathes et al., 2002; Voracek et al., 2006). Compared to women, men also tend to have a more positive, unrestricted, and promiscuous attitude towards casual sex (Carroll et al., 1985; Fisher et al., 1988; Hendrick et al., 1985; Schmitt, 2005; Townsend, 1995; Wilson, 1987), exhibit greater willingness to pursue extra dyadic involvement (Seal et al., 1994), and more often seek short-term and casual relationships (Blumstein & Schwartz, 1983; Eysenck, 1976; Laumann et al., 1994; Mathes et al., 2002; Wiederman, 1997). Furthermore, compared to women, men generally prefer more sexual partners (Buss & Schmitt, 1993; McBurney et al., 2005; Schmitt, 2005) and are more sexually aroused by novelty, hence the "Coolidge effect" (Buss, 2003, Symons, 1979). These differences not only appear to be universal (Schmitt, 2005), but are also found throughout the animal kingdom (Stewart-Williams, 2005).

From an evolutionary perspective, there are more reproductive benefits for a man to mate with multiple women within one year than vice versa. Consequently, the reproductive benefits from promiscuous sex are much higher for men than for women. In evolutionary terms, this exerts strong selective pressures on men's mating strategies to favour more unrestricted sexual practices and multiple partners (Buss & Schmitt, 1993).

This prediction is also in line with Trivers' (1972) empirically well-supported *Parental Investment Theory* stating that the gender that invests the least in the offspring typically shows 1) a higher eagerness to mate, 2) a tendency to engage in more indiscriminate mating with multiple partners, and 3) a willingness to invest more resources in intra-sexual competition and fights (Anderson, 1994; Clutton-Brock, 1991; Maynard-Smith, 1977). Thus, in humans cross-culturally, men generally show a higher eagerness to mate, less discriminative mating behaviour, more intra-sexual competition and fights, have more sexual partners and spend less time and effort in child raising behaviours than women (Hald, 2006; Low, 1989; Munroe & Munroe, 1997; Quinn, 1977; Voracek et al., 2006).

Women *do* engage in short-term interpersonal and sexual relationships. However, Sexual Strategies Theory predicts that they most often do so as a mean of mating with males, who have highly desirable traits, genes, positions and/or resources, or in order to attract or evaluate potential long-term partners (Buss & Schmitt, 1993). Nevertheless, both Social Structure Theory (Eagly & Wood, 1999; Wood & Eagly, 2002) and Strategic Pluralism Theory (Gangestad & Simpson, 2000; Simpson & Gangestad, 1991) predict that in large urban societies with high socio-political, economical and relational gender-equality, and under circumstances where the need for bi-parental care is limited or eliminated, women may be inclined to engage in more unrestricted sexual practices, including short-term and casual sexual relationships.

Furthermore, both Social Role Theory (Eagly et al., 2000) and the Cultural Contingency Perspective (Clark & Hatfield, 2001) predict that if social roles, sexual 'scripts', and present cultural expectations are similar for men and women, this will be reflected in more gender corresponding sexual behaviours. However, as Schmitt's (2005) cross-cultural studies indicate under these circumstances too, males are significantly more unrestricted and promiscuous in their sexual patterns than women.

One of the most cited and famous, yet controversial, (see also Clark & Hatfield, 2003) social field experiments investigating men and women's receptivity to sexual invitations is Clark & Hatfield's (1989) classic experiment 'Consent to Sex with a Stranger'. The experiment was conducted for the first time at Florida State University in the USA in 1978. In the experiment, a reasonably attractive confederate walked up to a member of the opposite gender and after a standard introduction asked one of the following three questions: 1) 'Would you go out with me tonight?', 2) 'Would you come over to my apartment tonight?', 3) 'Would you go to bed with me tonight?' The experiment showed large and significant gender differences in receptivity to sexual invitations, with men being significantly more receptive than women the more explicit the sexual invitation became. The same authors repeated the experiment in 1982 and 1990 with similar results. The aggregated results from these three experiments as reported by Clark & Hatfield (1989) and Clark (1990) can be found in Table 1.

Although one of the most famous experiments in social psychology and widely cited (Baumeister, 2001), the experiment has never been replicated (e.g., Voracek et al., 2005). Nonetheless, three publications are relevant to mention in this connection.

First, Clark (1990) reports from an altered experimental version of 'Consent to Sex with a Stranger'. In this version, 44 male and 44 female subjects were contacted by a close friend (confederate) who personally testified to the integrity of a stranger of the opposite gender coming to town. The confederate told the subject that the stranger was warm, sincere, trustworthy, and attractive, and then went on to ask the subject one of the following two questions, depending on the prior randomly assigned experimental condition of the subject: "Would you be willing to go on a date with him/her?" *or* "Would you be willing to go to bed with him/her?" The results showed that 91 % of females and 96 % of males ($p > .05$) would be willing to go on a date with the stranger, whereas only 5 % of females but 50 % of males ($p < .05$) would be willing to go to bed with the stranger.

Second, Voracek et al. (2005) report from the so-called Molzer Project. The Molzer Project was a non-scientific journalistic 'experiment' in which an attractive male confederate walked up to a hundred women across different locations and settings in three major cities in Germany in 2003 and asked "Do you want to sleep with me?". Following an affirmative response from a woman, the confederate verified the response by actually having sex with the woman. Although the project is of a non-scientific nature and differs from Clark & Hatfield's original experiment in important ways (see also Voracek et al., 2005), the results are noteworthy. In the Clark & Hatfield experiment, none of the female subjects consented to sleep with any of the male confederates across the three experiments. In contrast, in the Molzer Project 6 % of the female subjects accepted the invitation and went on to have sex with the male confederate - a statistical significant difference according to Voracek et al. (2005).

Third, Voracek et al. (2006) also report from an Austrian experiment in which 195 adult males and 179 adult females were asked to predict the rates of consent of men and women to the three different sexual invitations used in the Clark & Hatfield (1989) and Clark (1990) experiments, i.e. the 'date', 'home', and 'bed' conditions described above. The results showed that on average the consenting rates of male subjects were underestimated and that of female subjects overestimated (Voracek et al., 2006), as compared to the aggregated results of Clark & Hatfield (1989) and Clark (1990).

On the basis of 'Consent to Sex with a Stranger', Clark & Hatfield (1989) and Clark (1990) concluded that the results were in accordance with the 'traditional hypothesis', stating that women want love and men want sex from a relationship. Consequently, men will more readily than women agree to a sexual invitation from a member of the opposite gender i.e. consent to sex with a stranger (Clark & Hatfield, 2001).

Not only would it be interesting and important to replicate the original experiments by Clark & Hatfield (1989) and Clark (1990) in a different time and setting to see how the results compare with the original experiments, but also to do so using a non-American sample and a different research team. In the following we report and discuss the results from the Danish replication of 'Consent to Sex with a Stranger'.

Method

Ethics

Due to the potentially sensitive and controversial nature of the experiment and because of the fact that subjects were participating without giving prior informed consent, the study was reviewed and approved by the national appointed Ethical Regional Committee of Denmark before commencing.

Participants

A total of 96 subjects were included in the experiment, 48 of each gender. All subjects were on the campus of Aarhus University, Denmark during the first two weeks of May 2005 between 11 a.m. and 4 p.m. Thirty-two subjects, sixteen of each gender, were assigned to each of the three experimental conditions, i.e. the 'date', 'home', and 'bed' conditions. The average attractiveness for female subjects was 6.85 (range 6.25; 7.30) and for male subjects 6.94 (range 6.58; 7.67) as rated by confederates ($p > .05$) using a 9 point 'attractiveness scale' ranging from 1 (extremely unattractive) to 9 (extremely attractive).

A total of eight confederates were included in the experiment. Initially, the attractiveness of potential confederates who volunteered to assist in the experiment was rated by using a panel of psychology students ($n = 27$) and the 9-point 'attractiveness scale' described above. Only confederates of average attraction i.e. 2.5 - 7.5 on the attractiveness scale were eligible to assist in the experiment. From the initial pool, eight confederates, four of each gender, were randomly selected. The average attractiveness for female confederates was 5.71 (range 4.59; 7.41) and for male confederates 5.62 (range 5.00; 7.37) as rated by the panel ($p > .05$).

Male subjects were found to be significantly more attractive than female confederates ($t = 5.92$, $df = 71$, $p < .001$, Cohen's (d) = 1.41). Likewise, female subjects were found to be significantly more attractive than male confederates ($t = 5.25$, $df = 71$, $p < .001$, Cohen's (d) = 1.25).

Procedure

After careful instruction and training, confederates were divided into three mixed gender teams and positioned at one of three pre-selected outdoor locations on the campus of Aarhus University on a day with reasonably good weather, i.e. no rain or heavy winds. Aarhus University is the second largest university in Denmark and has approximately 35,000 students. On his/her own, each confederate was to approach an unknown member of the opposite gender and - after the following standard introduction: “Hi, my name is ‘N.N.’, I am sorry to disturb you like this but I have been noticing you around and find you very attractive” - ask one of the following three questions: Q1: ‘Would you go on a date with me tonight?’; Q2: ‘Would you come over to my place tonight?’; Q3: ‘Would you go to bed with me tonight?’. In this connection it should be noted that in the original study Clark & Hatfield (1989) used the term ‘go out with me tonight’ as opposed to ‘go on a date with me tonight’ used in the present study. This slight modification was necessary to convey the same message and intention in Danish as in English.

Only subjects who were on their own, i.e. not in a group or with someone else were to be approached. Only after having selected a subject was the confederate instructed as to which of the three questions (experimental conditions) to ask by way of a small booklet carried on site by the confederate. The booklet contained 12 pages, each page clearly indicating which experimental condition the confederate was to use next. For each booklet the order of the assignment of the experimental conditions had been randomized by using a dice by the first author prior to the selection of confederates.

Each confederate was instructed to approach a total of twelve subjects and record their answer (yes/no), initial verbal responses, and how attractive they found the subject by using the attractiveness scale described above. Confederates were instructed only to approach subjects that they would hypothetically be interested in dating, inviting home, or sleeping with. To ensure the anonymity of both confederates and subjects, all confederates used a standardized recording sheet, which subsequently was placed in the mailbox of one of the two authors in a standard white envelope without name. At the conclusion of the experiment, all subjects were debriefed and given an opportunity to ask questions about the experiment or contact the authors if needed. None of the subjects contacted the authors.

Results

Results[1] from the Present Study

No significant overall difference between male and female subjects in receptivity to sexual invitations was found ($\chi^2 = 0.80$, $df = 1$, $p > .05$). No significant gender differences in consenting to a date (χ^2 (Yates correction applied) = 1.47, $df = 1$, $p > .05$, Cohen's (d) = .44, post hoc power = .23) come home to (χ^2 (Yates correction applied) = 2.47, $df = 1$, $p > .05$, Cohen's (d) = .58, post hoc power = .35). or go to bed with (χ^2 (Yates correction applied) = .00, $df = 1$, $p > .05$, Cohen's (d) = .00, post hoc power = .05) the confederates were found. However, for both the 'date' and 'home' conditions effect sizes were found to be moderate to large (See also Table 1).

Statistical Comparisons between the Present Study and Clark & Hatfield (1989) and Clark (1990)

Overall comparison by gender and experimental condition between the present study and the aggregated results from Clark & Hatfield (1989) and Clark (1990) for female subjects showed a near significant difference for the 'date' condition only ($n = 64$, χ^2 (Yates correction applied) = 3.61, $df = 1$, $p = .057$, Cohen's (d) = .49, post hoc power = .48). Thus, for the 'home' and 'bed' conditions no significant differences were found. For male subjects, a significant difference was found for both the 'date' ($n = 64$, χ^2 (Yates correction applied) = 13.35, $df = 1$, $p < .001$, Cohen's (d) = 1.03) and 'bed' ($n = 64$, χ^2 (Yates correction applied) = 17.68, $df = 1$, $p < .001$, Cohen's (d) = 1.24) conditions. For the 'home' condition, no significant difference was found (see also Table 2).

[1] It should be noted that for the Chi square statistics (χ^2) when one or more cells in the 2 by 2 contigency table contained frequencies less than five, a Yates correction was applied. Although Yates correction provides a more conservative estimate Yates correction to the Chi square statistics (χ^2) or alternatively Fisher's exact test are recommended by many when one or more cells in a 2 by 2 table contains frequencies less than five (e.g. Mcdonald, 2008).

Table 1: Consenting rates of 1) Hald & Høgh-Olesen and 2) Clark & Hatfield (1989) and Clark (1990) - Aggregated Data.

Experiment	Experimental Condition/Type of Request 'Date'	'Home'	'Bed'
Present Study (N = 96)			
Male confederate, female subject	19%[c]	13%[c]	0%[c]
Female confederate, male subject	0%	44%	6%
Cohen's (*d*)[a]	.44	.58	.00
Post hoc power analysis[b]	.23	.35	.05
Clark & Hatfield & Clark (1990) (Agg. Data; N = 288)			
Male confederate, female subject	50%	7%*[c]	0%*[c]
Female confederate, male subject	56%	63%	71%
Cohen's (*d*)[a]	.13	1.39	2.07
Post hoc power analysis	.09	n/a	n/a

Note. For Hald & Høgh-Olesen, n = 16 for each experimental condition. For Clark & Hatfield (1989) and Clark (1990) - Aggregated Data, n = 48 for each experimental condition. [a] Effect size Cohen's (*d*). Standard evaluation criteria: < .20 = small; .50 = medium; > .80 = large; [b] Using the following criteria: Two-tailed test, alpha = .05, observed n, and χ^2 with Yates correction; [c] Yates correction applied; * $p < .001$

Table 2: Differences in Consenting Rates between 1) Hald & Høgh-Olesen and 2) Clark & Hatfield (1989) and Clark (1990) – Aggregated Data.

	Experimental Condition/Type of Request 'Date'	'Home'	'Bed'
Male confederate, female subject	$\chi2(1) = 3.61$[c]	$\chi2(1) = .07$[c]	n/a
Cohen's (*d*)[a]	.49	.21	n/a
Post hoc power analysis[b]	.48	.06	n/a
Female confederate, male subject	$\chi2(1) = 13.35$*[c]	$\chi2\ (1) = 1.73$	$\chi2\ (1) = 17.68$*[c]
Cohen's (*d*)[a]	1.03	.33	1.24
Post hoc power analysis	n/a	.26	n/a

Note. N = 384; n = 64 for each experimental condition.
[a] Effect size Cohen's (*d*). Standard evaluation criteria: <. 20 = small; .50 = medium; >. 80 = large; [b] Using the following criteria: Two-tailed test, alpha = .05, observed n, and χ^2 with Yates correction; [c] Yates correction applied; *p <. 001

Rejection Rate, 'Success Rate', and Subject Awareness

In order to examine if the relatively large number of rejections in the present study was due to subjects' relationship status i.e. having a partner *or* subject awareness, the initial verbal responses of each approached subject was noted by confederates and later quantified by two raters. In case of disagreement a third rater was used. If a subject rejected the offer with reference to having a partner or subject awareness e.g. "I am flattered, but no thanks I have a boyfriend" or "Is this some kind of bet or something?" this was quantified as a rejection due to having a partner or subject awareness respectively.

For female subjects, male confederates the results showed that out of the 43 subjects who rejected the sexual invitation, 18 subjects or 42 % did so with reference to already having a partner. For male subjects, female confederates the corresponding results showed that out of the 40 subjects who rejected the sexual invitation, 28 subjects or 70 % did so with reference to having a partner. The gender difference in rejection rate due to relationship status i.e. having a partner was found to be significant ($n = 83$, $\chi^2 = 6.64$, $df = 1$, $p < .01$, Cohen's $(d) = .59$). In order to further quantify this result, a 'success rate' for each gender was calculated. The success rate was calculated as a percentage based on the number of affirmative answers ('Yes' answers) relative to the number of subjects not rejecting the sexual invitation with reference to having a partner (Total number of 'No' answers minus 'No' answers due to having a partner). For female subjects, male confederates the success rate was 17 % (5 'yes' of (48-18 'no')). For male subjects, female confederates the success rate was 40 % (8 'yes' of (48-28 'no')). The gender difference in success rate was found to approach statistical significance and the effect size moderately large ($n = 50$, $\chi^2 = 3.40$, $df = 1$, $p = .065$, Cohen's $(d) = .54$, post hoc power = .45).

For female subjects, male confederates one subjects' verbal responses indicated subject awareness. For male subject, female confederates two subjects' verbal responses indicated subject awareness. Due to the relative low number of verbal responses indicating subject awareness across gender no further calculations on this matter were conducted.

Discussion

In our replication of Clark & Hatfield's (1989) and Clark's (1990) experiments 'Consent to Sex with a Stranger', we found that overall women did not respond favourably to sexual invitations by a reasonably attractive member of the opposite gender. Furthermore, women's consenting rates dropped the more explicit the sexual invitation became. For men, the response pattern showed a parabola-like relationship. That is, men did not respond favourably to sexual invitations of either too implicit ('date' condition) or too explicit ('bed' condition) a nature, but favourably to a sexual invitation fairly explicit in nature, yet still not 'in the box' ('home' condition).

At first glance the results of our study may somewhat seem to contradict those reported by Clark & Hatfield (1989) and Clark (1990). Across most experimental conditions, the results were either significantly different or the effect sizes substantially smaller than that of the aggregated results of Clark & Hatfield (1989) and Clark (1990). Nonetheless, the trend of results for the 'home' condition (both genders), and the 'bed' condition (male confederate, female subject only) approximated those of Clark & Hatfield (1989) and Clark (1990). Furthermore, the quantification of subjects' verbal responses indicated that male subjects rejected the invitation with reference to relationship status, i.e. having a partner, significantly more than female subjects. Further, when calculating a gender specific 'success rate' based on the number of affirmative answers relative to the number of subjects not rejecting the sexual invitation with reference to having a partner, a moderately large gender difference in effect size approaching statistical significance was found with male subjects more readily accepting the sexual invitation than female subjects. This indicates a greater overlap between the results of the present study and the original studies than might be evident at first glance.

Theoretically, the results from the present study may be explained by socio-cultural theory, evolutionary theory, or a combination of both. These theories in relation to receptivity to sexual invitations have already been elaborated upon in the Introduction. Given the results from the original experiments, as well as the moderate to large gender differences in effect size for all three experimental conditions and 'success rate', we believe the results best framed within a predominantly evolutionary framework.

Both in the present- and in the original studies, we believe that methodological artifacts constitute important moderating variables and limitations to the results. In particular, relationship status of subjects, attractiveness, mate value of confederates relative to subjects, subject awareness, and appropriateness of experimental conditions relative to the dating and mating culture of the two countries respectively. Further we believe that limited statistical power provides further limitations to the reliability and external validity of the study results.

The quantification of subjects' self-reported reasons for rejecting a sexual invitation due to relationship status and subsequent magnitude (effect size) of the gender difference in 'success rate' suggest that differences in results between the present study and the Clark & Hatfield (1989) and Clark (1990) studies may be due to sample differences in relationship status. As Danish college students are generally much older than their US counterparts (Koss et al., 1987; Statistics Denmark, 2005), it seems reasonable to assume that a higher proportion of either or both gender subjects may have been in a committed relationship in the present study as compared to subjects in the Clark & Hatfield (1989) and Clark (1990) studies. Unfortunately, testing of this hypothesis is not possible as relationship status was not assessed in the Clark & Hatfield (1989) and Clark (1990) studies.

Only reasonably attractive confederates were used. However, when comparing the level of attractiveness of confederates relative to subjects, for both genders it was found that subjects were rated as significantly *more* attractive than confederates. Further, for both genders the magnitude of this difference was found to be large as indicated by Cohen's (*d*). In addition, sexual invitations were made at various campus locations and thereby amongst individuals likely to possess highly valued mate characteristics, e.g. a (future) good education, high income potential, and high intelligence (Buss, 1990). Consequently, the mate value of confederates relative to subjects might very well have been perceived as low by subjects and hence reduced the likelihood of subject consent,. In Clark & Hatfield (1989), attractiveness of confederates was not measured. In Clark (1990), two different rating scales for confederates and subjects respectively were used. This makes problematic both the gauging of the mate value of confederates relative to subjects in the Clark & Hatfield (1989) and Clark (1990) studies as well as how these may compare with the present study.

Even though the quantification of verbal responses of approached subjects generally did not indicate subject awareness, subjects might very well still have been suspicious of their responses being watched or recorded as reality TV, YouTube videos, and candid camera are integrated parts of everyday life in Denmark today. We believe this not to have been the case to the same extent when the original experiments of 'Consent to Sex with a Stranger' were conducted in the USA, although we can only speculate on this issue as the Clark & Hatfield (1989) and Clark (1990) studies did not include any measurement of subject awareness. For the present experimental procedure, we believe that greater subject awareness will decrease rather than increase consenting rates.

Some of the experimental conditions may better accommodate a Danish subsequent American dating and mating culture for either or both genders. By the term 'dating and mating culture' we refer to the various ways in which people meet, interact or engage with each other with the prospect, intention, and/or explicit goal of a romantic and/or sexual encounter.

The concept of dating and asking someone out on a date has traditionally been more integrated into the American than the Danish dating and mating culture. For both genders this may have influenced consenting rates of the 'date' condition in the direction found. The 'home' condition more reflects a Danish dating and mating culture. That is, generally Danish college students initiate contact with each other through friends, in class, student bars, dating sites, etc. and usually only then, is an invitation made and often along the lines of the 'home condition', e.g. an invitation to come home for dinner, coffee, or a drink. However, although this might explain the high rate of male subject consent it does not explain the relatively low rate of female consent and hence the large gender differences for this condition in both the present and original studies. For the 'bed' condition, an affirmative response to a 'direct' sexual invitation made in public during daytime may be considered (more) inappropriate and socially stigmatizing for female than for male subjects (Hald, 2007). We also believe that for the 'home' and 'bed' conditions issues related to personal safety and the risk of falling victim to false criminal accusations, e.g. rape would lower consenting rates for either or both genders. Especially the latter has received a lot of attention in the public eye in Denmark in recent years and may well have influenced the willingness of males to engage situations likely to become sexual, especially with a complete stranger making a direct and unusual sexual request.

Considering the relative moderate to large, yet non-significant, effect sizes along with power analyses revealing low statistical power for many of the experimental conditions, the relative lack of statistical significant results may very well be due to a limited number of subjects in some of the experimental conditions. Thus, increasing N in future experiments is strongly recommended in order to reduce the risk of making type II mistakes i.e., in these analyses, wrongly accepting the null hypotheses for the various experimental conditions. In this regard, the authors recommend using the present study results as a guide for a priori power analyses under the various experimental conditions in future comparable studies and cultural settings.

Apart from potential methodological artifacts and limited statistical power the present study suffers from limitations related to the literalness of the replication.

Strictly speaking, it could be argued that the findings of any research can be applied only to the very particular conditions used in the original research. Clearly, insisting on such literalness would render any replication of social experiments nearly impossible. However, it still should be noted that in some important ways not already discussed the current replication differs from that of Clark & Hatfield (1989). In particular: age of confederates *and* subjects, location and time of the study, and the slight change of wording in the sexual invitation to accommodate a Danish context.

In conclusion it should be noted that the potential methodological artifacts, literalness of the replication, and limited statistical power may limit the overall external validity and reliability of the study and make a direct comparison to the original studies by Clark & Hatfield (1989) and Clark (1990) less suited. Also, studies in which the above problems are taken into consideration are under way.

References

Anderson, M. (1994). *Sexual selection*. New Jersey: Princeton University Press.

Baumeister, R. F. (2001). *Social psychology and human sexuality.* Philadelphia: Psychology Press.

Blumstein, P., & Schwartz, P. (1983). *American couples.* William Morrow, USA.

Buss, D. M. et al. (1990). International Preferences in Selecting Mates. *Journal of Cross-Cultural Psychology, 21,* 5-47.

—. (2003). *The evolution of desire: Strategies of human mating*. New York: Basic Books.

Buss, D. M., & Schmitt, D. P. (1993). Sexual Strategies Theory: An evolutionary perspective on human mating. *Psychological Review, 100,* 204-232.

Carroll, J. L., Volk, K. D., & Hyde, J. S. (1985). Differences between males and females in motives for engaging in sexual intercourse. *Archives of Sexual Behavior, 14,* 131-139.

Clark, R. D. (1990). The impact of aids on gender differences in willingness to engage in casual sex. *Journal of Applied Social Psychology, 20*, 771-782.

Clark, R. D., & Hatfield, E. (1989). Gender differences in receptivity to sexual offers. *Journal of Psychology & Human Sexuality, 2*, 39-55.

Clark, R. D., & Hatfield, E. (2001). Gender differences in receptivity to sexual offers. In R. F. Baumeister (Eds.), *Social psychology and human sexuality*. Philadelphia: Psychology Press.

Clark, R. D., Hatfield, E. (2003). Love in the afternoon. *Psychological Inquiry, 14,* 227-231.

Clutton-Brock, T. H. (1991). *The evolution of parental care*. New Jersey: Princeton University Press.

Eagly, A. H., & Wood, W. (1999). The origins of sex differences in human behavior: Evolved dispositions versus social roles. *American Psychologist, 54,* 408-423.

Eagly, W. O., Wood, W., & Diekman, A. B. (2000) Social role theory of sex differences and similarities: a current appraisal. In T. Eckes & H. M. Trautner (Eds.), *The developmental social psychology of gender.* New Jersey: Erlbaum.

Eysenck, H. J. (1976). *Sex and personality.* London: Open Books Publishing Ltd.

Fisher, W. A., Byrne, D., White, L. A., & Kelley, K. (1988). Erotophobia-erotophilia as a dimension of personality. *Journal of Sex Research, 25,* 123-151.

Gangestad, S. W., & Simpson, J. A. (2000). The evolution of human mating. Trade-offs and strategic pluralism. *Behavioral and Brain Sciences, 23,* 573-644.

Hald, G. M. (2006). Gender differences in pornography consumption among young heterosexual Danish adults. *Archives of Sexual Behavior, 35,* 577-585.

—. (2007). *Pornography Consumption – A study of prevalence rates, consumption patterns, and effects.* Aarhus, Denmark: Faellestrykkeriet for Sundhedsvidenskab, Aarhus Universitet.

Hendrick, S., Hendrick, C., Slapton-Foote, M. J., & Foote, F. (1985). Gender differences in sexual attitudes. *Journal of Personality and Social Psychology, 48,* 1630-1642.

Koss, M. P., Gidycz, C. A., & Wisniewski, N. (1987). The scope of rape: Incidence and prevalence of sexual aggression and victimization in a national sample of higher education students. *Journal of Consulting and Clinical Psychology, 55,* 162-170.

Laumann, E. O., Gagnon, J. H., Michael, R. T., & Michaels, S. (1994). *The social organization of sexuality*. Chicago: University of Chicago Press.

Low, B. S. (1989). Cross-cultural patterns in the training of children: An evolutionary perspective. *Journal of Comparative Psychology, 103,*

313-319.
Mathes, E. W., King, C. A., Miller, J. K., & Reed, R. M. (2002). An evolutionary perspective on the interaction of age and sex differences in short-term sexual strategies. *Psychological Reports, 90,* 949-956.
Maynard-Smith, J. (1977). Parental investment: A prospective analysis. *Animal Behavior, 25,* 1-9.
McBurney, D. H., Zapp, D. J., & Streeter, S. A. (2005). Preferred number of sexual partners: Tails of distributions and tales of mating systems. *Evolution and Human Behavior, 26,* 271-278.
McDonald, J. H. (2008). Handbook of biological Statistics. Baltimore: Sparky House Publishing.
Munroe, R. L., & Munroe, R. H. (1997). A comparative anthropological perspective. In J. W. Bery, Y. H. Poortinga, & J. Pandey (Eds.), *Handbook of cross-cultural psychology, V1* (pp. 171-213). Boston: Allyn & Bacon.
Quinn, N. (1977). Anthropological studies on women's status. *Annual Review of Anthropology, 6,* 181-225.
Schmitt, D. P. (2005). Sociosexuality from Argentina to Zimbabwe: A 48-nation study of sex, culture and strategies of human mating. *Behavioral and Brain Sciences, 28,* 247-311.
Seal, D.W., Agnostinelli, G., & Hannett, C. A. (1994). Extradyadic romantic involvement: Moderating effects of sociosexuality and gender. *Sex Roles, 31,* 1-22.
Simpson, J. A., & Gangestad, S. W. (1991). Individual differences in sociosexuality: Evidence for convergent and discriminant validity. *Journal of Personality and Social Psychology, 60,* 870-883.
Statistics Denmark. (2005). *Students by location of educational institution, education, age, and sex.* Retrieved December 19, 2007, from Statistics Denmark's on-line database.
Stewart-Williams, S. (2005). Fitting data to theory: The contribution of a comparative perspective. *Behavioral and Brain Science, 28,* 294-295.
Symons, D. (1979). *The evolution of human sexuality.* London: Oxford University Press.
Townsend, J. M. (1995). Sex without emotional involvement: An evolutionary interpretation of sex differences. *Archives of Sexual Behavior, 24,* 173-205.
Trivers, R. (1972). Parental investment and sexual selection. In B. Campbell (Ed.), *Sexual selection and the descent of man: 1871-1971* (pp. 136-179). Chicago: Aldine.
Voracek, M., Fisher, M. L., Hofhansl, A., Rekkas, P. V., & Ritthammer, N. (2006). I find you to be very attractive...Biases in compliance

estimates to sexual offers. *Psicothema, 18,* 384-391.

Voracek, M., Hofhansl, A., & Fisher, M. L. (2005). Clark and Hatfield's evidence of women's low receptivity to male strangers' sexual offers Revisited. *Psychological Reports, 97,* 11-20.

Wiederman, M. W. (1997). Extramarital sex: Prevalence and correlates in a national survey. *Journal of Sex Research, 34,* 167-174.

Wilson, G. D. (1987). Male-female differences in sexual activity, enjoyment and fantasies. *Personality and Individual Differences, 8,* 125-127.

Wood, W., & Eagly, A. H. (2002). A cross-cultural analysis of behavior of men and women: Implications for the origins of sex differences. *Psychological Bulletin, 128,* 699-727.

CHAPTER NINETEEN

INTERGROUP JEALOUSY: EFFECTS OF PERCEIVED GROUP CHARACTERISTICS AND INTRASEXUAL COMPETITION BETWEEN GROUPS

LIGA KLAVINA, ABRAHAM P. BUUNK AND JUSTIN H. PARK

Abstract: An important aspect of intergroup conflict is competition for mates, especially among men. Because different outgroups pose different levels of threat, the group membership of rivals can be a characteristic that evokes jealousy. Outgroups perceived to pose greater threat to one's mating resources should evoke more jealousy. Among Dutch students, we investigated the relationship between jealousy towards rivals of different ethnic groups and perceived characteristics of these groups. We found that jealousy was especially high for rivals from outgroups that were perceived to have a tendency to engage in short-term mating (e.g., being attractive, promiscuous and seductive), and to have a high status (e.g., being successful and confident). Men scored higher on a scale designed to assess intergroup intrasexual competition, and the higher the men scored on this scale, the more they tended to derogate outgroup rivals. This effect was not found for women, underlining the higher salience of intergroup intrasexual competition for men.

Substantial research has established that humans spontaneously categorize their social world into ingroups and outgroups, exhibiting favouritism toward the former and derogation of the latter (e.g., Brewer, 1979; Hewstone, Rubin, & Willis, 2002). Research on intergroup conflict has identified competition over tangible resources as an important contributor to such conflict (e.g., Sidanius & Pratto, 1999; Stephan, Ybarra, Martinez, Schwarzwald, & Tur-Kaspa, 1998). Remarkably, a potentially important type of resource that has been overlooked in the literature is mating opportunities. Given the evolutionary significance of obtaining mates, it seems plausible that when men form coalitional alliances in order to exploit other groups (Kurzban & Leary, 2001), women are an important and fought-over resource. Indeed, formation of male coalitions is more likely to evolve within species in which such coalitions aimed at fighting other groups may enhance males' chances of obtaining mates (Geary et al., 2003). Abundant evidence from the animal literature illustrates how mating-related goals may be an important aspect of between-group aggression in various species, including chimpanzees (Watts, 1998; Mitani & Watts, 2001; Wilson & Wrangham, 2003;), common marmosets (Lazaro-Perea, 2001), lions (Packer, Gilbert, Pusey, O'Brien, 1991; Grinnell, Packer & Pusey, 1995) and dolphins (Möller, Beheregaray, Harcourt & Krützen, 2001). Indeed, in wild chimpanzees a lack of females in one group and surplus of females in another group can elicit violent intergroup behaviours (Boesch, 2007).

Humans have a long history of intergroup violence and warfare, and a substantial part of these conflicts are marked by abduction and rape of outgroup females (see Wrangham & Peterson, 1996). In addition, although members of stigmatized groups are usually avoided, females of such groups are not: instead, they are often abused and exploited (Kurzban & Leary, 2001). Short-term mating with a member of another group can be beneficial for males as it allows access to more females and minimizes investment in reproduction. In intergroup encounters males can fertilize females and avoid providing further parental investment. Furthermore, they can lead away females from their groups to become long-term mates. In general, looking for a mate in other groups can be an advantageous mate searching strategy for both high- and low-status males. For high-status males, intergroup mating provides them with wider access to potential mates, as outgroup females might be more attracted to high-status outgroup males than to low-status ingroup males. For low-status males, intergroup mating offers increased mating opportunities compared to mating within their group, particularly if the outgroup has an even lower status. There is indeed evidence that engaging in intergroup aggression can be reproductively beneficial for men. For instance, in tribal societies, men who engage in intergroup warfare have greater reproductive success (Chagnon, 1988). In modern society, men in street gangs have been found to have more sex partners (Palmer & Tilley, 1995).

From the females' perspective, it is more costly to engage in intergroup mating, as males can leave them without providing any resources for them and their offspring. However, the potential benefit is that the males who engage in intergroup aggression are more likely to have good genes—that is, they tend to be healthy, physically strong and tall—therefore intergroup mating may provide an opportunity for females to obtain good genes for their offspring. This trade-off might be more beneficial for low-status females, who may have difficulty mating with the high-status ingroup males. Confirming evidence has been found in common marmosets (Lazaro-Perea, 2001), where observed intergroup copulations involved mainly subordinate females (especially, eldest), both during forays into other groups' territory and during intergroup encounters in ingroup's territory. Generally however, for females, intergroup mating is more costly, except on specific occasions when low-status females can confer genetic benefits from outgroup males.

An important implication of the foregoing is that, as males benefit more from intergroup mating, they should also be more concerned with outgroup members as a mating threat. Thus, not to compromise one's reproductive success, potential mates should therefore be considered a group resource that must be protected from potential invaders. Given the sexual asymmetries in parental investment (Trivers, 1972), it is not surprising that males are more concerned about outgroup mating threats than females. This might be part of the reason why men tend to be more intergroup oriented than women (e.g., Baumeister & Sommer, 1997; Van Vugt, De Cremer & Janssen, 2006; Pemberton, et al., 1996).

Recently, functional approaches to prejudice have proposed that qualitatively different threats posed by different kinds of people are likely to elicit different emotional reactions and manifestations of prejudice (Cottrell & Neuberg, 2005; Kurzban & Leary, 2001; Schaller, Park, & Faulkner, 2003). For instance, when confronted with hostile outgroup members, fear and avoidance are likely to be especially adaptive. With respect to mating-related threats, humans appear to have evolved a specialized mechanism—jealousy—that functions to alert individuals about threats to their relationships (Buss, 2000; Buunk, 1991). According to functional approach, the more threatening a potential rival is, the more jealousy should be triggered (Dijkstra & Buunk, 1998). The perceived mating threat depends on the characteristics of the rival—those with traits especially attractive to the opposite sex are considered more threatening. Dijkstra and Buunk (2002) found that men indicated stronger jealousy than women when the rival was more socially or physically dominant or had a higher status than themselves; yet, women indicated stronger jealousy than men when the rival was more physically attractive than themselves.

In an intergroup setting, a rival's group membership can function as a jealousy eliciting cue, as people attribute different characteristics to groups. Some of these perceived characteristics may evoke more jealousy than others. As interactions between groups in ancestral times were often short-term, and as the interacting individuals were more likely to pursue short-term mating strategies, we predict that the characteristics of the outgroup that are most likely to evoke jealousy are those indicating short-term mating proficiency, such as, attractiveness, seductiveness and promiscuity. In addition, we expect that, especially among men, characteristics indicating high social status, such as confidence and success of the outgroup will be positively related to jealousy in response to outgroup rivals. Thus, we investigated jealousy among Dutch respondents with respect to potential rivals from several countries, and we examined how the degree of jealousy was related to the short-term mating proficiency and status attributed to these ethnic groups. Furthermore, we examined if indeed such responses were related to individual differences to engage in intergroup intrasexual competition. To examine this, we developed a new scale that assesses this type of competition, i.e., to what extent people are threatened by and competitive towards outgroup rivals. It includes general questions about attitudes towards intergroup mating, reactions to outgroup men attempting to mate with ingroup women, and outgroup women attempting to mate with ingroup men.

Method

Participants and Design

One hundred and twenty-five students (85 men, 40 women; mean age M = 21.88, *SD* = 2.64) at the University of Groningen participated in exchange for €5. Responses from ten participants were excluded because they were not Dutch, and five participants were excluded because they failed to follow the procedure. The study was completed individually in separate cubicles.

Procedure

Respondents first engaged in a jealousy-evoking visualization task that has been used in previous research (e.g., Dijkstra & Buunk, 2002). During this task, respondents were asked to imagine their partner showing romantic interest in another person. Immediately following this visualization task, respondents were provided with a list of 10 different ethnic groups in alphabetical order and were asked to indicate how jealous they would feel if their rival was from that group. To increase variability in responses, respondents were asked to rank the 10 groups, so that 1 indicated most jealousy and 10 indicated least jealousy. The 10 ethnic groups included the ingroup (Dutch), several highly salient immigrant groups in the Netherlands (Moroccan, Indonesian, Turkish, Surinamese), and several other groups that may hold relevance for Dutch people with respect to intergroup relations (American, German, Italian, French, Polish).

Subsequently, respondents were asked to indicate the extent to which they believed that the average person from each of these ethnic groups possessed a number of characteristics. Responses were provided on a 5-point scale (1 = not at all; 5 = very much). Two scales were constructed on the basis of this. First, the degree of perceived short mating tendencies, consisting of the characteristics – attractiveness, seductiveness and promiscuity, (Cronbach's alfa calculated for all ethnic groups was .80) and second, a scale for perceived social status, consisting of the characteristics 'success' and 'confidence' (Cronbach's alfa calculated for all ethnic groups was.76).

Intergroup Intrasexual Competition Scale

The scale to assess intergroup intrasexual competition included general questions about attitudes towards intergroup mating, like, "Men and women with different ethnic origin have too different backgrounds to marry", "People who marry other foreigners are responsible for the decline of the Dutch ethnic population", questions about outgroup men "Seeing that foreign men come here and seduce our women, makes me want to interfere" and questions about ingroup women "I find it disgusting that foreign women flirt with Dutch men". Participants gave their responses on a scale from 1 (completely disagree) to 7 (completely agree). The original scale consisted of 29 items, and was internally very reliable (Cronbach's alpha = .95). However, this scale included items asking the same questions both for men and women, for example, "I am disgusted to see foreign men flirt with Dutch women" and "I am disgusted to see foreign women flirt with Dutch men". We felt it was more appropriate to have sex specific versions of the scale, with men responding to items about foreign men and women responding to items about foreign women. The items measuring general attitudes about intergroup mating were kept in both male and female versions. This resulted in a version of 15 items. We submitted this scale to a factor analysis. For men, this analysis produced a strong first rotated factor, eigenvalue = 7.47, and 1 other factor with eigenvalue of 1.11. For women, a strong first factor was detected with eigenvalue = 4.63, for women and, 2 other factors with eigenvalues of 1.89 and 1.24. For the final version, we chose items that loaded higher than .644, for men, and .531, for women, on the first factor. The final 12 items scale had reliabilities of .94 for men and .81 for women. The scores on the intergroup intrasexual competition scale were generally low (M = 2.14, SD = .90), possibly as a result of self-presentation concerns. As expected, men had on average higher scores (M = 2.15, SD = 1.22) than women (M = 1.59, SD = .56), $F_{1,76} = 6.87, p = .01$).

Results

Jealousy in Response to Outgroups

In the first set of analyses, the 10 ethnic groups served as the unit of analysis, and analyses were conducted on the mean values of the responses with respect to these 10 groups, separately for men and women. First, the mean level of jealousy towards rivals from each group was computed by taking the mean jealousy ranking given to each group. Because lower values (on the ranking measure) indicated higher levels of jealousy, these values were reversed so that higher values indicated higher levels of jealousy. Table 1 shows the levels of jealousy elicited by each group, separately for men and women. Male participants rated the Italians as highest on evoking jealousy, followed by the Dutch, the Americans and the French. Immigrant groups – the Moroccans, the Surinamese, the Turks and the Indonesians were rated relatively low. The Germans were also rated rather low and finally, the Polish were rated the lowest in evoking jealousy.

For women, the Dutch were rated the highest in evoking jealousy, followed by the French, the Italians and the Americans. The Germans, the Indonesians and the Polish received average ratings of jealousy. The immigrant groups, such as the Turks and the Moroccans were rated as the lowest in evoking jealousy. There were a number of differences between the ratings of men and women. We performed a one-way ANOVA analysis, which showed that men were more jealous of Moroccan, Surinamese and Turkish rivals than women, whereas women were more jealous of German rivals than men (see Table 1).

Table 1: The means and standard deviations of jealousy evoked by groups and the differences between men and women

Jealousy	Men		Women		Differences between men and women
	M	SD	M	SD	
Italian	7.45	2.35	7.03	2.59	n.s.
Dutch	7.28	3.18	7.92	2.92	n.s.
Americans	6.29	2.67	6.65	1.95	n.s.
French	6.25	2.29	7.06	2.45	n.s.
Moroccans	5.06	3.04	3.75	2.31	$F_{1,119} = 5.35, p = .02$
Surinamese	5	2.61	3.92	2.71	$F_{1,119} = 4.25, p =. 04$
Turks	4.84	2.9	3.78	2.5	$F_{1,119} = 3.64, p = .06$
Indonesians	4.56	2.37	4.72	2.64	n.s.
Germans	4.74	2.29	5.69	2.33	$F_{1,119} = 4.34, p =. 04$
Polish	3.75	2.19	4.5	2.38	n.s.

Outgroup Jealousy as Related to Perceived Characteristics of the Outgroup and Intergroup Intrasexual Competition

We were mainly interested in the extent to which the perceived short-term mating proficiency and perceived social status attributed to the various ethnic groups would predict the level of jealousy. We first calculated the perceived short-term mating proficiency and perceived social status attributed to the various ethnic groups. The mean values for perceived short-term mating characteristics and perceived social status were computed for each ethnic group (see Table 2)

Table 2: The means and standard deviations of perceived short-term mating proficiency and social status of groups and the differences between men and women

Group	Perceived short-term mating proficiency Men		Women			Perceived social status Men		Women		
	M	SD	M	SD	Difference men/women	M	SD	M	SD	Difference men/women
Italian	3.77	.59	3.64	.60	n.s.	3.59	.60	3.85	.57	F=4.72 p < .05
Dutch	2.97	.50	3.19	.62	F=4.3 p < .05	3.48	.62	3.82	.51	F=8.61 p < .01
Americans	2.80	.62	3.14	.63	F=7.47 p < .01	3.96	.71	4.11	.78	n.s.
French	3.26	.67	3.42	.72	n.s.	3.55	.55	3.60	.58	n.s.
Moroccans	2.62	.74	2.71	.71	n.s.	2.94	.57	2.87	.60	n.s.
Surinamese	2.78	.67	2.73	.71	n.s.	2.99	.53	3.13	.66	n.s.
Turks	2.47	.68	2.53	.68	n.s.	2.94	.64	2.79	.61	n.s.
Indonesians	2.45	.70	2.54	.64	n.s.	2.81	.56	2.61	.63	n.s.
Germans	2.37	.53	2.50	.57	n.s.	3.35	.64	3.60	.66	n.s.
Polish	2.03	.55	2.60	.85	F=19.46 p < .001	2.35	.57	2.68	.79	F=6.75 p < .01

Then—with the ethnic groups as a unit of analysis—we examined the relationship between the mean values of each characteristic and the level of jealousy by using Pearson's correlation coefficient. Short-term mating proficiency correlated significantly with jealousy, for males r = .88, *p* = .001 and for females r = .80, *p* = .006 (see Figure 1 and 2).

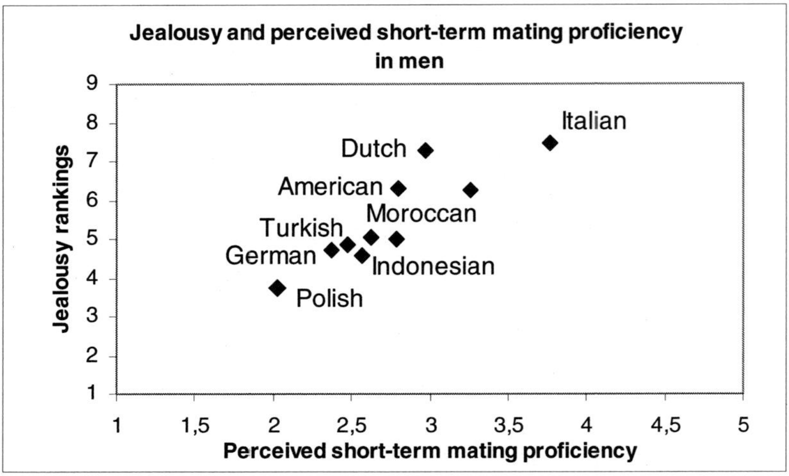

Figure 1. Relationship between jealousy and perceived short-term mating proficiency of groups in men

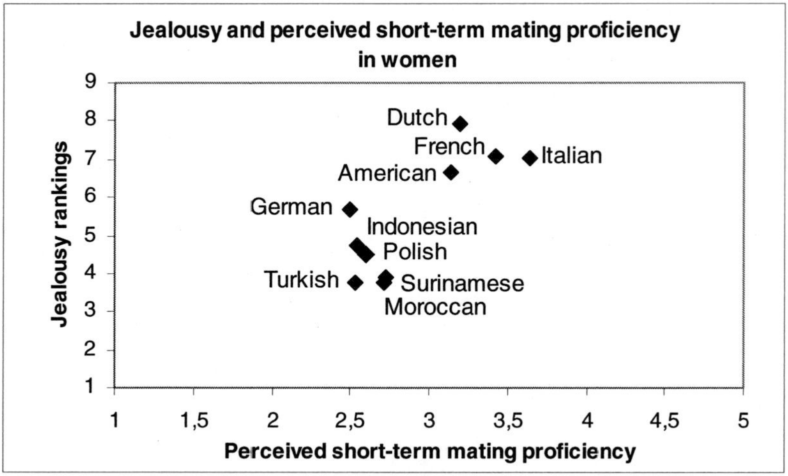

Figure 2. Relationship between jealousy and perceived short-term mating proficiency of groups in women

Social status correlated significantly with jealousy, for males r = .81, *p* = .004 and for females r = .85, *p* = .002 (see Figure 3 and 4).

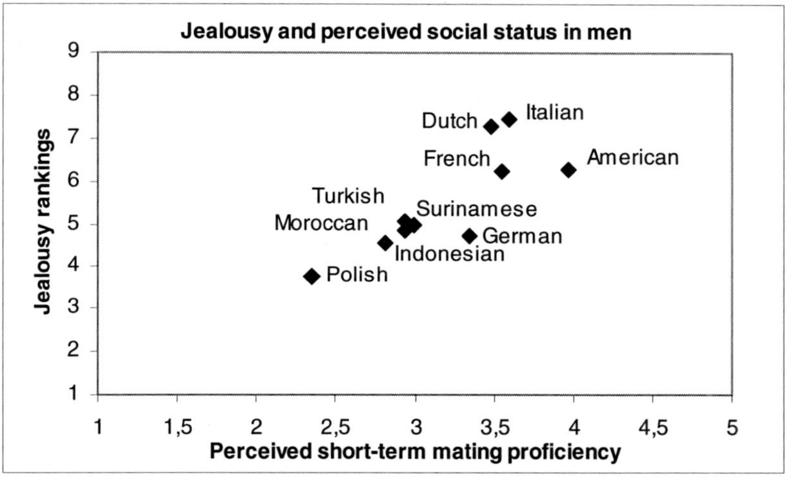

Figure 3. Relationship between jealousy and perceived social status of groups in men

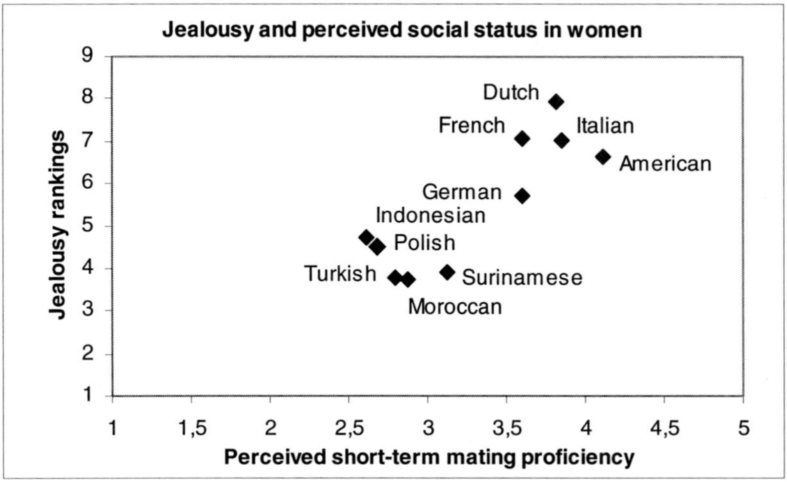

Figure 4. Relationship between jealousy and perceived social status of groups in women

Both factors correlated significantly with each other, r = .71, $p < .001$. When submitting both factors simultaneously to regression analysis, they significantly predict jealousy, with social status β = .51, t (19) = 3.2, p = .005 and short-term mating characteristics β = .45, t (19) = 2.83, p = .011 making independent contributions. Analyses for both sexes separately show that for men it is the short-term mating proficiency that predicts jealousy, controlling for social status (β= .63, t (9) = 3.31, p = .013), while social status is not such a strong predictor when controlled for short-term mating (β = .38, t (9) = 2.03, p = .081). Whereas for women both short-term mating proficiency and social status are significant predictors when entered separately (β = .80, t (9) = 3.75, p = .006) and (β =.85, t (9) = 4.54, p = .002), respectively, but when controlled for each they lose their significance. Overall, the more a group is perceived to be proficient in short-term mating, and the higher the social status of the outgroup, the more jealousy a member of the outgroup evokes. While for men short-term mating proficiency is an independent contributor to jealousy, for women it seems to be the combination of social status and short-term proficiency that contributes to jealousy.

Next, we explored the relationship between intergroup intrasexual competition and the perceived short-term proficiency and social status of the outgroups (this could not be done for jealousy, as these were rank order scores). Men who scored high on intergroup intrasexual competition perceived outgroups as being low on traits that constitute a potential short-term mating threat (r = -.38 p = .02). For women the relationship was not statistically significant. This may seem a somewhat surprising result, as one could expect that they might tend to overestimate the potential threat. However, in this case, the opposite seems to be true - they tend to derogate outgroups on this dimension. For women, intergroup intrasexual competition did not correlate with the tendency to perceive certain characteristics in outgroups. Thus men concerned with outgroup mating threats tend to derogate outgroups that possess traits relevant for short-term mating. Both for men and women, intergroup intrasexual competition was related to perceiving outgroup as having lower social status; however, the relationship was not strong enough to reach statistical significance.

Discussion

Research on intergroup processes has recognized competition over resources as an important source of intergroup conflict. However, one type of group resources – potential mates - has not received considerable attention. The long evolutionary history of human coalitional competition, including competition for reproductive benefits, seems to have created vigilance towards outgroups as a threat to one's mating opportunities. Functional approaches to prejudice maintain that qualitatively different threats elicit different emotional reactions (Cottrell & Neuberg, 2005; Kurzban & Leary, 2001; Schaller, Park, & Faulkner, 2003). Humans seem to have evolved a specialized alarming mechanism that alerts them to mating related threats – jealousy (Buunk, 1991). People are likely to be jealous of members from groups that are perceived to pose a threat to their relationship.

We investigated jealousy among Dutch respondents with respect to potential rivals from several countries, and the extent to which perceived characteristics of these ethnic groups predict the level of evoked jealousy. The results indicate that for both men and women, the Italians, the French and the Americans were the outgroups that caused the most jealousy. Interestingly, the Dutch as the ingroup was rather high in evoking jealousy. For women Dutch rivals were the number-one in jealousy rankings and for men they were the second in jealousy rankings. This may be explained by the fact that people are mostly surrounded by other ingroup members. However, while women seemed primarily concerned with threats from ingroup rivals, men were most concerned about the potential threats from a particular outgroup – the Italians. In addition, men and women differed in their jealousy rankings. While women were more jealous of German rivals than men were, men were significantly more jealous than women of Moroccans, Surinamese and Turks – all three visible immigrant groups in The Netherlands. Perhaps because men compared to women tend to use more direct mating strategies, the regular presence of immigrant groups makes men more threatening as rivals than women.

In general, our data clearly suggests that the rivals from outgroups were particularly threatening as they possessed high short-term mating potential and high social status. Namely, we found strong evidence that on the group level, a perceived inclination to engage in short-term mating was an important factor associated with jealousy in response to outgroup rivals. Groups that were perceived to be more open to short-term mating, in terms of being attractive, seductive and promiscuous, evoked high levels of jealousy for both, men and women. This is in line with our assumption, that a typical feature of many interactions between groups is their short-term nature, and that therefore interacting individuals in intergroup settings are more likely to apply short-term mating strategies. In addition, our prediction that the perceived status of the group would be related to jealousy was also substantiated. In interpersonal setting, traits indicating high social status and social dominance have been found to induce jealousy, particularly for men (Dijkstra & Buunk, 2002). In the intergroup setting in the present study, both men and women were more jealous of groups that were perceived to be high in social status and dominance. This may be due to the fact that social status and socially dominant behaviours are signals that can be noticed very quickly. As interactions between groups in our ancestral past would usually have been short-term based, a high status socially dominant rival would have more chance to succeed in attracting the attention of the opposite sex than someone who lacked those qualities. Particularly, it seems to be the combination of social status and short-term mating characteristics that drive jealousy, especially for women. While for men short-term mating proficiency contributes to jealousy independently. Thus an outgroup male that has the qualities to engage in short-term mating is perceived as a high threat, regardless of other factors.

What refers to jealousy of ingroup rivals, women perceive other Dutch rivals as significantly more short-term mating proficient, and of a higher status than men perceive other Dutch rivals. This finding is consistent with the preposition that women are more ingroup-oriented, in this case, also what refers to evaluating rivals in mating domain. Men, however, seem to be more concerned with outgroup mating threats than women.

The results found, with our newly developed scale for intergroup intrasexual competition, were generally in line with our theoretical perspective. First, men scored significantly higher in this type of competition than women, which implies that they are more concerned with outgroup mating threats than women. We can speculate that the origins of this difference lie in the long history of male coalitional alliances and their competition for resources, including mates. Thus for men, the link between mating competition in general and intergroup competition might be evolved as more salient. Second, we found that intergroup intrasexual competition was negatively related to the tendency to perceive groups as possessing traits indicating short-term mating ability (attractiveness, seductiveness and promiscuity) only among men. These traits were the ones that most clearly related to the ratings of jealousy. It therefore seems that men high on intergroup intrasexual competition derogate outgroups on this dimension. This might be one of the competition strategies, especially if the outgroups can be derogated in the presence of women. We did not find any relationship between intrasexual intergroup competition and perceptions of short-term mating ability for women. This confirms that for men intergroup mating competition might be more relevant and elaborate, resulting in more evolved mating competition strategies.

This study had several limitations. First, participants had to rate the outgroups on how much jealousy they evoke, consequently some of the differences might be exaggerated by the constraints of a rating scale. However, to assess the relationship between perceived characteristics a rating scale of jealousy seemed to provide the most information. The results are correlational as we did not manipulate mating threats or other psychological constructs. Future studies may manipulate mating threat and compare it with other types of outgroup threats, like, resource threat, territorial threat and disease transmission threat, to tease apart the particular effect of mating threat on intergroup conflict and prejudice. Additionally, more traits, including those suggesting perceived cognitive ability, like intelligence and creativity, and prosocial traits, like altruism and generosity should be included to extend the type of characteristics that predict jealousy in intergroup context. Despite these limitations, the current findings indicate that a functional approach, taking into account the evolutionary history of coalitional alliances in intrasexual competition and cost-benefit analysis of intergroup mating can illuminate how a perceived threat from outgroups can affect jealousy and, potentially, also further prejudice at an intergroup level.

Acknowledgements

This work was supported by a grant from the Royal Netherlands Academy of Arts. We thank Karlijn Massar, Shelli Dubbs and Ashley Hoben for their assistance.

References

Baumeister, R.F., & Sommer, K.L. (1997). What do men want? Gender differences and two spheres of belongingness. *Psychological Bulletin,* 122, (p.38-44).

Boesch, C. (2007, August 17). Intergroup Dynamics in Wild Chimpanzees – from Territoriality to Primitive Warfare. Paper presented at 2007 Human Mind Human Kind conference. Abstract retrieved September 15, 2007, from http://www.psy.au.dk/en/hk/programme/keynotes/boesch

Brewer, M. B. (1979). In-group bias in the minimal intergroup situation: A cognitive-motivational analysis. *Psychological Bulletin,* 86, (p.307-324).

Buss, D. M. (2000). *The Dangerous Passion: Why Jealousy is as Necessary as Love and Sex.* New York: Free Press.

Buunk, B. P. (1991). Jealousy in close relationships: An exchange theoretical perspective. P. Salovey (ed.), *Psychological Perspectives on Jealousy and Envy* (p. 148-177). New York: Guilford Publications.

Chagnon, N. A. (1988). Life histories, blood revenge, and warfare in a tribal population. *Science,* 239, (p.985–992).

Cottrell, C. A., & Neuberg, S. L. (2005). Different emotional reactions to different groups: A sociofunctional threat-based approach to "prejudice." *Journal of Personality and Social Psychology,* 88, (p.770-789).

Dijkstra, P., & Buunk, B. P. (1998). Jealousy as a function of rival characteristics: An evolutionary perspective. *Personality and Social Psychology Bulletin,* 24, (p.1158-1166).

Dijkstra, P., & Buunk, B. P. (2002). Sex differences in the jealousy-evoking effect of rival characteristics. *European Journal of Social Psychology*, 32, (p.829-852).

Geary, D. C., Byrd-Craven, J., Hoard, M. K., Vigil, J., & Numtee, C. (2003). Evolution and development of boys' social behavior. *Developmental Review*, 23, (p.444-470).

Grinnell J., Packer, C., & Pusey A. E. (1995). Cooperation in male lions: Kinship, reciprocity or mutualism? *Animal Behaviour*, 49, (p.95-105).

Hewstone, M., Rubin, M., & Willis, H. (2002). Intergroup bias. *Annual Review of Psychology,* 53, (p.575-604).

Lazaro-Perea, C. (2001). Intergroup interactions in wild common marmosets, Callithrix jacchus: Territorial defence and assessment of neighbours. *Animal Behaviour*, 62, (p.11-21).

Kurzban, R. & Leary, M. R. (2001). Evolutionary origins of stigmatization: The functions of social exclusion. *Psychological Bulletin,*127, (p.187-208).

Mitani, J. C., & Watts, D. P. (2001). Why do chimpanzees hunt and share meat? *Animal Behaviour*, 61, (p.915–924).

Möller, L. M., Beheregaray, L. B., Harcourt R. G., & Krützen, M. (2001). Alliance membership and kinship in free-living male bottlenose dolphins. *Proceedings of the Royal Society of London: Biological Sciences, 268*, (p.1941–1947).

Packer, C., Gilbert, D. A., Pusey, A. E., O'Brien, S. J. (1991). A Molecular Genetic Analysis of Kinship and Cooperation in African Lions. *Nature,* 351, (p.562-565).

Palmer, C. T., & Tilley, C. F. (1995). Sexual access to females as a motivation for joining gangs: An evolutionary approach. *Journal of Sex Research,* 32, (p.213-217).

Pemberton, M. B., Insko, C.A., & Schopler, J. (1996). Memory for and experience of differential competitive behavior of individuals and groups. *Journal of Personality and Social Psychology,* 71, (p.953–966).

Schaller, M., Park, J. H., & Faulkner, J. (2003). Prehistoric dangers and contemporary prejudices. *European Review of Social Psychology,* 14, (p.105-137).

Stephan, W. G., Ybarra, O., Martinez, C. M., Schwarzwald, J., & Tur-Kaspa, M. (1998). Prejudice toward immigrants in Spain and Israel: An integrated threat theory analysis. Journal of Cross-Cultural Psychology, 29, 559-576.

Trivers, R. (1972). Parental investment and sexual selection. In: B. Campbell (ed.), *Sexual Selection and the Descent of Man* (p. 136-179). Chicago: Aldine.

Van Vugt, M., De Cremer, D. & Janssen, D. P. (2006). Gender differences in cooperation and competition: The male warrior hypothesis. *Psychological Science,* 18,1.

Watts, D. P. (1998). Coalitionary mate guarding by male chimpanzees at Ngogo, Kibal National Park, Uganda. *Behavioral Ecology and Sociobiolgy,* 44,1.

Wilson M. L., & Wrangham R. W. (2003). Intergroup relations in chimpanzees. *Annual Review of Anthropology,* 32, (p.363-392).

Sidanius, J., & Pratto, F. (1999). Social dominance: An intergroup theory of social hierarchy and oppression. New York: Cambridge University Press.

Wrangham, R., & Peterson, D. (1996). Demonic males. New York: Houghton Mifflin Company.

CHAPTER TWENTY

HUMAN CHARACTERISTICS. DIFFERENCES AND SIMILARITIES BETWEEN HUMAN AND NON-HUMAN ANIMALS

HENRIK HØGH-OLESEN

Abstract: Summarizing and supplementing perspectives.
The following four central fields of activity are surveyed and comparatively analyzed: 1) complex symbolic behaviours, 2) tool making and tool use, 3) culture and social transmission, and 4) sociality and morality. Further, a tentative outline is made of humans' species-specific behaviours, social practices and psychological processes.

So, tell me if you can, "what makes man a man?"

Mapping the Rubicon between man and beast is a perennial endeavour, but it is time to summarise and supplement the perspectives presented, and to highlight some of the unique behaviours, social practices and psychological processes that make us human.

It is muddy waters that we are mapping, and the general continuity between us and the rest of the animal world, makes the contours even more blurred; but – at least to this researcher – some structures and demarcations seem to appear. Additionally, the task for the rest of this book is to single out and outline these elusive forms to the reader.

Aesthetic preferences may be functional to others than humans (cf. Stange, Chapter 1, Nadal et al., Chapter 2), but the areas of art and complex symbolic behaviours may still be a good place to track down some of the human characteristics.

Similarly today, we know that animals communicate and use tools (Nørager, Chapter 6). Complex mental processes, such as "mental time travelling"(Oswath, Chapter 3) and some form of "theory of mind" may also be found outside the human species, as may "moral inclinations" such as altruistic and empathic behaviours, but still these capacities may be substantially more advanced on the human level (Kingo & Krøjgaard, Chapter 4).

By looking at social cognition, it is obvious, that the human brain is not simply an enlarged chimpanzee brain (Smedt, Cruz & Braeckman, Chapter 8, Toft, Chapter 12), instead natural selection seems to have favoured a more collaborative kind of sociality in our species (Ingram, Piazza & Bering, Chapter 10). Our complex religious cults and rituals (Levi, Chapter 15), our symbolic demarcation of social groups, and our propensity to radical self-sacrifices and even martyrdom (Lieberoth, Chapter 14), for abstract ideas and ideological mega-groups, we identify with, are unique human features as well, and we could go on. However, simply piling up facts will not help us to obtain a clearer picture; instead we have to put the details into perspective.

In the following, four central fields related to:

1) Complex symbolic behaviours.
2) Tool making and tool use.
3) Culture and social transmission.
4) Sociality and Morality

will be surveyed and comparatively analysed for similarities and differences. Supplementary data, to the ones already presented, from a broad range of sciences, will be included, hopefully contributing to the puzzle of what makes us human.

On paper this may look like plain sailing; but experience shows that this is a neck-breaking and hazardous endeavour.

As late as in 1962, one of the leading "New synthesis" biologist, T. Dobzhansky, laid down as a fact that: "*Homo sapiens is not only the sole tool-making animal and the sole political animal, he is also the sole ethical animal*" (Dobzhansky 1962, p 339), just to find himself corrected in everyone of these areas a few years later by primatologists such as Goodall (1969,1986) and De Waal (1982, 1996), and a similar fate most likely awaits my own claims.

So why persist in these follies?

Because these demarcations simply have to be drawn once and again. They focus attention, make us wonder, and direct and stimulate research, because they provoke and challenge other researchers to take up the glove and prove us wrong.

So here we go again, and we start with a comparative look at our symbolic behaviours, from language and art to cults and rites.

Symbolic Capacities

Language

Language is a system for translating meanings into signals and signals into meanings (Hurford 2007, p.3), and for a long period it was cherished as a species-specific marker common to humans only, but also this bastion has lately been challenged.

We have long known that a wide range of earth, air, and sea-living animals use sounds and calls in order to attract mates or conspecifics, or to signal danger (Eibl-Eibesfeldt, 1970), but so far, referential communication and not least syntax (i.e. the ability to create new meanings through the rearrangement of discrete vocal elements) has been viewed as a strictly human hallmark.

Nevertheless, studies of several species (e.g. domestic fowl (*Gallus gallus)* (Evans & Evans, 1999), black-capped chickadees (*Poecile atricapillus) (*Templeton et al., 2005*)*, vervet monkeys (*Cercopithcus aethiops)* (Cheney & Seyfarth, 1990), surricats (*Suricata suricata)* (Manser, 2001), California ground squirrels (*Spermophilus beecheyi)* (Leger et al., 1980)) have documented qualitatively and quantitatively different alarm calls in response to different predators, specifying, for instance, whether the danger was aerial or terrestrial.

Other studies have shown that putty-nosed guenon monkeys (*Cercopithecus nictitans*), as well as gibbons (*Hylobatidae*), use combinations of calls, and rearrange their cries in order to relay new and specific information to one another. Different sounds were combined to create more complex sequences specifying e.g. whether a threat was a leopard, a serpent, or an eagle, and the responses of conspecifics demonstrated that these syntactic differences were meaningful (when somebody shouts "leopard" climb up a tree, when somebody shouts "eagle" hide under a bush) (Arnold & Zuberbühler, 2006; Clarke, Reichard, & Zuberbühler, 2006).

Some species may even incorporate a range of more descriptive information about a predator besides its spatial whereabouts. Prairie dogs (*Cynomys gunnisoni, Cynomys ludovicianus*) encoded information into their alarm calls specific to the colour of clothes (grey, green, orange, blue) and general shape (male/female, tall/small, slim/round) of humans acting as potential predators in their territory. When some of these humans, in another set of experiments, started to either feed the prairie dogs or fire a gun, these individual references were also encoded in distinct calls, whenever these persons later appeared (Slobodchikoff et al., 1991, Frederiksen & Slobodchikoff, 2007).

Of course this is not speech in the human sense, but it certainly is a close forerunner, and one could argue that these species may have some sort of "lexical items" or referential sounds related to central life-categories, that one can point to or act out. Something equivalent to our nouns ("eagle"/"dangerous air predator"), verb phrases ("watch out for…") and simple specifying adjectives related to physical properties.

Apparently no grammatical items (prepositions, conjunctions, articles or auxiliary verbs like "can", "may", "will" and "should") are found, in sharp contrast to human language, where up to half of the words are purely grammatical items, with no physical referent that one can point to (Diamond 2006, p.153).

Language trained chimpanzees (*Pan troglodytes)* and bonoboes *(Pan paniscus)* are able to understand semantic content as well as grammatical syntax, and they are quite good at utterances with present time imperative force ("Give me this or that"), but considerably more reticent than humans at giving information about events that happened yesterday or earlier. Information related to the future beyond the next few minutes seems to be more or less closed land as well, although the renowned Kanzi has been known to remind Sue Savage-Rumbaugh to give him the treat that she promised him yesterday (Hurford 2007, p.50, Savage-Rumbaugh, 1986).

Captive, language trained dolphins (*Tursiops truncates)* easily understand and react to complex instructions like: "Put the ball on your left, in the basket on your right" the first time they are presented, just as they do not react to semantically meaningless instructions, but looks expectantly at the instructor for further information. Referential gestures like pointing are spontaneously understood, and they are able to see and react to symbolic representations as TV exposures (even to carry out orders presented through TV), just as they are able to recognize themselves in mirrors and TV presentations (Herman, 2006).

These are specially trained animals, so what we see here are well stimulated potentials for syntactical, grammatical and semantic understanding, rather than natural capacities. But if we mix these insights with the natural capacities found in the free-living species presented before, a picture begins to emerge. This picture questions the Chomsky´an "Big bang theory of language" (first there was nothing, and then - after a major mutation in the human line sometime in the Pleistocene period - language emerged) (Chomsky, 1972,1995), suggests that some of the building blocks of human language and speech can be found in other species (even species outside the primate line), and consequently must have a much longer and more continuous evolutionary history than classical Chomsky´an theory presumes[1].

Human language may be unique in its grammatical and syntactic complexity, as well as, in its ability to signify abstract matters and relations involving unobservable mental and physical causes, that one cannot point to or act out.

Likewise, we may be the only species to work up and reinterpret our first order perceptions into abstract higher order principles and categories, as Penn et al. (2008) maintains. But, at least to this writer, these symbolic and abstract capacities rest on a staging of more basic linguistic and cognitive forerunners, that we share with some other species (just take a look at Pepperberg's (1987,1988, 2006) grey parrot studies). This, however, may not be the case for all our symbolic behaviours.

If we take a closer look at some of our more complex, referential and symbolic actions, something uniquely human seems to appear. Unlike all other species, we produce art, we bury our dead, and we conduct complex religious ceremonies directed at unseen powers. Something exceptional happens.

Art, Cult and Rite

In a period, some 40.000 years ago, where we, according to Maslow (1970) and all sensible people, should be fully occupied with satisfying our most basic needs, we went out of our way and into deep inaccessible caves to adorn and decorate these from floor to ceiling! And it was not even the game that we survived on that was painted. These Palaeolithic hunters had reindeer and grouses in their stomachs, and horses, bisons, and bears on their mind, as Leakey (1995) put it.

And why bury our dead? No other animal does that (elephant graveyards are only found in Tarzan books, should anyone be in doubt), and also inside the human line this practice is a rather late achievement. Excavations from Atapuerca in Spain suggest that archaic *Homo sapiens*, some 300.000 years ago, may have tugged dead tribesmen into caves and collectively thrown them into deep shafts where scavengers could not reach them. But individualized interments, where a deceased is placed in a grave, are first known from about 100.000 years ago (Eliade, 1995 v. 1).

---

[1] In his later writings, Chomsky maintains this saltational view, but specifies, that it is only the faculty of language in its narrow sense (FLN), i.e. the core computational systems that governs the recursive functions of mapping, allowing us to create sentences of potentially unlimited complexity, that is uniquely human (Hauser, Chomsky & Fitch 2002).

Where there are burials, there is awareness of death and of self, and where there are burial rites and offerings (in the form of weapons, food, goods, and gifts) this awareness is further intensified. Death is no longer just a physical fact – something that happens – it is conceived as a transmission from one mode of being to another. Now a reflective and narrative mind with a natural propensity to causality perception – and projection of meaning, purpose, and mind into the meaningless, coincidental, and mindless (Heider & Simmel, 1944; Mischotte , 1963; Boyer, 2002) – has started to seriously ponder the human conditions and to answer the perennial questions of who we are, why we are here, and what will happen in the afterlife. In other words: the quest for meaning, origin and ontological status, so typical to our species, has now become an all-pervading part of human existence.

These rites also suggest that a higher sense of individuality, self-identity, and self-consciousness, together with a deeper kind of bonding, may have arisen. The deceased must have had a very strong personal presence – he must have been "somebody" to someone – to survive as a lasting impression among the living, and he must have been *alive* to such a degree that he simply cannot die, but has to live on in another sphere (Figure 1. sums up the differences).

In short: With the rise of modern Homo sapiens, a new and more complex mind seems to have seen the light of day, and there is more evidence to support this presumption.

Behavioural Field 1: *Complex Symbolic Behaviour*

- Grammatical items with no physical referent
- Syntactic complexity – Ability to signify abstract matters, unobservable mental and physical clauses
- Art
- Burial rites
- Cults and religions
- Quest for origin, meaning, ontological status

Mental preconditions: *Self-reflective, narrative mind with the ability to abstract and symbolize higher order categories*

Figure 1: Human Characteristics – Behavioural Field 1.

Technology

Many animals use and produce simple tools (Barnard, 2004), but here a few examples will suffice. New Caledonian crows (*Corvus moneduloides*) are consummated toolmakers and users, and may, for instance, bend a piece of wire into a hook in order to pull food out of a pipe (Weir et al., 2002; Hunt, 2000). Recently they have even been shown to spontaneously solve a demanding meta-tool task in which a short tool is used to extract a longer tool that can be used to obtain meat (Taylor et al., 2007).

Meta-tool use, as e.g. the use of simple stone tools to make more complex tools, was one of the major innovations in our own history, and to many an indication of the "cognitive leap" that initiated technological evolution in hominines. This capacity, so far only found in humans and apes, has now also been documented outside the primate line.

Dolphins and whales may practise "sponge-carrying" in their foraging tactics, which involves finding and tearing off conical marine sponges from the substrate, placing the sponge in the rostrum and using it to ferret fish from the sea floor (Smolker et al., 1997).

Elephants may show the highest frequency and diversity of tools outside the primate line. They brake off branches, leaves and sticks in order to make flyswatters, fans and backscratchers and throws stones and sticks at antagonists. In a 100-hour field study of African savannah elephants (*Loxodonta africana*), Chervalier-Skolnikoff & Liska (1993) observed as many as 9 types of tool use, with an average of 1.45 acts per hour.

Chimpanzees (*Pan troglodytes*) have special tool kits, i.e. repertoires of tools used in special areas (hygiene, agonism, extractive foraging, etc.), and sometimes mash leaves to manufacture a sponge, throw rocks and sticks at antagonists (Goodall, 1986), strip and bend branches to fish termites (Sanz, Morgan & Gulick, 2004), or even produce 'spears' with sharpened points to hunt bush babies (*Galago senegalensis*) hiding in tree holes (Pruetz & Bertolani, 2007). Several of these tools are not only made but also reused and transported from one place to another. Moreover, they have tool sets, i.e. tools used in a necessary functional sequence to achieve goals and to solve problems (McGrew, 2004) , such as the "termite tool set" which includes puncturing sticks (that are often kept and carried and specially used for subterranean nests) followed by modified twigs with brush-tip fishing probes made in one end (Sanz, Morgan & Gulick, 2004).

The complete chimpanzee tool kit may be as complex as the simplest tool kits known to modern humans, such as the one used by the aboriginal

Tasmanians (McGrew, 1992), but in spite of its relative complexity something is conspicuously missing: So far, no animal has been seen to spontaneously combine two items to make one new tool, and what is even more conspicuous: Apparently none of our hominid ancestors did that either!

If we look into the toolboxes of *Homo habilis, Homo erectus*, and archaic *Homo sapiens* (c. 2 mill. - 150,000 BC), no multi-component tools, no special-purpose tools, and no socially marked tools are found, just as we do not find any manifestations of art or decoration in their remains. In fact, almost nothing happens between the invention of the drop-shaped hand axe 1,4 million years ago, and until modern Homo sapiens set off the Palaeolithic revolution less than 60,000 years ago. Then the axe is connected with a wooden shaft, the spear with a spear thrower and the bone hook with a line and a stick. Our technical skills are fused with our knowledge of the natural world in the production of special-purpose tools and weapons that take into account e.g. that the skin of elephants and rhinos are sterner than the skins of deer and birds. These things become beautifully decorated and hereby loaded with social information about the possessor. And, as suggested by Mithen (1998), all these material differences may very well indicate that a new, functionally integrated and generally reflective mind or brain has developed with the ability to mentally combine information (social, technical and natural historical) stored in different brain systems, previously not in communication with each other (Figure 2 sums up the differences).

Yet the fact that all these complex cultural patterns have only been found in our species does not imply that culture as such is a strictly human phenomenon. That depends entirely on how culture is defined.

Behavioural Field 2: *Complex Combinatory Tool Making*

- Multi-component tools
- Special-purpose tools
- Socially marked tools

Mental preconditions: Functionally integrated, generally reflective mind combining information stored in various brain systems

Figure 2: Human Characteristics – Behavioural Field 2.

Culture

If you, as De Waal (2001, p. 30-31), simply define culture as *"the non-genetic spreading of habits"*, and insist that the rest is nothing else than *"embellishment"*, or as Laland & Hoppit (2003, p. 151) maintain that culture is *"those group-typical behaviour patterns shared by members of a community that rely on socially learned and transmitted information"*, then a number of species, from birds and whales to elephants and primates can be granted a membership to "the culture club".

In a review of social transmission of foraging behaviour, Lefebvre & Palameta (1988) documented 97 examples of socially learned variation in animals as diverse as fish, lizards, sparrows and baboons. One study gathering data from six long-term chimpanzee studies from across Africa reported 39 different cultural features (Whiten et al., 1999), and recently Zapolsky (2006) has convincingly documented how a distinctive social tradition of low aggression and high affiliation was established, passed on to the following generations, and socially transmitted to newcomers in a group of wild anubis baboons (*Papio anubis*).

However, other researchers (Zihlman & Bolter, 2004) insist that we maintain a strict dividing line between tradition and culture, and that the transmission of knowledge through symbols with an abstract reference (which allows cultural learning to take place without the physical presence of living models) is the defining basis of culture proper. Yet others (Bergman, 2006) suggest that we leave out socially functional and ecologically functional behaviours, and hereby the patterns that appear as responses to demographic and environmental contingencies.

Personally, I fail to see what is gained by this exclusivity. Why reserve the concept of culture to the highest order of human complexity? Such restrictive definitions not only exclude inquiries by biologists and comparative psychologists into the question of whether or not other species have culture; but also preclude an exploration of the evolutionary roots of human culture and of humans' place in nature generally! And besides, who would insist that e.g. Indian or Egyptian culture is independent on such ecological factors as the monsoon, the desert, or the Nile?

The environmental forces both evoke (Tooby & Cosmides 1992), afford and select for (cultural, as well as, natural selection) specific adaptive behaviours and cultural strategies, that are passed on through generations, and I fail to see why these specific group patterns, should be less cultural than other patterns?

A much more fertile approach would be, to start out with a few descriptive and common denominators that most of us would agree upon, and which would allow us to compare human cultural patterns with discrete and group specific patterns found in other species. Something like the following (inspired by Potts, 2004):

Culture: Descriptive common denominators

1) Culture is a system of non-genetic information transfer between generations and among individuals of the same generation.

2) Culture is manifested as discrete forms (i.e. readily identifiable behaviours, artefacts, mental outputs, or social organisations) distinctive of particular populations.

3) These distinct forms show geographic and temporal differentiation inside a species.

4) There is a potential for change across generations.

Cumulative Culture

So other species may have different cultural features, but a distinctive feature of human culture is its cumulative character. Behaviours and artefacts that are transmitted are modified and added on over many generations leading to more complex artefacts and behaviours. Even an instrument as simple as a spear consists of several components and demands several meta-tools such as scrapers, wrenches, knives and hammers for its manufacture, and such complex innovations are not invented by individuals, but evolve gradually over many generations.

Evidence of cumulative culture in non-human animals remains minimal and controversial (Richerson & Boyd, 2005, Marshall-Pescini & Whiten, 2008). Our complex symbolic capacities allow us to store and stockpile vast repositories of social information, and another important factor may be our ability to acquire novel behaviours through observational learning and imitation. The last two abilities are also found in a range of other species, but much more seldom and in less sophisticated forms (Richerson & Boyd, 2005). A well described incidence of cultural accumulation occurred when the famous potato-washing Japanese macaques (*Macaca fuscata*) also began to wash wheat and rice, and later in the following generation started to dig waterholes, when the tide was low, so that they did not have to walk too far in order to wash and salt their groceries (Matsuzawa, 2003). But until now, the scientific registration of such incidences has been limited, and experimental approaches with, for instance chimpanzees, indicate that they, like most other animals, are conservative and hesitant in adopting new habits.

Marshall-Pescini & Whiten (2008) showed that young chimpanzees (age 2-8 years) where able to learn and copy a honey-dipping technique (where they had to use their index finger to slide open a trap door, whilst – with their other hand - inserting a rod to reach the honey) demonstrated by a familiar human. But when they were later shown a more efficient probing technique, built cumulatively on the first one (using the same rod to poke a bolt in the side of the honey pot, allowing the lid to go off so that not only more honey but also peanuts could be reached), yet nobody upgraded to this more efficient technique despite excessive series of demonstrations (exceeding 200). Instead they seemed to be stuck on the fishing technique they initially learned, inhibiting cumulative social learning, and possibly constraining the species′ capacity for cumulative cultural evolution.

The jury is still out on this question awaiting more evidence, but so far complex cumulative culture, where an artefact through the endless modifications of generations is transformed e.g. from a spear to a missile carrying spacecraft, appears to be a uniquely human feature, and there are other differences.

Culture is sharing by learning, and human culture differs from the cultures of non-human species, in what it is that can be shared, and a substantial part of human culture (from religion and marriage to the latest hip thing in fashion) consist of "social constructions", that cannot be reduced beyond the common intentional states and agreements of the individuals making up these forms (Plotkin 2007). A sum of Euros is not just a unit of currency, it is also an agreement among a group of individuals to use it as the basis for trading relationships, and such imagined worlds and symbolic tokens, made real by collective agreements only, are not seen in other species.

Culture's single most adaptive feature is, that it allows the gradual, cumulative assembly of adaptations over generations. Adaptations that no single individual could evoke on his own (Richerson & Boyd, 2005). But it is difficult to pin down, exactly what it is, that causes this difference between us and other species in cultural ability?

At least four interrelated factors may be involved. Basically, it is the same functionally integrated brain and mind, that allows us to picture and produce multi-component tools, that makes it possible for us to generate cultural accumulations. Other cognitive derivates of this mind, related to our ability to abstract and synthesize higher order categories, may be in play as well.

Then humans have languages, and hereby complex symbolic capacities for spoken and written sign-systems, that allow us to store and stockpile vast repositories of information, and to socially transmit knowledge, without the physical presence of living models, and that has to count for something as well. And last, but not least: Cumulative culture may be a derived side-effect of our ultra-social human mind (see next paragraph).

Humans have "theory of mind"(ToM), i.e. the ability to attribute intentional mental states to others, and simply cannot help joining forces, sharing perspectives and taking part in each others' projects, and that is imperative for the spreading and accumulation of knowledge, as well as for collaborative ventures (Figure 3 sums up the perspectives).

Behavioural Field 3: *Culture and Social Transmission*

- Cumulative culture
- Social constructions made real by collective agreements
- Vast repositories of information stored in sign-systems
- Social transmission without the physical presence of living models

Mental preconditions: Complex symbolic capacities. Cognitive ability to abstract and synthesize high-order categories. Ultra-social mind with ToM

Figure 3: Human Characteristics – Behavioural Field 3.

Sociality and Morality

Collaboration and Shared Intentionality

In a headline making study, comparing chimpanzees (*Pan troglodytes*), orangutans (*Pongo pygmaeus*) and 2.5 years old human children on a comprehensive battery of cognitive tests, Herrmann et al. (2007) showed that children and apes had very similar cognitive skills for dealing with the physical world, but that children had more sophisticated skills, than either of the ape species, for dealing with the social world.

Other researchers (Boesch, 2007, De Waal, Boesch, Horner & Whiten, 2008) have questioned these findings and objected, that when a human experimenter provides the social cues, the apes are at a disadvantage (and this may be especially crucial in the social tasks), and on top of this, they point out that the differences between the setups for children and apes in these tests are manifold, making comparisons difficult.

This is undoubtedly true, but it does not change the fact that a growing number of studies indicate that important features of human sociality are not to be found in the rest of the animal world, not even among our closest primate relatives.

Tomasello and colleagues have launched a theory stating that the decisive difference between apes and us is that only humans have "shared intentionality" – and that we are the only species that are: 1) biologically adapted to participate in collaborative activities involving shared goals and socially coordinated action plans, and 2) are motivated to share emotions, experiences, and activities with conspecifics (Tomasello et al., 2005, p. 676).

Indeed, among apes and monkeys there are very few triadic interactions or communications about third entities or topics, and very few signals serve a declarative or informative motive. They usually do not point, show, or actively offer things to conspecifics, and generally they do not engage in simple tasks, such as carrying or constructing things together.

In the mother-infant dyad there does not seem to be any of the shared dyadic face-to-face engagements, or 'proto conversations' with turn-taking, which is so common to humans and their infants (Trevarthen, 1974; Hobson, 2002), and there are other subtle but crucial differences in their interactive behaviours.

Warning and food calls left aside, apes and monkeys communicate almost exclusively with the purpose of getting others to do what they want. Human infants, in addition, also make gestures and talk in order to share information with others, just as they share their emotions freely, as when an infant points to a running dog to show his/her mother and squeals with delight. This unprompted information sharing reveals a helpful, interested other-directedness that seems natural to humans, but almost absent in other species.

So, it may be fair to conclude that "shared intentionality" and socially coordinated actions are significantly more widespread and cognitively more elaborated in our species than among our non-human relatives. But is it also fair to conclude, as Tomasello does, that no convincing signs of shared intentionality are found outside the human line?

Many would disagree and throw incidences in strong defence of chimpanzees (*Pan troglodytes*) engaging in collaborative hunting with role division, coordinating territorial patrolling and raids behind enemy lines (Boesch, 2001, 2003), soliciting support during coalition formation (De Waal & van Hooff, 1981), mediating reconciliation (De Waal & van Rosmalen, 1979), or performing various helping behaviours such as holding a "ladder" for others to climb (Menzel, 1972; De Waal, 1982), etc.

Lately, even Tomasello and collaborators have presented evidence that young human-reared chimpanzees may volunteer to help familiar or unknown humans in obtaining an object beyond their reach, or assist a conspecific in gaining access to a room with food, by removing a chain that only they can control (Warneken et al.2007), and other species follow suit.

Rooks (*Corvus frugilegus*) cooperate in pulling in a heavy platform loaded with food, by pulling both ends of a string simultaneously, and sharing the food afterwards (Seed, Clayton & Emery, 2008), and species as diverse as hawks (*Parabuteo unicinctus*), social cetaceans (humpbacked whales, killer whales, dolphins) and lions (*Pantera leo*) hunt in groups of up to ten individuals with coordinated movements and division of roles (some drive or flush, some block and surround and some ambush), resembling the collaborative hunting seen in chimpanzees (Bednarz, 1988, Gaydos et al., 2005, Rendell & Whitehead, 2001, Connor, Heithaus & Barre, 1999, Byrne, 1994, Boesch, 2003).

Sceptics may object, that what we see here are acts of opportunistic mutualism, in which everybody does their thing in the spur of the moment – a case of adequate reactive improvising - rather than cooperative endeavours governed by a shared plan, and this may be the case in some incidences. In other cases it seems difficult not to see these complex, coordinated and collaborative acts as expressions of shared intentions. But there are other differences.

Humans not only engage in complex large scale cooperation, unparalleled in the rest of the animal world outside the eusocial insects, involving tens, hundreds or even thousands of participants, united in super-groups, performing almost like "super-organisms", pursuing common goals through elaborate individual specialization and division of labour, and sometimes involving radical self-sacrifices for the benefit of the group. But most peculiar: we do so with non-kin and total strangers, just as we trade systematically and exchange goods and services, accepting symbolic tokens such as money and stocks, based on a system of mutual trust.

Such large scale and high-risk engagements with non-kin can only work if we can detect cheaters, social parasites and non-reciprocators, and if defection is collectively sanctioned, i.e. not only by the ones cheated, but also by third-parties watching or hearing about the deed.

Indeed, people do seem to be 1) endowed with cognitive and emotional cheater detection mechanisms that react to even minor indications of cheating (Cosmides 1989, Frank et al., 1993, Brown, Palameta & Moore, 2003), 2) to keep score of (and gossip about-) other people's behaviours , 3) to use these reputation-scores to gauge the quality of potential partners, and most spectacularly, 4) to practise third-party altruistic punishment and rewards ("strong reciprocity") towards group members, as well as, total strangers, that they will never meet again, just to enforce the social norms that they abide by (Gintis, 2000; Fehr, Fishbäcker & Gächter, 2002; Fehr & Fishbäcker, 2004). We even have evidence, that social and generous acts, spontaneously inspire good will and reciprocal generosity, not only in the beneficiary, but also in unrelated third-parties (Weedekind & Milenski, 2000).

Many species, from fish to non-human primates, are scorekeepers, and register others' behaviours towards them in order to predict future behaviours (Silk, 2002; Dugatkin, 2006). We even have evidence that chimpanzees may learn the reputation of strangers by observing third-party interactions, and later use this information in their own interactions with the stranger (Subiaul et al., 2008). However, these scores are still limited to the individual's personal perceptions and experiences, whereas humans have language that allows them to share the scores of others, and this makes it possible for large groups of non-kin, without day-to-day interactions, to reap the benefits of cooperation by monitoring each others' behaviours (via gossip), shunning or punishing cheaters, and rewarding team players. Besides, chimpanzees may be recruited to take part in conflicts and attacks involving kin or allies, but generally they take a self-centred and opportunistic stance in relation to the on-goings of group-life, and we still have no evidence of them rewarding a good deed done to others. Moreover, experimental findings indicate that this should not be expected either.

Sharing and "Other Regarding Preferences"

As humans we simply are each other's means to the common goals, and the *human niche* – i.e. the physical and social environment that we have adapted to throughout history – is a niche characterized by considerable inter-dependence, collaboration, closely bound up relationships, intensive gift-giving, reciprocity, extended sharing of resources and services, and radical self-sacrifices (Rochat, 2006, Høgh-Olesen, 2006).

Comparatively speaking, our capacities to share are quite unique. For instance, as humans we often take pleasure in simply giving something to others and take delight in their joy. And when flooded by beggars on holidays in less privileged parts of the world, we sometimes have to install cognitive counter strategies – such as only carrying a certain amount of coins in our pockets – simply to restrain our inclinations to share. In other species sharing may be endured, but it is hardly enjoyed, and they most certainly do not need cognitive counter strategies in order to keep things for themselves.

Chimpanzees do share, but it is first and foremost passive sharing as a response to intense begging – it is rarely unsolicited and never egalitarian as in our hunter-gatherer societies (Nissen & Crawford 1936, De Waal, 1989, 1997; Boesch, 2001a; Høgh-Olesen, 2009). Therefore, a prey killed alone and without the other's knowledge will not be shared but eaten on the spot (Knauft, 1991), whereas the human impulse clearly is to *actively* share what we have with others, and a prey killed by a hunter is brought to the common home base and divided among the tribesmen (Barret et al., 2002; Hawkes, 1991).

Likewise, reciprocity comes in simple and complex forms, and with the norm of active sharing, humans move from the simple, responsive, "tit for tat reciprocity", found among our *Pan* relatives (Packer, 1977; Seyfarth, 1980; De Wall & Luttrell, 1988, De Waal, 1989, 1997; Brosnan & De Waal, 2002 ; Melis, Hare & Tomasello, 2008), towards a more complex, "golden rule reciprocity", that may be unique to our species, in which the sharing initiative is clearly placed upon the individual: ***You** have to give in order to receive!* Moreover, hardwired neuro-endocrine feed-back systems seem to reinforce this behavioural normativity, since the centres associated with reward in the brain (*nucleus accumbens, caudate nucleus, ventromedial frontal and orbitro frontal cortex, rostal anterior cingulated cortex),* simply light up, when we choose to help or cooperate, as Rilling et al.´s (2002) neurological studies have shown. This means that a good deed "does well" both outside and inside the organism.

Experimental studies further develop these differences. Several studies indicate that generally chimpanzees (*Pan troglodytes*) are naturally motivated to act competitively rather than cooperatively in food related situations (Hare et al., 2000, 2001), even if individual differences in relation to e.g. helpfulness appear (Melis et al., 2006a, 2006b).

Later experiments by the Leibzig lab, using a novel twist on the object-choice paradigm, further consolidated these findings. Hare & Tomasello (2004) made their chimpanzee subjects familiar with two human experimenters: one was cooperative, shared food and showed positive emotions when the chimpanzees found food, while the other acted competitive with the animals, kept the food for himself and expressed negative reactions when they found food. In the actual experiment, the competitive experimenter made a reaching gesture towards one of two containers hiding food, while the collaborative experimenter, in another set of experiments, made a pointing gesture towards the container hiding food. In the competitive situation, the chimpanzees interpreted the reaching movement quite well, and tried to grab the food container before the competitors reached it. However, in the collaborative situation, the chimpanzees did not understand it and so only chose the correct container at a near chance level. The difference between their predominantly correct responses to a competitor and their random responses to a human co-operator were highly significant (P=0.009). Later replications, comparing a human co-operator with a chimpanzee competitor showed the same differences.

The lessons learned seem to be, that even if chimpanzees do seem to be able to read the intentions of others, they do not expect those intentions to be cooperative. Instead these kinds of intentions seem to be a more or less closed book to them. So even if this line of experiments presents experimental evidence of "Theory of Mind" in chimpanzees, it is a special (restricted) kind of ToM that anticipates that others – at least in situations related to food – will have competitive rather than cooperative intentions. More generally: a number of studies indicate that their "other-directedness" may be quite different to ours.

When chimpanzees are given a choice between a food reward for themselves or the same reward for themselves and a visible, well-known but unrelated conspecific in an adjoining cage, they show no other-regarding preferences (Silk et al., 2005, Jensen et al., 2006). They are highly observant and motivated to receive the food reward themselves, but seem indifferent to the opportunity to deliver cost-free rewards to others, even when these rewards could be released, after they have consummated their own reward, and after intense begging gestures and sounds from the other chimpanzee (Vonk et al., 2008).

Similar experiments with young children show that their other-regarding preferences develop strongly between the ages of 3 and 8 and take the form of inequality aversion. At age 3-4 most children behave selfishly, and fail to choose the other-regarding possibility above chance level, whereas most children at age 7-8 prefer resource allocations that remove disadvantageous, as well as, advantageous inequality, and this even more so, if the other player is an in-group member (Fehr, Bernhard & Rosenbach, 2008).

Capuchin monkeys (*Cebus apella*) and chimpanzees (*Pan troglodytes*) may at times display an aversion to disadvantageous inequity (Brosnan, & De Waal, 2003; Brosnan, Schiff & de Waal, 2005; Fletcher, 2008, but see also Bräuer, Call & Tomasello, 2006; Jensen, Call & Tomasello, 2007, for critical replications), but so far only humans have been shown to actively avoid advantageous inequality!

The benchmark test for examining other-regarding preferences and sensitivity to fairness is the ultimatum game. When this game is played among humans, the proposer typically offers 40-50 % of the windfall, and offers below 20-30% are routinely rejected by the responder (Camerer 2003, Henrich et al., 2005), suggesting that humans have an aversion to inequitable outcomes and punish offers that are too low. When an adapted mini version of this game was played among chimpanzees, in which the proposer can choose between two pre-set offers, which the responder can then accept or reject (by collaborating or not in drawing in two food trays with different numbers of raisins), chimpanzees, in contradiction to humans, did not –

A) take outcomes affecting the responder into account
B) make fair offers, when given the opportunity
C) nor, refuse to accept unfair offers from conspecifics.

In short: did not show any other-regarding preferences or sensitivity to fairness and behaved according to traditional economic models of self-interest and rational maximization (Jensen, Call & Tomasello, 2007).

Other-regarding preferences and inequality aversion are decisive for the ability to achieve and maintain cooperation in large groups of genetic strangers. If an individual cares for the welfare of other group members, this individual is more likely to refrain from free-riding in cooperative projects, and if he is endowed with an aversion to inequality he is more likely to monitor and punish free-riding in others, and these inter-species differences are probably important in explaining why humans are so exceptionally cooperative - so "ultra-social" - compared to other primates.

New evidences may still modify or change this picture, but so far our data suggest that chimpanzees are considerably more competitive than humans and that other-regarding preferences (and perhaps also reputational concerns) are derived properties of humans that have evolved after *Homo* and *Pan* diverged.

Morality - from "Gut-feelings" to "Ought-scenarios"

Sociality and morality are communicating vessels, and the latter stems from the fact that we, as social creatures, are each others' means to common goals, and hence rely on a support system for our survival. A solitary organism has no need for moral rules, nor does a creature that lives among others without mutual dependency.

As students of Plato and Kant, we may have learned that morality is a strictly human phenomenon, depending on the faculty of pure reason and founded on higher cognitive and cultural processes, and consequently: That we become moral creatures only by opposing our basic nature. Today comparative studies and modern neuroscience show that many of our highest moral, altruistic, and sympathetic inclinations may be evolutionarily anchored in mammalian sociality, and based on pre-programmed, emotional and neuro-endocrine feedbacks – operating as spontaneous "gut-feelings" - rather than pure reason, and just as selected for, as our more selfish impulses! (De Waal, 1996, 2006; Rilling et al., 2002; Greene & Haidt, 2002).

Hamilton's (1964) kin selection theory (genes for altruism can evolve if altruism is targeted at kin) explains why we, and many other species, help, care and make costly sacrifices for those genetically close to us, and Trivers' (1971) reciprocal altruism theory (genes for altruism can evolve if altruism and vengeance are targeted at those who do and do not return favours) explains why so many social exchanges in humans (and a variety of other species from worms (*Ophrocha diadema*), sea bass (*Hypoplectrus nigrans*), and sperm whales (*Physeter macrophalus*), to vampire bats (*Desmodus rotondus*), impalas (*Aepyceros melampus*) and primates in general) are governed by a reciprocity programme (Høgh-Olesen, 2006, Dugatkin, 2000, Mesnic, Evans & Taylor, 2003, Wilkinson 1984, De Waal, 1989, 1997).

But as several researchers (Richerson & Boyd, 2005, Haidt, 2007) have noted, these basic mechanisms cannot explain the extraordinary degree to which people cooperate with strangers they will never meet again and sacrifice for large groups of non-kin. Reciprocity may explain the cooperation of smaller social units with day-to-day interactions over longer periods, but for large scale cooperation among strangers to take place, some of the above-mentioned mechanisms related to indirect reciprocity (score-keeping, reputation gossiping) and third-party altruistic punishments and rewards, found in humans only, have to play an important role. This has allowed the same researchers to assume that group selective processes in humans, may have evolved a kind of "tribal overlay" in our species, i.e. a coevolved set of cultural practices and moral intuitions, that are not about how to treat other individuals (kin selection and reciprocal altruism take care of that), but about how to be a part of a group competing with other groups (Ibid.) These generally cooperative dispositions were superimposed onto human psychology, without eliminating those older social instincts that favour friends and kin, and this renders us susceptible to an inherent conflict between other-regarding and public-minded concerns, and more selfish and nepotistic dispositions, that other species are spared (Richerson & Boyd, 2005). And there are other moral dilemmas that appear to be distinctly human.

That many of our highest moral impulses, rather than "pure reason", are anchored in our mammalian and primate sociality, and based on spontaneous, emotional intuitions and "gut-feelings", does not exclude the possibility, that some of our moral choices may be both exclusively human and strictly rational. Take for instance the famous trolley dilemma (see Fig. 4).

Most people choose without hesitation to press a shift that will lead a runaway trolley from one track, where it would kill five innocent people, to another track, where only one person will be killed. In other words: Sacrificing one to save five.

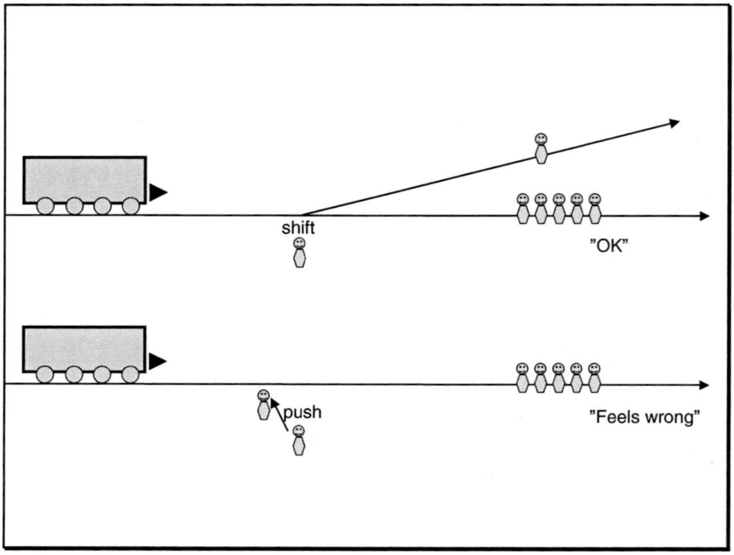

Figure 4: The Trolley Dilemma.

However, when they have to save the five by actually pushing this person onto the tracks themselves, a clear majority refuses flatly to do so. The reason given is that it simply does not feel right, and FMRI scannings of their brain activity show that a number of old mammalian brain areas (more specific the *Precuneus* area) are activated during this kind of problem solving (Greene & Haidt, 2002), lending support to the above mentioned idea that substantial parts of our morality may be anchored in mammalian sociality.

But what about the few rationalists who ignore their gut-feelings and still decide to sacrifice one to save five by pushing one onto the tracks?

FMRI scannings show that these subjects: 1) take a longer time to decide, and 2) show more activity in prefrontal brain areas, and hereby more cognitive processing before making the decision. And this morally rational choice, in which higher cognitive processes apparently overrule more basic programmes, could be an example of a moral choice, which no other social animal than us would be able to make!

Let us widen the perspective slightly. It is possible for humans to have an organising, "top-down" impact on some of their basic constituents - e.g. when a child chooses to inhibit the spontaneous reflex of crying when hurt, until others who can comfort are within sight, when a monk chooses celibacy, or when a mountaineer overrules basic needs for body temperature, oxygen and food in order to climb Mount Everest. And this seems to be an essential difference.

Thanks to this ability, humans can be motivated by "an ought to" or "a should", and decide to "row a canoe against the stream", metaphorically speaking. To lead lives governed by principles and values so important, that they are worth dying for, and to strive to be impartial, courageous, responsible, honest, unselfish, and so on. Just as we can be tormented by complex, self-reflective emotions such as shame and guilt, when we fail to live up to our standards, and that makes up for another life, than a life ruled by instincts, desires, needs, and emotions, even if such a life can also be a social and intelligent life.

Apes may sometimes be helpful, caring and brave, but that is not because they think they should be, and that is a crucial difference (Korsgaard, 2006). Our neo-cortical superstructures have given us the ability to consider and reflect on any number of different scenarios for the future, and that gives us, as probably the only creatures, the ability to see before our mind's eye the world as it *could* be. Figure 5 summarises the human particulars.

Behavioural Field 4: *Sociality and Morality*

- Increased inclination to share and sacrifice
- Proactive "golden rule reciprocity"
- Other-regarding preferences
- Third-party altruistic punishments and rewards
- Large scale, high-risk collaboration with non-kin and strangers
- Moral choices governed by "pure reason" and principles
- Moral projects governed by "ought-scenarios" for the future

Mental preconditions: Ultra-social mind: Hardwired programmes of moral intuitions, active sharing and reciprocity. Higher cognitive (prefrontal) top-down organization of basic programmes

Figure 5: Human Characteristics – Behavioural Field 4.

Concluding Remarks

There is continuity between us and the rest of the world, but each species has its species-specific traits, and if we are going to track down some of the human particulars, a good place to start could be in the above-mentioned fields and capacities.

However, things are progressing rapidly, and this applies especially to the areas of culture and sociality/morality. Research on tool use, collaboration, "theory of mind" and "mental time travels" is likewise gaining new momentum, thanks to the fascinating studies on rooks, crows and scrub jays conducted in Cambridge (Dally, Emery & Clayton, 2006, Seed, Clayton & Emery, 2008). While you are reading these lines, studies - currently under review, that are able to challenge or introduce light and shade onto the suggestions presented here - may already be pilling up. But that is just part of the game, and it shall not prevent us from summing up the conclusions reached so far (See Figure 6).

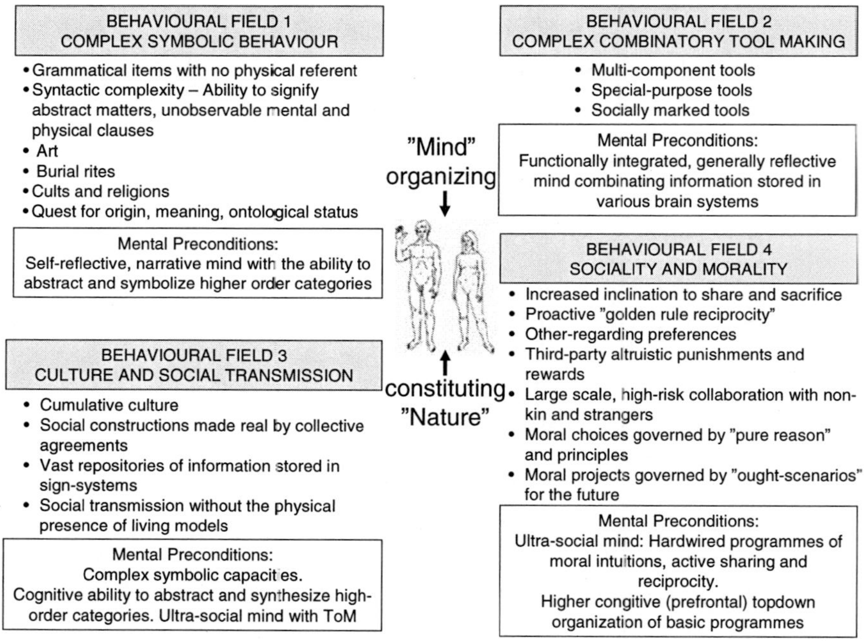

Figure 6: Behaviours and Capacities Apparently Unique to *Homo sapiens sapiens.*

The complex symbolic activities of art, cult and rite, listed in behaviour field 1, still seem to be wholly unprecedented outside the line of modern *Homo*, but this may be due to an almost total absence of relevant comparative studies in this area. Apart from a few dubious examples of captive chimpanzees and elephants "painting" pictures (see Stange Chapter 1.), and anecdotal incidences of chimpanzees performing so-called "rain dances", in the form of stylized power displays, when confronted with noisy natural phenomenon's, like waterfalls and heavy showers (sometimes interpreted as an indication of awe (Goodall, 1995), this area is more or less *terra incognita.* So, to the inventive experimental or naturalist developer among the readers: Please enlist here! And why not start with taking up the experimental challenge presented by Stange in the first chapter of this book? Aesthetic preferences may be functional to others than humans, but do other species show aesthetic sensibility and appreciation? We simply do not know, but the quest goes on.

References

Arnold, K. & Zuberbühler, K. (2006). Semantic combinations in primate calls. *Nature, 441*, 303.

Barnard, C. (2004). *Animal behaviour.* England: Pearson Education.

Barret, L. Dunbar, R., & Lucett, J. (2002). *Human evolutionary psychology*. New York: Palmgrave.

Becker, H. (1956). Man in reciprocity. New York: Praeger.

Bednarz, J. (1988). Cooperative hunting in Harris´hawks. *Science 239,* 1525-1527.

Bergman, T. J. (2006). Comments. *Current Anthropology, 47*(4), 648-649.

Boesch, C. (2001). Chimpanzee hunters! Chaos or cooperation in the forest? In L. Dugatkin, *Model systems in behavioral ecology* (pp. 453-465). Princeton University Press.

—. (2001a). Cooperative hunting roles among Tai chimpanzees. *Human Nature, 13,* 27-46.

—. (2003). Complex cooperation among Tai chimpanzees. In F. B. M. De Waal & P. L. Tyack, *Animal social complexity* (chapter 4). Boston: Harvard University Press.

—. (2007). What makes us Human (*Homo sapiens*)? The challenge of cognitive cross-species comparison. *Journal of Comparative Psychology 121, No. 3,* 227-240.

Boyer, P. (2002). *Religion explained.* London: Vintage.

Bräuer, J., Call, J. & Tomasello, M. (2006). Are apes really inequity averse? *Proceedings of the Royal Society of London, B, 273, 3123-3128.*

Brosnan, S.F. & De Waal, F.B.M. (2002).Empathy: Its ultimate and proximate bases. *Behavioral and Brain Sciences, 25, 1-72.*

Brosnan, S. F. & De Waal, F. B. M. (2003). Monkeys reject unequal pay. *Nature, 425,* 297-299.

Brosnan, S. F., Schiff, H. C. & De Waal, F. B. M. (2005). Tolerance for inequity may increase with social closeness in chimpanzees. *Proceedings of the Royal Society of London, B, 1560, 253-258.*

Brown, W. M., Palameta, B. & Moore, C. (2003). Are there nonverbal cues to commitment?. *Evolutionary Psychology. Human Nature. Comp/EP-2003, 1,* 42-69.

Byrne, R. (1994). The evolution of intelligence. In P. Slater & T. Halliday. *Behaviour and evolution* (pp. 223-265). Cambridge University Press.

Camerer, C.F. (2003). *Behavioral game theory – Experiments in strategic interaction*. Princeton NJ. Princeton University Press.

Cheney, D. L. & Seyfarth, R. M. (1990). *How monkeys see the world.* Chicago. University of Chicago Press.

Chevalier-Skolnikoff, S & Liska, J.(1993). Tool use by wild and captive elephants. *Animal Behaviour, 46, 209-19.*

Chomsky, N. (1972). *Language and mind.* New York. Harcourt Brace Jovanovich.

—. (1995). *The Minimalist Programme*. Cambridge, MA. MIT Press.

Clarke, E., Reichard, U. H., Zuberbühler, K. (2006). The syntax and meaning of wild gibbon songs. Plos one 1(1): E73, DOI: 10.1371/Journal. Pone .0000073.

Connor, R. C., Heithaus, M. R. & Barre, L. M (1999). Superalliance in bottlenose dolphins. *Nature, 371*, 571-572.

Cosmides, L. (1989). The logic of social exchange. Has natural selection shaped how humans reason? *Cognition, 31*, 187-276.

Dally, J. M., Emery, N. J. & Clayton, N. S. (2006). Food-caching western scrub-jays keep track of who was watching when. *Science, 312*, 1662-1665.

Darwin, C. (1859). On the origin of species. Camb. Mass.: Harvard University Press (1966).

Darwin, C. (1874). The descent of man (2nd ed.). New York: Prometheus Books (1998).

De Waal, F. B. M. & van Hooff, J. A. (1981). Side-directed communication and agonistic interactions in chimpanzees. *Behaviour, 77,* 164-198.

De Waal, F. B. M. & van Roosmalen, A. (1979). Reconciliation and consolation among chimpanzees. *Behavioral Ecology and Sociobiology, 5,* 55-66.

De Waal, F. B. M. (1982). *Chimpanzee polics.* London: Jonathan Cape.

—. (1989). Food sharing and reciprocal obligations among chimpanzees. *Journal of Human Evolution, 18,* 433-459.

—. (1996). *Good natured – The origins of right and wrong in humans and other animals.* Camb. Mass.: Harvard University Press.

—. (1997). The chimpanzee's service economy: Food for grooming. *Evolution and Human Behaviour, 18,* 375-386.

—. (2001). *The ape and the sushi master.* New York: Basic Books.

—. (2005). Our inner ape. London: Grania Books.

—. (2006). *Primates and philosophers.* Princeton University Press.

De Waal, F.B.M., Boesch, C., Horner, V. & Whiten, A. (2008). Comparing social skills of children and apes. *Science 319,* 569.

De Wall, F. B. M & Luttrell, L. M. (1988). Mechanisms of social reciprocity. *Ethology and Sociobiology, 9*, 101-118.

Diamond, J. (2006). *The third Chimpanzee.* New York. Harper Perenial.

Dobzhansky, T. (1962). *Mankind evolving: The evolution of the human species.* New Haven: Yale University Press.

Dugatkin, L. (2000). *Cheating monkeys and citizen bees.* Cambridge, MA. Harvard University Press.

Dugatkin, L.A. (2006). Trust in fish. *Nature, 441*, 937-938.

Eibl-Eibesfeldt, I. (1970). *Ethology – The biology of behavior.* New York: Holt, Rinehart, & Winston.

Eliade, M. (1995). *De religiøse ideers historie V. 1.* København: Gyldendal.

Evans, C. S. & Evans, L. (1999). Chicken calls are functional referential. *Animal behaviour, 58*, 307-319.

Fehr, E. & Fishbacker, U. (2004). Third-party punishment and social norms. *Evolution and Human Behaviour, 25, 63-87.*

Fehr, E., Fishbacher, U. & Gächter, S. (2002). Strong reciprocity, human cooperation and the enforcement of social norms. *Human Nature, 13*, 1-25.

Fletcher, G.(2008). Attending to the outcome of others: Disadvantageous inequity aversion in male Capuchin monkeys. *American Journal of Primatology, 70, 901-905.*

Frank, R. H., Gilovicht, L. & Reagan, D. T.(1993). The evolution of one-shot cooperation: An experiment. *Ethology & Sociobiology, 14,* 247-256.

Frederiksen, J. K. & Slobodchikoff, C. N. (2007). Referential specificity in the alarm calls of the black-tailed prairie dog. *Ethology, Ecology & Evolution, 19*, 87-99.

Gaydos, J., Raverty, S., Baird, R. & Osborne, R. (2005). Suspected surplus killing of harbour seal pups (*Phoca vitulina*) by killing whales (*Orcinus orca*). *Northwestern Naturalist, 86*, 150-54.

Gintis, H. (2000). Strong reciprocity and human sociality. *Journal of theoretical biology, 206*, 169-179.

Goodall, J. (1969). *My friends the wild chimpanzees.* Washington DC: National Geographic Society.

—. (1986). *The chimpanzees of Gombe.* Boston: Bellknap Press.

—. (1995). *The new chimpanzees*. USA. National Geographic DVD.

Gouldner, A. Q. (1960). The norm of reciprocity. *American Social Revue, 25*(2), 161-178.

Greene, J. & Haidt, J. (2002). How (and when) does moral judgement work? *Trends in Cognitive Science, 16*, 517-523.

Haidt, J. (2007). The new synthesis in moral psychology. *Science, 316*, 998-1002.

Hamilton, W.D.(1964). The genetical evolution of social behaviour. *Journal of Theoretical Biology, 7*, 1-52.

Hare, B. E, Call, J. & Tomasello, M. (2001). Do chimpanzees know what conspecifics know? *Animal Behaviour, 61*(1), 139-151.

Hare, B. E. & Tomasello, M. (2004). Chimpanzees are more skilful in competitive than in cooperative tasks. *Animal Behaviour, 68*, 571-581.

Hare, B. E., Call, J., Agnetta, B. & Tomasello, M. (2000). Chimpanzees know what conspecifics do and do not see. *Animal Behaviour, 59*, 771-785.

Hauser, M., Chomsky, N. & Fitch, W.T.(2002): The faculty of language: what is it, who has it, and how did it evolve ? *Science, 298, 1569-1579.*

Hawkes, K. (1991). "Showing off": Test of another hypothesis about men's foraging goals. *Ethology and Sociobiology, 11*, 341-361.

Heider, F. & Simmel, M. (1944). An experimental study of apparent behaviour. *American Journal of Psychology, 57*, 243-254.

Henrich, J., Boyd, R. & Bowles, S. et al. (2005). "Economic Man" in cross-cultural perspective. Behavioural experiments in 15 small-scale societies. *Behavioral and Brain Sciences, 28*, 795-855.

Herman, L. M. (2006). Intelligence and rational behaviour in the bottlenosed dolphin. In *Hurley, S & Nudds, M. (Ed.). Rational animals.* Chapter 20. Oxford University Press 2006.

Herrmann, E., Call, J., Hernandez-Lloreda, M.V., Hare, B. & Tomasello, M. (2007). Humans have evolved specialized skills of social cognition: The cultural intelligence hypothesis. *Science, 317*, 1360-1366.

Hobson, R. P. (2002). *The cradle of thought*. New York: Macmillian.

Høgh-Olesen, H. (2006). The sacrifice and the reciprocity-programme in religious rituals and in man's everyday interactions. *Journal of Cognition and Culture, 6*(3-4), 499-519.

—. (in press 2009). The will to sacrifice – Sharing and sociality in humans, apes, and monkeys. In A. Gertz (Ed.), *Origins of religion, cognition and culture*. London: Equinox.

Hunt, G. R. (2000). Tool use by the New Caledonian crow. *EMU, 100*, 109-114.

Hurford, J. R (2007). *The origins of meaning*. Oxford University Press.

Jensen, K., Call, J. & Tomasello, M. (2007). Chimpanzees are vengeful but not spiteful. *PNAS: Proceedings Natl. Acad. Sci. USA, 104, 1346-1350.*

Jensen, K., Hare, B. E., Call, J. & Tomasello, M. (2006). What´s in it for me? Self-regard precludes altruism and spite in chimpanzees. *Proceedings of the royal Society of London, B, 273,* 1013-1021.

Knauft, B. M. (1991). Violence and sociality in human evolution. *Current Anthropology, 26*(2), 223-240.

Korsgaard, C. M. (2006). Morality and the distinctiveness of human action. In F. B. M. De Waal, *Primates and philosophers* (pp. 103-123). Princeton University Press.

Laland, K. & Hoppit, W. (2003). Do animals have culture? *Evolutionary Anthropology, 12,* 150-159.

Leakey, R. (1995). *The origin of humankind*. New York: Basic Books.

Lefebvre, L. & Palameta, B., (1988). Mechanisms, ecology, and population diffusion of socially-learned food-finding behaviour in feral pigeons. In T. Zentall & J. B. G. Galef (Ed.). *Social learning, psychological and biological perspectives* (pp. 141-165). Hillsdale, NJ. Lawrence Erlbaum Associates.

Leger, D.W. & Owings, D.H. & Gilfand D.L. (1980). Singel note vocalizations of California ground squirrels. *Zeitschrift für Tierpsychologie, 52*, 227-246.

Linnét, C. von (1758). Systema naturae. Stockholm: Rediviva (1976).

Manser, M.B (2001). The acustic structure of suricate alarm calls varies with predator type and the level of urgency. *Proceedings of the Royal Society of London (B)* 268, 2315-2324.

Marshall-Pescini & Whiten, A. (2008). Chimpanzees (*Pan troglodytes*) and the question of cumulative culture: an experimental approach. *Animal Cognition 11*, 449-456.

Maslow, A. H. (1970). *Motivation and personality (2nd ed.)*. New York: Harper & Row.

Matsuzawa, W. (2003). Koshima monkeys and bossou chimps: Long-term research on culture in non-human primates. In F. B. M. De Waal & P. L. Tyack, *Animal social complexity* (chapter 14). Boston: Harvard University Press.

McGrew, W. C. (1992). *Chimpanzee material culture*. Cambridge. Cambridge University Press.

—. (2004). *The Cultured Chimpanzee*. Cambridge, UK. Cambridge University Press

Melis, A. P., Hare, B. E. & Tomasello, M. (2006a). Chimpanzees recruit the best collaborators. *Science, 311*, 1297-1300.

Melis, A. P., Hare, B. E. & Tomasello, M. (2006b). Engineering cooperation in chimpanzees: Tolerance constraints on cooperation. *Animal Behaviour 72*(2), 275-286.

Melis, A.P., Hare, B. & Tomasello, M. (2008). Do chimpanzees reciprocate recieved favours? *Animal behaviour,76, 951-962.*

Menzel, E. (1972). Spontaneous invention of ladders in a group of young chimpanzees. *Folia Primatologica, 17,* 87-106.

Mesnic, S. L., Evans, K., & Taylor, B. L. (2003). Sperm whale social structure. In F. B. M. De Waal & P. L. Tyack, *Animal social complexity* (chapter 6). Camb. Mass.: Harvard University Press.

Mischotte, A. E. (1963). *The perception of causality.* London: Methuen.

Mithen, S. (1998). *The prehistory of the mind.* London: Phoenix.

Nissen, H. W. & Crawford, M. P. (1936). A preliminary study of food-sharing behavior in young chimpanzees. *Journal of Comparative Psychology, 22*, 383-419.

Packer, C. (1977). Reciprocal altruism in *Papio anubis. Nature 265,* 441-443.

Penn, D. C., Holyoak, K. J., Povinelli, D. J. (2008). Darwins Mistake: Explaining the discontinuity between human and nonhuman minds. *Behavioral and Brain Sciences, 31*, 109-178.

Pepperberg, I. M. (2006). Ordinality and inferential abilities of a Grey parrot. *Journal of Comparative Psychology, 120*, 205-216.

—. (1987). Acquisition of the same/different concept by an African grey parrot (Psittacus erithacus). *Animal Learning and Behavior, 15*, 423-32.

—. (1988). Acquisition of the concept of absence by an African grey parrot. *Journal of the Experimental Analysis of Behavior, 50,* 553-564.

Potts, R. (2004). Sociality and the concept of culture in human origins. In R. W. Sussman & A. R. Chapman (Eds.), *The Origins and nature of sociality* (chapter 12). New York: Aldine de Gruyter.

Plotkin, H. (2007). The power of culture. In R.I.M Dunbar & L. Barrett (Eds.), *The Oxford Handbook of Evolutionary Psychology* (chapter 2). Oxford, Oxford University Press.

Pruetz, J. D. & Bertolani, P. (2007). Savanna chimpanzees, *Pan troglodytes verus,* hunt with tools. *Current Biology, 17*(1-6), March 6.

Rendell, L. & Whitehead, H. (2001). Culture in whales and dolphins. *Behavioral and Brain sciences, 24*, 309-382.

Richerson, P. J. & Boyd, R. (2005*). Not by genes alone*. Chicago. Chicago University Press.

Rilling, J. K., Gutman, D. A., Zeh, T. R., Pagnon, G., Berns, G. S., & Kilts, C. D. (2002). A neural basis for social cooperation. *Neuron, 35,* 395-405.

Rochat, R. (2006). What does it mean to be human? *Journal of Anthropological Psychology, 17,* 100-107.

Sanz, C., Morgan, D. & Gulick, S.(2004). New insights into Chimpanzees, tools, and termites from the Congo Basin. *The American Naturalis, 164* (5), 567-581.

Savage-Rumbaugh, E. S. (1986). *Ape language: From conditioned response to symbol.* New York. Columbia University Press.

Seed, A. M., Clayton, N. S. & Emery. N. J.(2008). Cooperative problem solving in rooks (*Corvus frugilegus). Proceedings of The Royal Society B 275, No.1641,* 1421-1429.

Seyfarth, R. (1980). The distribution of grooming and related behaviours among adult female vervet monkeys. *Animal Behaviour, 28, 798-813.*

Silk, J. B. (2002). Kin selection in primate groups. *International journal of primatology, 23,* 849-75.

Silk, J. B., Brosnan, S. F. & Vonk, J. et al.(2005). Chimpanzees are indifferent to the wellfare of unrelated group members. *Nature, 437,* 1357-1359.

Simmel, G. (1950). *The sociology of George Simmel.* Free Press, Glencoe Ill.

Slobodchikoff C. N., Kiriazis, J. & Creef, E. (1991). Semantic information distinguishing individual predators in the alarm calls of Gunnisons prairie dogs. *Animal Behaviour, 42*, 713-719.

Smolker, R. A., Richards, A. F., Connor, R.C. & Mann, J. et al. (1997). Sponge-carrying by Indian Ocean bottlenose dolphins. *Ethology, 103*, 38-69.

Subiaul, F., Vonk, J., Okamoto-Barth, S. & Barth, J.(2008). Do chimpanzees learn reputation by observation. *Animal Cognition, 11,* 611- 623.

Taylor, A. H., Hunt, G. R., Holzhaider, J. C. & Gray, R. D. (2007). Spontaneous Metatool use by New Caledonian Crows. *Current Biology, 17*, 1504-1507.

Templeton, C. N., Greene, E. & Davis, K. (2005). Allometry of alarm calls. Black-capped chickadees encode information about predator size. *Science, 308*, 1934-1937.

Tomasello, M., Carpenter, M., Call, J., Behne, T., & Moll, H. (2005). Understanding and sharing intentions! The origins of cultural cognition. *Behavioral and Brain Science, 28,* 675-735.

Tooby, J. & Cosmides, L. (1992). The psychological foundations of culture. In J. Barkow, L. Cosmides & J. Tooby (ED). *The Adapted Mind.* 19-136. New York. Oxford University Press.

Trevarthen, C. (1974). Conversations with a two-month-old. *New Scientist, 62,* 230-235.

Trivers. R. L. (1971). The evolution of reciprocal altruism. *The Quarterly review of Biology, 46*, 35-57.

Vonk, J., Brosnan, S. F., Silk, J. B. & Henrich, J. et al. (2008). Chimpanzees do not take advantage of very low cost opportunities to deliver food to unrelated group members. *Animal Behaviour 75,* 1757-1770.

Warneken, F., Hare, B., Melis, A.P.& Tomasello, M. (2007). Spontaneous altruism by chimpanzees and young children. *PloS Biol 5(7): e184,doi:10.1371/journal.pbio.0050184.*

Weedekind, C. & Milenski, M. (2000). Cooperation through image scoring in humans. *Science*, *288*, 850-852.

Weir, A. A. S., Chappell, J. & Kacelnik, A. (2002). Shaping of hooks in New Caledonian crows. *Science, 297,* 981.

Whiten, A. J., Godall, W., McGrew, T., Nishida, V. et al. (1999). Culture in chimpanzees. *Nature, 399,* 682-685.

Wilkinson, G. S. (1984). Reciprocal food sharing in the vampire bat. *Nature, 308,* 181-184.

Zapolsky, R. M. (2006). Social cultures among nonhuman primates. *Current Anthropology, 47*(4), 641-656.

Zihlman, A. L. & Bolter, D. R. (2004). Mammalian and primate roots of human sociality. In R. W. Sussman & A. R. Chapman (Eds.), *The*

Origins and nature of sociality (pp. 23-52). New York: Barnes & Noble.

List of Contributors

Bering, Jesse M.	Institute of Cognition and Culture, Queen's University, Belfast, Northern Ireland
Bertelsen, Preben	Department of Psychology, Aarhus University, Denmark
Braeckman, Johan	Department of Philosophy and Ethics, Ghent University, Belgium
Brickman, Harry R.	UCLA Geffen School of Medicine, USA
Buunk, Abraham P.	Department of Psychology, University of Groningen, the Netherlands, Royal Netherlands Academy of Arts and Sciences
Capò, Miquel	University of the Balearic Islands, Palma de Mallorca, Spain
Cela-Conde, Camilo José	University of the Balearic Islands, Palma de Mallorca, Spain
De Cruz, Helen	Centre for Logic and Philosophy of Science, Free University of Brussels, Belgium
De Schrijver, Jelle	Ghent University, Belgium
De Smedt, Johan	Department of Philosophy and Ethics, Ghent University, Belgium
Dubbs, Shelli L.	Department of Psychology, University of Groningen, the Netherlands
H.-Olesen, Henrik	Department of Psychology, Aarhus University, Denmark
Hald, Gert Martin	Clinic of Sexology, Copenhagen University Hospital, Denmark
Ingram, Gordon P. D.	Institute of Cognition and Culture, Queen's University, Belfast, Northern Ireland
Inoue, Mariko	University of the Sacred Heart, Tokyo, Japan
Kennair, Leif Edward O.	Department of Psychology, Norwegian University of Science & Technology, Norway
Kingo, Osman S.	Department of Psychology, Aarhus University, Denmark
Klavina, Liga	Department of Psychology, University of Groningen, the Netherlands

Krøjgaard, Peter	Department of Psychology, Aarhus University, Denmark
Levi, Sergio	Università degli Studi di Milano, Dipartimento di Filosofia, Milano, Italy
Lieberoth, Andreas	Centre for the Study of Religion, University of Southern Denmark
Nadal, Marcos	University of the Balearic Islands, Palma de Mallorca, Spain
Nissen, Gorm	Department of Media, Cognition and Communication, Philosophy Section, University of Copenhagen, Denmark
Nørager, Rune	Department of Psychology, Aarhus University, Denmark
Osvath, Mathias	Lund University, Sweden
Park, Justin H.	Department of Experimental Psychology, University of Bristol, United Kingdom
Peters, Avanti	University of the Balearic Islands, Palma de Mallorca, Spain
Petersen, Michael Bang	Department of Political Science, Aarhus University, Denmark
Piazza, Jared R.	Institute of Cognition and Culture, Queen's University, Belfast, Northern Ireland
Shibata, Reiko	Sugiyama Jogakuen University, Tokyo, Japan
Stange, Ken	Nipissing University, Ontario, Canada
Takahashi, Keiko	University of the Sacred Heart, Tokyo, Japan
Toft, Kristian Høyer	Department of Media, Cognition and Communication, Philosophy Section, University of Copenhagen, Denmark
Tønnesvang, Jan	Department of Psychology, Aarhus University, Denmark
Yamakawa, Kayoko	Sugiyama Jogakuen University, Tokyo, Japan

The *Human Mind - Human Kind* Conference
preceding this publication
was sponsored by:
Arla Foods, The Danish Research Council (FKK)
and Videnssamfundet, Aarhus University, Denmark.

INDEX